Philosophy and Medicine

Volume 129

More information about this series at http://www.springer.com/series/6414

Jamie Carlin Watson • Laura K. Guidry-Grimes
Editors

Moral Expertise

New Essays from Theoretical and Clinical Bioethics

Editors
Jamie Carlin Watson
Department of Medical Humanities and Bioethics
University of Arkansas for Medical Sciences
Little Rock, Arkansas, USA

Laura K. Guidry-Grimes
Department of Medical Humanities and Bioethics
University of Arkansas for Medical Sciences
Little Rock, Arkansas, USA

ISSN 0376-7418 ISSN 2215-0080 (electronic)
Philosophy and Medicine
ISBN 978-3-030-06510-2 ISBN 978-3-319-92759-6 (eBook)
https://doi.org/10.1007/978-3-319-92759-6

Printed on acid-free paper

This Springer imprint is published by the registered company Springer Nature Switzerland AG
The registered company address is: Gewerbestrasse 11, 6330 Cham, Switzerland

Contents

Chapter 1
Introduction

Jamie Carlin Watson and Laura K. Guidry-Grimes

1.1 The Problem of Moral Expertise

When an attending physician needs specialized assistance with a patient's end-stage renal disease, she requests a consult from a nephrologist—an expert on kidneys. When a fellow has a question about the effects of depression on a patient's decisional capacity, she calls for a psychiatric consult—an expert on mental conditions. And so, if a physician needs information about the moral implications of a treatment plan, why shouldn't she turn to someone competent in the subject of ethics—a moral expert?[1] While the former requests are regarded as prudent and commendable, the latter strikes many as scandalous.[2] And yet the role of ethicists in professional decision-making in fields such as business, research, and medicine is steadily increasing.

Moral concerns about how patients and research participants are treated by medical staff and researchers led to the subfield of bioethics. In the United States, accreditation by The Joint Commission requires hospitals to have a mechanism for

[1] Some scholars attempt to maintain a distinction between morality and ethics, but here we follow the majority of academic philosophers in using them interchangeably to refer to the study of the related concepts of good, bad, right, wrong, permissible, impermissible, and obligatory.

[2] Richard Zaner expresses the timidity many ethicists feel embarking upon the task of ethics consulting: "[M]any of us felt acutely out of place and recoiled in shock and dismay" (1988: 5). Giles Scofield excoriates the notion, arguing that "medical ethics consultants neither know nor agree on what they do for a living, much less what one needs to know and what skills one needs to do whatever it is they do for a living (2008: 96). And Julia Driver notes that this sentiment extends fairly widely, since most of us are even more willing to accept aesthetics experts than ethics experts, "displayed by a willingness to be guided by the advice of art critics as to what movie we ought to see, and what art exhibit is the most worthwhile" (2006: 619).

J. C. Watson (✉) · L. K. Guidry-Grimes
Department of Medical Humanities and Bioethics, University of Arkansas for Medical Sciences, Little Rock, Arkansas, USA
e-mail: lguidrygrimes@uams.edu

J. C. Watson, L. K. Guidry-Grimes (eds.), *Moral Expertise*, Philosophy and Medicine 129, https://doi.org/10.1007/978-3-319-92759-6_1

addressing ethical concerns.[3] In most cases, this takes the form of an ethics committee, but in some cases also includes an ethics consultation service. Motivated independently of professional regulation, many large hospitals now offer ethics consultation services.[4] A 2007 report shows over 300 clinical ethics mechanisms in German hospitals.[5] And in 2011, the UK Clinical Ethics Network (now, UKCEN) website included over 100 UK hospitals claiming to have clinical ethics mechanisms.[6]

What's more, a large number of active and thoughtful scholars—working with or through organizations like the American Society for Bioethics and Humanities, the Association for Practical and Professional Ethics, and UKCEN—have developed and tested strategies for enhancing practical moral decision-making in clinical contexts, putatively improving the quality of patient care and reducing the emotional and psychological effects of morally charged situations.

There remains, however, a raw uneasiness with the idea that ethicists might contribute something meaningful to clinical decision-making. This uneasiness is not new. The idea that one can justifiably tell others what they morally ought to do has been controversial at least since the fifth century BCE, when Heraclitus reportedly said, "Of all those whose accounts I've listened to, none gets to the point of recognizing that which is wise…."[7] In the early 300 s BCE, Plato tells us that Socrates attempted to convince Meno and Protagoras that virtue cannot be taught.[8] In contrast, Plato's pupil Aristotle argued that *phronesis* (practical wisdom, which includes wisdom regarding virtue) not only can be attained, but that a dialectical process might allow us to identify and trust those who have attained it.[9] These discussions have become more sophisticated over the succeeding centuries, informed by ever more nuanced debates over moral theory,[10] yet it remains contentious whether anyone can speak authoritatively on moral matters.

Of course, apart from the philosophical debates, we all share a deep sense that, at least in some instances, we *know*[11] the morally preferable thing to do, from decisions about lying to loved ones to voting for public policies. Further, assuming there is some moral reality, we cannot avoid making decisions that have moral implications:

[3] Joint Commission (1992). The Joint Commission is the independent, not-for-profit accrediting body for hospitals in the United States (formerly known as the Joint Commission on Accreditation of Healthcare Organizations (JCAHO)).

[4] Fox et al. (2007). Fox, et al. estimate that every hospital over 400 beds has an ethics consultation service, but there are concerns with their sampling methods, and there is disagreement over how broadly they construe "consultation service."

[5] Döries and Hespe-Jungesblut (2007).

[6] UK Clinical Ethics Network (2011).

[7] Heraclitus: *Fragments*, Fr. 108. T. M. Robinson, trans.

[8] Plato, *Meno* (1997a), *Protagoras* (1997b).

[9] See Khan (2005: 49–51).

[10] For more on the history of this debate, see the first seven chapters of Rasmussen (2005), which include discussions of Socrates, Aristotle, David Hume, J. S. Mill, Josiah Royce, John Dewey, and G. E. Moore.

[11] Or that we at least have *well-justified beliefs* about the morally preferable thing to do.

we *will* make decisions about lying to loved ones and public policies. And in medical contexts, such decisions have higher moral stakes, greater complexity, and more urgency (see Shuster 2014). Unless there are no objective moral truths, it seems at least plausible that some of our beliefs about morality are justified. For instance, it is generally uncontroversial that slavery is morally impermissible, that kindness towards animals and humans is morally good, and that parents have a moral obligation to feed their children. Setting aside the difficult theoretical debates on this topic, we will assume, for the sake of this volume, that there is an objective moral reality of some sort, and that, in some cases, we really do have justified beliefs (perhaps even knowledge) about what we ought morally to do.

But even if we can have justified moral beliefs about *our own* moral decisions, there is still a question of whether we could form such beliefs about others' decisions, and, of significance for this volume, whether we could *expertly* advise others regarding their moral decisions. The prominent—perhaps dominant—answer of philosophers has been *no*. The standard interpretation of Immanuel Kant is that moral authority can derive only from one's own will.[12] John Locke famously doubted whether we can learn much through testimony, writing, "we may as rationally hope to see with other [people's] Eyes, as to know by other [people's] Understandings."[13] In the twentieth century, A. J. Ayer argued that it is mistaken to look to philosophers for moral guidance.[14] Gilbert Ryle argued that no one can be a moral expert because there is nothing in morality for anyone to be an expert about.[15] C. D. Broad argued that moral philosophers have no "special information" about ethics that is not available to everyone else, and so "it is no part of the professional business of moral philosophers to tell people what they ought or ought not to do."[16] And Bernard Williams says it is a *notorious* fact that there are no experts in ethics, singling out medical ethics as a prominently implausible attempt at developing them.[17]

This skepticism, combined with the growing number of ethicists serving in professional contexts, leads to an apparent conflict. If the very idea of someone's

---

[12] See Wolff (1970) and Zagzebski (2012). Some scholars, like Immanuel Kant and J. S. Mill, can be interpreted as defending the idea that part of what makes your knowledge of morality *knowledge* (in a strong sense) is that it is a result of reflecting on your own decision-making process. In other words, it is a function of your autonomy as a rational agent. Mill writes, "If a person possesses any tolerable amount of common sense and experience, his own mode of laying out his existence is the best, not because it is the best in itself, but because it is his own mode" (2002: 69). Of course, even Kant doesn't rule out the possibility of checking your reasoning against the informed opinion of others. See Zagzebski (2012: 23–26) for an excellent discussion of how Kant regards testimony in his *Anthropology*

[13] Locke (1979).

[14] Ayer (1954).

[15] Ryle (1958). However, see Arjo in this volume (Chap. 2) for a discussion of Ryle's account of knowledge.

[16] Broad (1952: 244).

[17] Williams criticizes the field of medical ethics in two places: "Who needs ethical knowledge?" (1993) and "Truth in Ethics" (1995).

speaking authoritatively in matters of moral decision-making is problematic, what goals might these scholars, committees, and consultants aim to achieve? Are they a sort of "ethics police," bringing abstract philosophical concepts to bear on the concrete realities of medical care? Are they conflict-mediators in disguise? Do they simply ensure compliance with professional codes and legal standards? Are any claims of ethics expertise legitimate? If so, who could claim it, and what would it imply for non-experts? Should non-experts defer to these experts, as they would to their accountants and physicians?

There are, to be sure, an increasing number of educational degrees, programs, fellowships, conferences, and scholarly works aimed at helping ethics committee members and consultants develop the ability to recognize and respond to difficult moral situations. Such initiatives suggest the possibility of establishing and conferring credentials on participants, similar to other professional organizations. But given classic concerns about having moral authority, the question remains as to what these initiatives are accomplishing: What sort of expertise are they credentialing?

In the first half of this collection, contributors explore theoretical debates over the possibility, nature, and implications of moral expertise in general. In the second half, contributors explore practical debates over purported moral expertise of clinical ethics consultants (CECs). This volume should appeal to moral philosophers, academic bioethicists, CECs, clinicians, hospital administrators, ethics committee members, and others who want to gain a deeper understanding of the potential scope of moral expertise in real-world contexts.

1.2 The Role of Clinical Ethicists

There is a rough consensus among scholars of clinical ethics consultation that the goal of ethics committees and consultants is to help clinical staff, patients, and families navigate a variety of moral and non-moral features of difficult situations so they can make better moral decisions.[18] Non-moral features include professional responsibilities and constraints, political and cultural expectations, institutional policies, legal responsibilities and constraints, and the cultural-historical backdrop of medical practice.[19] Moral features include respect for (and threats to) patients' well-being and autonomous interests, which include risks regarding pain and death, respect for an aspect of identity (such as gender, sexuality, race, religion),

[18]Puma et al. (1995); ASBH (1998) and (2011); Aulisio (2003); Jonsen et al. (2010); Dubler and Liebman (2011); Hester and Schonfeld (2012). See Scofield (2008) for a critique of these stated goals.

[19]To be sure, each of these has moral implications. The point is to simply highlight that neither laws nor institutional policies are, in themselves, moral statements, even if motivated by moral concerns with moral implications for care. For instance, a federal law requiring that all new employees have the legal right to work in the country is not itself a moral requirement. Similarly, an institutional policy that all team members wash their hands has clear moral implications, but not every instance in which that policy is violated is a moral infraction.

marginalization, and unjust treatment. Threats to such interests also include patients' *perception* of medical treatment. In some cases, patients do not recognize threats that may ultimately compromise their moral interests. In other cases, they perceive threats to their well-being or autonomy where there are none. And clinicians are not immune from these blind spots; their perceptions can be underinformed, unnuanced, or simply mistaken. This fallibility suggests that even when a clinician perceives that a patient's expressed preference is "irrational,"[20] he should take that preference seriously in decisions regarding patient care.

Consider two common moral situations clinicians face:

- A 45-year-old patient has been refusing above-the-knee amputation for a gangrenous leg, even though the surgery would likely extend his life. The infection is now imminently threatening his life, and his capacity has become unclear with his condition worsening. The patient's wife says the patient is not in his right mind and really wants to live, and she begs the surgical fellow to perform the amputation against the patient's wishes. The patient's son says the patient has very strong opposition to being "crippled" or "disabled" and would prefer to "die with his boots on." The nurses on the unit are expressing increasing discomfort with letting a patient die from a treatable illness. The surgical fellow cannot get in touch with the attending physician and calls for an ethics consultation.
- A 60-year-old patient with unclear capacity and no surrogate decision-maker requests to stop kidney dialysis, which will result in her deterioration and eventual death on this admission to the hospital. The dialysis is expected to keep her relatively stable for years to come, and she could probably be discharged back to her nursing home. The attending physician believes that requesting to stop a life-saving treatment is tantamount to requesting suicide, so the physician refuses to accept the patient's request. The physician treats the patient's increasing agitation with Haldol before the next dialysis session, but the patient still physically resists. A nurse working with the patient anonymously calls for an ethics consultation.

Other controversial cases include questions regarding "dignified death," maternal-fetal conflict, "medical futility," the appropriateness of unilateral DNR, and requests for posthumous ovum or sperm retrieval. While some of these moral complexities will be rare, moral red flags and problems arise on a daily basis in a hospital. As emphasized by Mark Aulisio, healthcare settings are increasingly ripe for ethics consultation: "Even as the complexity and number of choices facing patients, families, and providers multiplies, the contemporary health care environment is increasingly less conducive to good decision making."[21] Standard care in a hospital frequently involves many specialties and sub-specialties, imperfect hand-off among physicians, the rise of managed care with less time spent per patient, length of stay pressures, and economic considerations.[22] From all of this, it is evi-

---

[20] See MacIntyre (1988) for the influence of different accounts of rationality on decision-making.

[21] Aulisio (2003: 5)

[22] Many of the items on this list come from Aulisio (ibid.).

dent that ethical difficulties will abound in any healthcare setting. If a hospital reports having relatively few ethics consults (or none), this is not evidence that ethical issues are few and far between; rather, this is evidence that the health care professionals do not have a robust ethics consult service, do not know of the existence of the service, and/or do not trust in the service.

What might ethicists contribute to these complex ethical decisions? Is there a recognized and widely accepted subject matter the ethicist can apply to such decisions, as an internist or psychiatrist might in their fields? Should ethics consultants offer moral advice on how to resolve such conflicts, or should they only clarify putatively competing values? Should committees make recommendations to medical teams about what is preferable (as, say, a nephrologist or oncologist might), or should they only talk through the morally salient considerations, "supporting a critical inquiry of moral convictions and moral questions"?[23] And if they do make recommendations, to what degree, if any, are those morally binding on the participants' decisions?

The answers to these questions depend on the nature and scope of moral expertise, whether there are any moral experts, and if there are, what their expertise implies for the recipients of their testimony. These are the topics addressed by the contributors to this collection, and the chapters that follow are relevant not only for potential ethics consultants, but for clinicians who engage with ethics committees and consult services, ethics committee members, policy committee members, board members, and academic bioethicists.

In the remainder of this introduction, we sketch the major outlines of contemporary debates on these topics. We begin with some basic concepts and distinctions common to the debate, namely: different ways of conceiving expertise, different conceptions of moral expertise, the distinction between political and epistemic authority, and the distinction between situational authority and expert authority. We then review debates over what moral experts could tell us if there were any, and we discuss some important developments since Lisa Rasmussen's important 2005 collection *Ethics Expertise: History, Contemporary Perspectives, and Applications*. We close this introduction with an outline of the contributions to this volume.

1.3 Concepts and Distinctions

1.3.1 What Is Expertise?

Whether it is plausible to be or to justifiably believe someone is a moral expert depends heavily on what it means to be any sort of expert. Unfortunately, the increasingly rich literature on expertise is often neglected in discussions of moral

[23] Stolper et al. (2010: 151).

expertise.[24] A brief review of two prominent accounts of expertise highlights the importance of this literature for moral expertise.

The most prominent account of expertise is veritism, from the Latin *veritas*, which means "truth." According to veritists about expertise,[25] experts have authority because their beliefs are "truth-tracking," that is, they have a "substantial fund" of reliably true beliefs in a subject matter (see Goldman 2001). Some veritists require that experts have more reliably true beliefs *than false in a subject matter* (Goldman 2001). Others argue, more modestly, that experts have more reliably true beliefs *than others in their epistemic*[26] *community* (Fricker 2006; Coady 2012).

The term "reliably" is important because it indicates that experts do not have their true beliefs accidentally (because you happened to overhear your physician mother talking as you grew up) or through some arbitrary means (memorizing the claims in a medical textbook). Experts have more true beliefs because they gained access to the relevant content by a means that allowed them to understand and apply evidence relevant to those claims, and they used those claims in a way that developed competence.

A moral expert, on this account, would be someone who has substantially more reliably true moral beliefs than others. This raises important questions for scholarship on moral expertise: Can someone have more reliably true moral beliefs than someone else? How might we know? And are there true and false moral beliefs, or is morality substantially different from other types of subject matter? The implication is that which account of expertise we presuppose affects how we evaluate the possibility and plausibility of moral expertise.

An alternative prominent account of expertise was developed by Hubert Dreyfus and Stuart (1980; 1986). This account focuses on expertise as skill acquisition rather than merely knowledge acquisition: someone is an expert in a subject matter if they can *do* something competently in that subject matter, even if the doing requires a substantial degree of knowledge. The Dreyfuses view expertise as the upper end of a competence continuum, starting with novice, progressing through advanced beginner, competence, and proficiency, and culminating in three categories that allow one to practice and eventually teach: expertise, mastery, and practical wisdom.

The benefits of this account highlight the roles of teaching and culture in the acquisition of expertise. Different teachers have different styles of practice, and while copying a style may be sufficient for basic expertise, a master recognizes that she must adapt her teacher's style into her own. Further, the language, communica-

---

[24] Exceptions include Hopkins (2007) and Priaulx, et al. (2014).

[25] Those who defend some version of veritism include Alvin Goldman (2001), who coined the term, Elizabeth Fricker (2006), Jimmy Alfonso Licon (2012), and David Coady (2012).

[26] The term "epistemic" refers to concepts associated with knowledge or justified belief. A person's "epistemic community" is the group of people closest to the person in terms of what they are interested in knowing, how questions are framed about that subject, and the relevant evidence and strategies for answering those questions. For instance, the international community of scientists would be members of a chemist's epistemic community.

tion techniques, attitudes toward our subject matters, and even the structure of a profession are determined in large part by that field's culture. As one masters the content of a subject along with that field's culture, they come to embody the practical wisdom associated with that field. All of this suggests that a large part of being an expert is *tacit* rather than *explicit*, that is, it is less about acquiring and applying knowledge claims and more about being an active member of a field.

A recent prominent version of performative expertise was developed by Harry Collins and Robert Evans (2007). Like the Dreyfuses, Collins and Evans view expertise on a continuum from low levels of expertise to high levels. They begin with *ubiquitous expertise*, low-level skills that everyone can master (such as native language use) and proceed to higher-level skills of *interactional expertise*, which allows an expert to engage with a subject matter on its own terms (such as a science writer for the *New York Times*). If one pursues specialized experience in a discipline, one may be able to achieve the highest level of competence, *contributory expertise*, which is the competence to participate in a subject matter as a contributor (e.g., a respected astrophysicist). Also like the Dreyfuses, they focus on tacit knowledge as the primary feature of expertise.

Unlike the Dreyfuses, Collins and Evans ground their continuum in a number of social science experiments, and based on that research, they develop a rich set of conceptual distinctions that help explain many of our intuitions about expert authority. For example, they argue that interactional experts rely heavily on contributory experts, that contributory expertise entails interactional expertise, and that both require one to develop the social capacities they call "interactive ability" and "reflective ability."

Crucial to their account is a set of meta-criteria that help explain how one becomes an expert and also how non-experts can recognize experts. As you might expect, these criteria include the putative expert's credentials, experience, and track record. Interestingly, these meta-criteria also admit of meta-expertises, which are the continuum of skills necessary for using these criteria well. These distinctions have important implications for evaluating the role and authority of clinical ethicists, especially considering the recent debate over whether anyone could be credentialed as a moral expert (see Vogelstein in this volume).

We will not delve deeper into this account of expertise except to note that it was introduced into the discussion of moral expertise in 2014 by Nicky Priaulx, Martin Weinel, and Anthony Wrigley. They argue that the Collins/Evans account entails that moral expertise is not only possible, but plausible. If we all have basic moral competence, we share what the Collins/Evans model deems ubiquitous moral expertise. And it seems that those who have studied moral philosophy may achieve interactional and contributory expertise in the subject matter of ethics. Further, even those who haven't studies moral philosophy may, nevertheless, achieve specialized expertise in "robust moral judgment."

Unfortunately, Priaulx, et al. stop at robust moral judgment and do not discuss whether this judgment is restricted to personal moral decisions or whether it could be extended to helping others make better moral decisions. While it seems uncontroversial that a moral philosopher could achieve interactional moral expertise in the academic study of ethics, it is less clear whether someone could achieve it in helping others. Nevertheless, Priaulx, et al. have demonstrated how the concept of expertise can enrich debates over moral expertise.

While veritism and performative expertise are the most widely discussed accounts of expertise, there are others that may be helpful in discussions of moral expertise (see Hopkins 2007, Watson 2018, and Quast 2018). This growing literature on general expertise has remained largely disconnected from discussions of moral expertise. Moral expertise may be more or less plausible depending on which account of expertise is most plausible. Further, the sorts of skills and knowledge one may need to become a moral expert may be different depending on which account of expertise is plausible.

1.3.2 What Kind of Moral Expertise?

Moral expertise refers to several distinct concepts. In order to restrict our focus to the moral expertise relevant to the clinical context, it will be helpful to compare and contrast these notions. Perhaps the most common type of moral expertise is that of the scholar who studies and teaches moral philosophy. It is uncontroversial to regard someone who has achieved advanced degrees in moral philosophy and actively participates in academic discussions through publication and conferences as an expert in moral philosophy. Such a person tends to have the ability to explain the details and implications of a variety of metaethical and normative moral theories, detail contemporary controversies surrounding those theories, and help others understand the significance of those debates. We will call this sort of expertise *academic moral expertise*.

Yet, academic moral experts might be incapable of expertly advising others about what morally they ought to do (see Butkus in this volume). Although many academic moral experts teach applied or practical ethics, they often abstract from the complexities of scenarios in which some of those decisions must be made.[27] For example, in discussions of abortion, moral philosophers will often talk about the nature of personhood, individual rights, and types of moral standing. All of these questions are relevant to practical moral decision about abortion, but they are certainly not the only relevant features in any actual clinical case. Further, in academic discussions, philosophers can assume certain variables are fixed in order to test intuitions and isolate specific moral concerns.

---

[27] Burch (1974) puts this point eloquently: "In the typical moral problem, the ethically relevant features are tricky to specify and extremely difficult to weigh with respect to one another. Moreover, there is no given short-list of possible actions to be decided upon; instead there looms before the person deciding what do to an open field of infinitely diverse actions, shading into one another in countless, different ways. To be or not to be is hardly ever the moral question, but rather when, where, how, for whom, how much, and in what respect to be or not to be. A moral problem calls not for a mechanical response, but rather for a creative act" (655).

By contrast, in actual cases where abortion is considered, decision-makers face questions of decisional capacity, religious commitments, legal constraints, relational autonomy in a variety of cultural contexts, actual or potential risks, insurance limitations, and a wide variety of medical contingencies. In some cases, answers to crucial questions are not available to decision-makers. In other cases, crucial answers are disregarded by decision-makers as irrelevant. Further, clinical ethicists do not have the luxury of time and distance. And yet, in all these cases, decisions must be made.

None of these variables can be suspended or held fixed in the clinic as they can in the classroom. Even when certain stakeholders (such as physicians, patients, or family members) hold beliefs that strike us as irrational or contentious, no stakeholder's perspective is irrelevant. The student and the CEC therefore stand in a different relationship to the ethical problem or conflict. And so, as important as academic moral expertise is, it is not obvious that it is sufficient for helping clinicians, patients, and families make better medical decisions. At least some clinical knowledge and experience making decisions in complex, non-hypothetical cases are necessary.

A second type of moral expertise is the competence to reliably make good moral decisions for oneself. That is, one might be a moral expert if one were competent at living a morally good life. Describing this sort of competence, Robert Burch (1974) writes that ethicists are:

> … good at discerning what is right and wrong, and in doing something about it. … [They have][28] the capacity of discerning details and the knack of penetrating beneath the surface of convention and idle talk. [They have] insight into the ways one can twist or blunt moral issues, and [they have] competence in stiffening [their] wills so that [they do] not always take the easy way out (652).

We will call the ability to act as a morally good person, regularly or perhaps more often than not, *performative moral expertise*.[29] This type of expertise is more controversial. How might someone achieve it? How might they demonstrate it? For a contemporary defense of moral expertise as performative, see Hulsey and Hampson (2014).

Again, however, even if one could achieve performative moral expertise, this is not obviously the sort of expertise one would need to serve as an ethics consultant. One may perform well as a moral agent without being able to help others achieve that competence. Many of us have had classes with exceptional scholars who could not communicate clearly or effectively regarding their subject. The question of whether one can expertly navigate one's own moral path is independent of whether she can help others expertly navigate theirs.

This is not to say that the sort of moral experts we are interested in should (or would) have no ability to make good moral decisions for themselves. In fact, it

---

[28] The brackets in this paragraph replace masculine pronouns with plural pronouns.

[29] There is no widespread consensus on this terminology. Cheryl Noble (1982) might call this "moral wisdom," and Bruce Weinstein (1994) calls this "expertise in living a good life."

would be difficult to trust someone who did not act at least in the spirit of the advice they give (see Cholbi 2007 for an argument along these lines).[30] The point is simply that, the competence to advise someone in a type of decision-making does not entail that one regularly incorporates such advice into her own decision-making.[31]

And, of course, even if one were motivated to do so, one might not have achieved mastery in this for any number of reasons, including weakness of will, lack of opportunity to practice making such decisions for themselves, or simply being too close to some decisions to view them objectively. This latter concern is precisely why academics and researchers value processes like peer review. Plausibly, one aspect of becoming (and continuing to be) an expert involves regularly putting oneself in a position to receive feedback from other experts. But one can explain the problem of, say, confirmation bias, along with strategies for avoiding it, even if one does not avail himself of those strategies.

The type of expertise in which we are interested, then, is the competence to help others make better moral decisions. As noted, this likely entails some degree of academic and performative moral expertise, but it is conceptually distinct, consisting primarily in the authority to speak (in a sense yet to be explicated) on moral matters within the scope of a certain subject matter. This type of expert understands both sophisticated moral philosophy and the concrete complexities of a particular subject matter, whether business, or research, or medicine (and in some cases, subfields, like end-of-life care and pediatrics).

In medicine, these complexities might include the risks to a patient who will not likely comply with the post-op care for a medically indicated surgery, the risks associated with discharging a patient to an unstable environment, the undue influence of an overbearing family member, the paternalistic stance of a physician toward any patient with a psychiatric history, or the seemingly irrational fears of a patient. This type of moral expert can draw informed distinctions between the moral and non-moral features of a case, work with clinicians and families to weigh the conflicting and complex features against one another (recognizing that how much weight some considerations have depends on how much they give it), and form a moral judgment about a morally preferable plan of care. This moral judgment would then need to be conveyed in a manner and language useful to the clinicians, patients, or families who receive that judgment. To make her advice useful, the moral expert would need skills for translating rich moral notions into the practical realities of the context at

[30] Julia Driver offers a humorous example: "Satan could well be an example of a being with superior moral knowledge, but it would be unwise to defer to Satan's judgment on what to do. I might be confident in his ability to know, but not confident in his accurate transmission of that knowledge, because I view him to be deceitful" (2006: 630).

[31] Dale Miller (2005) notes that some, like J.S. Mill, hold that there is no "intrinsic connection between moral beliefs/knowledge and moral motivation" (a view known as moral externalism), which means that knowing the right thing to do does not entail that one will feel any motivation to act on that belief. "This implies that while those with greater moral expertise might be able to lay claim to greater moral knowledge, … it would be a mistake to assume that they are automatically more virtuous…than anyone else" (83).

hand. We will call this competence to speak authoritatively about moral matters, *practical moral expertise*.

To be sure, practical moral expertise may not be possible. Some, for instance, argue that there is not a single moral judgment that answers a moral question, but instead, only opportunities to learn more about others' meanings and perspectives.[32] But even if it is possible, it might be extremely rare. And even if it is not rare, clinical ethics consultants may not need *this sort* of expertise to contribute to high quality patient care (see Rasmussen 2011). Iltis and Sheehan (2016) argue that the field would benefit from dropping the term altogether. As will be illuminated in the chapters that follow, clinical ethics consultants (CECs) may be viewed as skilled professionals who can offer important contributions even if they are not, or should not be regarded as, practical moral experts (see Fiester in this volume). Nevertheless, if some CECs are plausibly moral experts, this has important social and professional implications for bioethics, clinical education, and a variety of clinical relationships.

1.3.3 Political vs. Epistemic Authority

Calling someone an expert has a number of social connotations. Someone who is an expert has *authority* to speak on certain matters. And the concept of authority is associated with everything from the right to coerce (as in cases of political or legal authority) to religious and civil practices of including and excluding ("I baptize you…"; "By the power vested in me, I pronounce you…"), to certain representative capacities (as in the case of ambassadors or executors). These examples also highlight that the notions of "authority" and "rights" are closely related: the *right* to do something, the *right* speak about a subject. These latter are so closely associated, in fact, that some philosophers have conflated them.

Hannah Arendt (1961) associated the notion of authority exclusively with the political relationship between leaders and citizens, in particular, the power to command. If the relationship holds, citizens obey a leader's commands without needing an explanation (the leader has authority). If a political figure must exert coercion to enforce a command, then authority has failed (the leader has no authority to command). If a political figure uses reason and citizens follow because of the reasons given, authority is not invoked in the command; the citizens follow because of what they understand, not because the leader has commanded (in this case, she says, "authority is in abeyance").

Robert Paul Wolff (1970) agrees with Arendt that authority must be distinguished from the ability to coerce (power) and persuasive argument, and concludes that authority "resides in persons; they possess it—if they do at all—by virtue of what

[32] Widdershoven and Molewijk (2010).

they are and not by virtue of what they command."[33] But Wolff then argues that authority—in this sense—is inconsistent with *autonomy*, which he says is the moral obligation of each of us to take responsibility for our own actions. The concept of authority entails, according to Wolff, that someone cannot accept or follow the testimony of any putative expert on the basis of their testimony alone without also relinquishing her autonomy: "by refusing to engage in moral deliberation, by accepting as final the commands of the others, he forfeits his autonomy."

Yet, while this might be an apt way to describe authority in certain political structures, it is not obviously what is implied when an expert speaks about her special subject matter, and to assume without argument that the implications of political authority apply *mutatis mutandis* to non-political conceptions is controversial, at best.

When, say, a geneticist, speaks about the relationship between alleles and gametes, there is a sense in which others *should* take her seriously. We *ought* to accept her testimony (at least *prima facie*); she has authority in her field. We might even say she has a *right to be listened to*. But the sense of "right" here is weaker than we find in political contexts. No one would worry that in accepting an expert's testimony they are forfeiting their autonomy or giving up their independence of judgment. We wouldn't call someone *immoral* who refused to accept a scientist's or say that they violated that scientist's rights. We might, of course, say such a person is *irrational*. But whether it is rational to listen to someone is distinct from whether they have a strong or political right to be heard.

Let us assume, then, that expert authority, at minimum, places a default *rational* demand on the recipients of that expert's testimony to accept that testimony, to defer one's judgment to the expert's. This type of authority is known as *epistemic authority*. To avoid confusion, then, we will not say an expert *has a right to be listened to*, but instead that an expert *should be taken seriously*, leaving the implications of "seriously" open to interpretation. The reason to leave it open at this stage is that there are wide-ranging disagreements about what epistemic expertise implies normatively, one of which we will briefly review briefly in the remainder of this subsection. But even on a fairly strong view of what it means to defer one's judgment, epistemic authority is distinct from political authority.

[33] 1970: 6. George Agich (1995: 274) calls this the "command-obedience" model of authority, which is grounded in political structures. He contrasts this with "social role authority," according to which someone accepts a person's testimony based on a set of complex, informal social relationships. For example, "a teacher does not order students, except when he behaves as a disciplinarian and then does so as a school official in charge of conduct. Teaching as such involves complex processes of communication that bind student and teacher into an authority relationship where teaching and learning occur. A scientist interacting with peers might rightly take their word on a particular scientific point over that of a layman. Such trust is based not simply on other scientists' power or position, though that might to some degree contribute to the initial acceptance, but also on their common commitment to methods of work and modes of demonstration" (276). In subsequent paragraphs, we call social role authority "epistemic authority."

1.3.4 Taking Experts Seriously

In considering how one should receive authoritative testimony, one might take a strong view, that one should defer completely to an expert, irrespective of any other reasons one might have. Alternatively, one might take a weak view, that an expert's testimony is one piece of evidence among several—albeit a rather weighty piece—that must be evaluated in light of one another.

Linda Zagzebski (2012) defends a version of the strong view: the obligation to take an epistemic authority seriously entails that one accepts the authority's testimony about a subject *in lieu of any other reasons* one might have about that claim.[34] According to Zagzebski, an expert's authority "preempts" or "replaces my other reasons relevant to believing *p* and is not simply added to them."[35] The strongest case for this view appeals to empirical evidence that people who defer to experts do better and have more true beliefs than people who do not.[36]

Thus, in listening to an expert, one is more likely to form a true belief than if he tried to evaluate the reasons for that belief himself, and the expert's judgment would most likely stand up to his own efforts to evaluate those reasons (in other words, he would still regard the expert as reliable after what Zagzebski calls "conscientious reflection").[37] If this right, then we couldn't do better than listen to authorities when they declaim on a subject. And if we couldn't do better, we have a normative reason to defer to their judgments, irrespective of our other reasons.

In saying that an authority's testimony replaces or preempts one's reasons, Zagzebski does not mean that one ignores the other reasons she may have, nor does she mean that one actively decides to let an authority's testimony override her other reasons. She means simply that an authority's testimony constitutes a sufficient and overriding reason to believe that testimony.

Jennifer Lackey (2018) identifies a number of problems for this strong view of epistemic authority, and we will briefly highlight two. One problem, Lackey argues, is that Zagzebski's account *"fails to provide the resources for rationally rejecting an authority's testimony when what is offered is obviously false or otherwise outrageous."*[38] Experts are not only fallible; they sometimes say patently false things. Lackey gives the example of a pastor who may be highly regarded as a moral

---

[34] Zagzebski draws heavily from Joseph Raz's (1986) account of authority, but for simplicity we will focus on Zagzebski here.

[35] Zagzebski, 107.

[36] In a suggestive study that Zagzebski cites by Mlodinow (2008), when animals discern that one choice is better a majority of the time, they choose that option every time. And thus, they choose the better choice most of the time; they are outcome-maximizers. Humans, on the other hand, are probability-matchers. If a choice is better about 75% of the time, humans will choose that option about 75% of the time, making it very likely that they will almost always choose the better option less than 75% of the time. (Zagzebski, 2012: 115)

[37] 110–111. By "conscientious reflection" Zagzebski means, "[u]sing our faculties to the best of our ability in order to get the truth" (2012: 48).

[38] (2018: 234).

expert but nevertheless makes a claim that women are morally inferior to men. If you are screening off other reasons, you should accept the pastor's testimony without question. And yet, even if you weren't a moral expert, you would likely have substantial reasons to challenge the claim.

A related concern Lackey raises is that "*it is unclear how the testimony of an authority can even strike one as clearly false or outrageous, given that all of one's other relevant evidence has been normatively screened off*."[39] To be sure, one can experience a certain amount of outrage at a claim; Zagzebski's account doesn't make claims on our psychological states. Nevertheless, if you have normative reasons to set aside all other evidence, then you have normative reasons to dismiss your outrage as misplaced or misguided. But if this is right, how strongly should we trust the initial evidence we used to adopt the expert in the first place? And to what evidence would we appeal if two experts disagreed?

In contrast to the strong view, Lackey offers a weak version of expertise that she calls the "expert-as-advisor" view.[40] Lackey argues that it is far more plausible to view experts as advisors rather than authorities in Zagzebski's strong sense. Unlike authorities, advisors offer guidance, that is, their testimony counts as evidence for believing something. Lackey says that an expert witness at a trial is a paradigmatic example of an expert-as-advisor:

> No one would tell the jurors that the testimony of a given expert is authoritative or provides preemptive reasons for belief. Indeed, jurors themselves would be superfluous in many ways if experts functioned authoritatively. Instead, competing expert testimony is often presented from both sides—the prosecution and the defense—with jurors needing to evaluate the full body of evidence in reaching a verdict. The experts here are, then, advising the jurors rather than dictating to them what they ought to believe.[41]

To keep our terminology consistent, we regard this type of expert competence as a type of authority, but a type weaker than that advocated by Zagzebski. This weaker, advisory view of epistemic authority takes seriously the fact that experts are not only fallible, but that they can make audaciously false claims. Further, it takes seriously the fact that experts disagree over claims in their own fields, even fields as highly revered as medicine and physics. And it takes seriously that the relevance of expertise is often contingent on a number of contextual and decisional factors, some of which might not be available to the expert. She even names CECs explicitly as a case of expert advising:

> An ethics consultant serving at a hospital will be effective largely by helping doctors, patients, and their families navigate through difficult medical decisions. Sure, her reliably offering true testimony is important, but equally important are her abilities to clearly explain the terrain, to listen attentively and receptively to the concerns and values of those around her, and to answer questions in a thoughtful and constructive way.[42]

---

[39] Ibid., p. 235, italics hers.

[40] Ibid., pp. 238ff.

[41] Ibid., p. 239.

[42] Ibid.

By maintaining that expert testimony is one source of evidence among many, according to Lackey, we have far more opportunities for forming and enhancing beliefs in a subject. "We can evaluate the arguments proffered on behalf of a particular view, we can assess how able the expert in question is at enhancing our understanding of the matter, we can determine how effective the expert is at being an advisor, and so on."[43]

Although Lackey's view has intuitive force, it does not undermine Zagzebski's empirical argument that we would do better, overall, if we simply deferred to experts. The question, then, would be whether CECs could exemplify this stronger type of authority. The debate over strong and weak epistemic authority is more complex than we can pursue here. But this brief segment highlights its relevance for moral expertise. If couched strongly, moral expertise is less plausible and likely rare. If couched weakly, moral expertise is more plausible but perhaps more difficult to identify in training and practice.

1.3.5 Situational Authority vs. Expert Authority

A related and no less crucial distinction is between the authority to speak about a particular claim in a particular context, which we call *situational epistemic authority*, and *expert epistemic authority*, which is the epistemic authority to speak about a range of claims in a subject matter. One might have epistemic authority in some cases without being an expert and vice versa. For example, if you don't know what time it is and you ask someone with a watch, you have good reasons to trust their testimony regarding the time even if that person is not an expert on watches or the concept of time. And the person with the watch is in a better position to justifiably believe what time it is than you are[44] even if *you are* an expert on time or watches.

This phenomenon is explained by a person's access to evidence relevant for believing a claim. This access is called *epistemic positioning*.[45] An epistemic position refers to the relationship someone has to evidence. If we are on different sides of a brick wall, we have access to different visual evidence. If there is a dog on your side, then, other things being equal, you are in a better position to judge that there is a dog there. Many factors mitigate your position, such as how tired you are, whether you are on certain medications, whether I have reliable testimony that there is a dog there or a video feed of the dog. When such factors don't render our epistemic positions equal, that is, when one person is in a *better* epistemic position than another, that person has an *epistemic advantage* over the other person.

---

[43] Ibid.

[44] Whether the person is in a better position to *know* the time (instead of merely *having a justified belief* about the time) is a more complicated question, leading to questions about the reliability of watches, the proper functioning of that person's watch, etc.

[45] See Elizabeth Fricker (2006).

In standard cases, experts are better positioned than non-experts with respect to evidence about claims in their subject matter. So, even though we have all had experience with time and watches, someone who is an expert about the concept of time has an epistemic advantage over the rest of us regarding the relation of time to space, the logic of temporal language, competing arguments regarding the structure of time, etc. This distinction becomes important in cases where non-experts have situational epistemic authority over experts with claims relevant to that expert's judgment. This can occur when some piece of information relevant for an expert's judgment is not directly accessible to her; it is contingent on someone else to provide it.

A prime example of a case where a non-expert has situational authority over an expert comes from the growing literature on *patient expertise*.[46] Patients, according to some, have insights into their medical conditions and personal values that physicians could not have, even in virtue of their medical expertise. And thus, patients have a certain degree of epistemic advantage over some claims regarding their treatment plans that physicians should take seriously.[47]

In the case of morality, too, we might recognize that sometimes people are in a better position than we are to know what is morally good, irrespective of whether they are experts. For example, parents tend to have situational epistemic authority over young children regarding basic moral heuristics—do not lie, cheat, steal, or cause pain—even if they cannot reliably discern when to do or not do those things in their complicated adult lives. Further, a minister or counselor might have some authoritative moral insight into a particular decision based on their experiences working with people who faced similar decisions. Further still, in circumstances where someone is emotionally overwhelmed or experiencing decisional fatigue, she might benefit from the judgment of someone else who is more distant from the situation and can therefore process all of the relevant considerations.

Expert authority, in contrast, refers to a more robust epistemic position. Experts do not simply have authority to speak about particular claims in particular contexts. Rather, they can speak authoritatively about a wide range of claims in a subject. They can answer questions about the terms, claims, and arguments in that subject, and they can use them to the satisfaction of others in that field. They can explain how those terms and claims came to be part of that subject, and they can apply those terms and claims to novel cases. To be sure, expertise comes in degrees, and advanced laypersons may engage with a subject in ways indistinguishable from new experts; nevertheless, the development of these traits is widely regarded as constituting one's epistemic expertise.

So, whereas most of us can add 237 to 458 without any trouble, an expert in mathematics can explain how to derive the rule that allows one to perform that func-

[46] See, for example Civan and Pratt (2007), Heldal and Tjora (2009), and Hartzler and Pratt (2011).

[47] It is controversial whether patient authority is plausibly regarded as "expertise." Given that any particular medical condition involves extensive subject matter outside the patient's competence, we have categorized this authority as situational with respect to evidence only the patient could have.

tion from Peano's axioms. Similarly, a practical moral expert, if there were one, could do more than simply state a moral aphorism, for example, that we should not violate a patient's autonomy interests. They could also explain why autonomy is morally valuable relative to other moral interests, identify when a decision would violate autonomy interests, and evaluate whether such a violation is nonetheless justified by competing interests.

1.3.6 Is Morality Unique?

A further relevant distinction is that between the content of ethics and the content of other subjects, such as mathematics or medicine. Could one have an expert-level grasp of the content of morality the way we assume one can have of calculus? As we saw in the discussion of authority, there is a long history of concern that morality is unique among subject matters. Whereas the empirical facts about nephrology or oncology are the special province of trained and certified people, we are all moral agents—we all have a sense of what is good and bad, right and wrong.

Immanuel Kant, for instance, argued that morality makes rational demands of us precisely because it is something to which we all have access. Kant argues that anyone who can reason can figure out what our moral obligations are in any given situation.[48] And if we all have equal access to it, it may seem unclear what rational demands one person could make of another person regarding her beliefs about that subject matter. This leads to the conclusion, as Giles Scofield puts it, that "either all are experts or none are" (1994: 420). We call this the *uniqueness problem* for moral expertise.

David Archard argues that universal access to moral content implies that *no one* is a moral expert. He argues as follows:

> A claim of moral expertise is a claim to command knowledge in respect of the making of normative judgments not commanded by others. But moral philosophers see themselves as required to construct moral theory on the foundations of common-sense morality. The latter is the set of moral maxims of which ordinary people have knowledge and of which they make use in their quotidian lives. These maxims comprise basic judgments of what is morally right and wrong. Thus by their own lights moral philosophers do not have command – in respect of the making of normative judgments – of knowledge lacked by nonphilosophers. Moral philosophers cannot, consistent with their own commitments to common-sense morality, claim moral expertise (123).

What Archard means by "common morality" is a technical matter (see Beachamp and Childress (2013) for a full analysis), but a rough distillation is "the minimal core set of ethical precepts that can be observed to be shared by all conscientious humans who seek to live their lives morally" (Archard, 124). And so, what ethicists study is precisely what all of us are already committed to. Archard takes this to

[48] His famous defense of this is in *Groundwork for the Metaphysics of Morals* (1785/1997).

imply they do not have a command of it that others do not, and thus—by his definition of expertise—no one is a moral expert.

Christopher Cowley argues, alternatively, that universal access implies that *everyone* is a moral expert, and thus, no one can effectively act as an expert: "*we are all ethical experts*, and so effectively none of us are [sic]" (2005: 276, emphasis his). Why might someone hold that universal expertise nullifies any social benefit of expertise? The presupposition seems to be that expert authority serves only non-experts. As Michael Cholbi puts it: "One's own expertise obviates the need to seek out other experts in the first place. Experts don't need the expertise of other experts" (2007: 324).

There are a number of important questions raised by these arguments. Is Archard's account of expertise plausible? Is it true that one's own expertise obviates the need for other experts? And are moral philosophers actually committed to respecting common morality, and, if so, whether this prevents them from being able to stand in a position of expertise relative to other moral agents?[49]

Another defense of the idea that morality is a distinct subject-matter—and thus that there can be no moral experts—is that, unlike other subjects, morality *can only be learned first-hand*, and, therefore, cannot be acquired through testimony. Charles Hendel (1958) puts it this way:

> [T]o allow of any possible role for authority in the moral life of [people] is to take away its properly ethical character, no matter whether the authority be divine or regal, because morality consists in actions of an individual's own authentic choice, choice in the light of [their] own knowledge, appraisal, and conviction, without any external inducements or sanctions (7).

The idea seems to be that, in order for a moral decision to be *authentic*, it cannot be supplemented with moral advice. One must understand and evaluate all the moral reasons relevant to a decision for oneself.

Christopher Cowley (2005) also offers a more recent version of this argument. He contends that, even if there were moral experts, moral decisions are personal in a way that other decisions are not. Even if you were to receive competent moral advice, "[you] cannot abdicate the decision to someone else in a way that would shift responsibility and blame onto that person, in a way that [you] can to the dentist or cartographer" (278). In cases of moral decision-making, an advisor must give you reasons, and then you must decide whether to accept those reasons are your own. If you do, then they are no longer the advisor's reasons.

Cowley defends the distinctness of morality by noting that much of the evidence for our moral beliefs is "direct and intuitive, without any plausible validating reasons that could be given." Further, the evidence is not only direct, it is often emotional, and therefore, "nonrational." Consider that, if you don't already experience a certain revulsion to an action, it is unlikely that you could be convinced that it is

---

[49] See John-Stewart Gordon (2014) for an argument that moral philosophers are not, contra Archard, committed to respecting common morality. And Dale Miller (2005) argues that J. S. Mill views the role of moral philosophers as going beyond common morality, critiquing and improving it.

wrong, even by an expert (277–278). Robert Hopkins (2007) calls this objection the *Unusability Argument* against moral testimony. The idea is that, in order to have a right to a moral belief—unlike a belief about physics or mathematics—one must meet a sufficiency requirement: "having the right to a moral belief requires one to grasp the moral grounds for it" (630). If one must understand the relevant grounds for oneself (feel a "natural repugnance," as Cowley puts it), and one cannot acquire that understanding from others, then moral testimony is *unusable*.

Interestingly, even if Hendel and Cowley are right about the distinctness of morality, it is not clear that this diminishes or undermines the role of moral experts. If a moral expert provides you with reasons you had not previously considered, then even if those reasons become your own, you may be better positioned to make good moral decisions. Elizabeth Anscombe (1981) and Karen Jones (1999) addressed the uniqueness problem—and its corollary, the unusability argument—along these lines. Anscombe contends that there is at least one sense in which morality is not unique: "[O]nly a foolish person thinks that his own conscience is the last word… about what to do. … [A]ny reasonable man knows that what one has conscientiously decided on may later conscientiously regret" (46). And Jones writes that, "just as borrowing scientific knowledge can enhance our capacity to discover truths about the nonmoral world, borrowing moral knowledge can enhance our capacity to understand the world of value" (56).

Further, it might be that moral expertise does not require or presume to transmit distinctly moral knowledge by testimony. If the primary expertise of CECs is in bioethical mediation (see Dubler and Liebman, 2011), the primary content of their testimony may not be moral. Or, if the primary testimony of CECs is an "all things considered judgment" regarding the complex of moral and non-moral features of a medical decision (see Rasmussen 2016), then one might not need distinctly moral knowledge to provide epistemically authoritative advice.

1.3.7 What Is Moral Expertise About?

A final important debate is whether moral expertise requires that experts have *objective moral knowledge* or simply a *facility with moral reasoning*. As in the early cases of Broad (1952) and Ryle (1958), many critiques of moral expertise attack the idea that philosophers have privileged access to moral knowledge. In 2012, Martin Hoffman concluded that, "The scope of ethics expertise is limited to working out and explicating the logical structure and the empirical conditions of intricate moral problems. But it does not consist in genuine moral expertise, which allows the expert to have an epistemic access to esoteric moral knowledge" (305). And as recently at 2014, Edward J. Bergmann and Autumn Fiester wrote that "No ethicist possesses the moral wisdom to objectively rank the values at play in [conflicts where parties hold incommensurate values]" (703).

In response, many defenders of practical moral expertise have argued that ethicists do not, and need not, make any strong claim about having moral knowledge (see Yoder 1998). What ethicists are really experts about, they purport, is moral *reasoning*.[50] In 1972, for example, Peter Singer made a case for the idea that ethicists were, at minimum: "more than ordinarily competent in argument and the detection of invalid inferences," able to understand and clarify "moral concepts and the logic of moral argument," and able to dedicate her full-time professional life to the study of moral issues (189). Singer concluded that a person with this background and skill set "may reasonably be expected to reach a soundly based conclusion more often than someone who is unfamiliar with moral concepts and moral arguments and has little time" (189).

What does Singer mean by a "soundly based conclusion"? He may be referring to deductive soundness, in which case, he is committed to the idea that moral experts have objective moral knowledge. Alternatively, he may be using sound in a more colloquial way, referring to conclusions that are well-reasoned and carefully inferred. It seems reasonable to think someone could draw a conclusion in the latter sense without arriving at objectively true beliefs. For example, when Olympic judges assign scores to an athletic performance, the scores are not "objectively true" (that's not the sort of thing Olympic scores are), but they are, nevertheless, based upon the judge's expertise in evaluating a variety of subtly technical features.

Of course, when it comes to subjects that are comprised of claims that are true or false, reasoning alone is not the only relevant goal. For instance, someone could devote extensive time to understanding the concepts and research practices of creation science and could, thereby, reach a soundly based conclusion (in some sense of "soundly based") within that subject. But such competence doesn't strike us as particularly useful for scientific decision-making. R. G. Frey (1978) criticized Singer along these lines, arguing that the whole point of reasoning well is to obtain more true beliefs: "[Q]uestions of moral expertise are not concerned with the skills that go to comprise the critical examination of moral issues but with the outcome of the use of those skills in terms of particular moral judgments" (48–49). This concern has clear affinities with the unusability argument against moral testimony: Why should ethics experts be content with their competence in moral reasoning if that reasoning is not arriving at conclusions that are likely to be true? Why would they offer testimony in the first place, and who should accept it?[51]

On the other hand, if they *are* arriving at conclusions that are likely to be true, why shy away from saying that experts have objective knowledge of morality? Can moral philosophers make such a claim? Bernward Gesang (2010) defends a qualified "yes." Gesang distinguishes stronger and weaker versions of competence in moral reasoning. He begins by distinguishing the ways experience in used in different fields. In science, for example, scientific theories must be adapted to fit

[50] See also David Adams, this volume (Chap. 12).

[51] Martin Hoffman (2012) draws a distinction between "ethics expertise" and "genuine moral expertise," arguing that, while moral philosophers might be competent to apply moral concepts to complex situations, it is a mistake to think that it gives them privileged access to "esoteric moral knowledge (304–306). This suggests that ethics experts might be trusted for their epistemic virtues even if they cannot dispense moral truths.

observation. "Moral experience," however, "is influenced by the concrete subject, its preferences, traditions, and so on" (156). The idea is that the role of theory and experience is flipped in the case of morality: theory helps to shape our moral judgments about experience. The opposite is true in science, where theory is shaped by experience and experimental results. This difference between science and ethics is supported by the fact that moral experience is not "intersubjectively reproducible," rendering it weaker than scientific experience.[52]

This does not, however, mean that moral philosophers do not have *any* epistemic advantage over anyone else. Gesang admits that moral philosophers will likely have beliefs that are "better founded and more likely to be right" than non-philosophers. Nevertheless, because their moral intuitions "are not intersubjectively reproducible" they cannot make strong claims to expertise like physicists. They cannot, for example, say to anyone "Your opinion is false" the way a physicist can" (158).

If Gesang is right, the fact that moral judgments are weaker than scientific judgments means that ethics consultants could not be experts in the same way that scientists can. They can, however, be "semi-experts." Moreover, this might help explain the reticence to regard CECs as moral experts, thus suggesting new ways of explaining and acknowledging the role that CECs should have in medical decision-making. Gesang's conclusion, however, may render CECs' judgments too weak. If a patient's family member were to say, "Cutting off limbs is never ethically permissible under *any* circumstances," it would be strange to think an ethics consultant could not justifiably respond (with some tact), "False!" Nevertheless, much work is needed on identifying precisely what sort of judgments CECs can offer and which are necessary for practical moral expertise.

1.4 Recent Developments

In 2005, Lisa Rasmussen published a collection of essays addressing a number of these topics.[53] Contributors to that volume pushed the debate forward in a number of ways, clarifying historical perspectives, exploring the public implications of moral expertise, and rendering more palatable the idea that ethics consultants can

[52] Gesang's conclusion depends on adopting what he calls the "coherence theory of moral justification," which he contrasts with the "deductive theory." We won't rehearse these details here but will simply note that whether one adopts the coherence theory affects the plausibility of Gesang's conclusion. See Cowley (2012) for a critique of Gesang.

[53] Rasmussen (2005). Rasmussen (2011) distinguishes between "ethics expertise" and "moral expertise" as a heuristic to help distinguish the sorts of epistemic authority CECs might possess. Though there is no widely accepted account of the sorts of recommendations that ethicists can make, one may think that CECs can make decisive recommendations that effectively and objectively resolve moral dispute (what she calls "moral expertise"). She argues that this is not the sort of expertise a CEC could plausibly have, and argues, instead, that they have "ethics expertise," the authority to offer "non-normatively binding recommendations grounded in a pervasive ethos or practice within a particular context" (650).

speak as authorities in a variety of roles. In addition to the need to follow-up on these conversations, a number of important shifts have occurred since Rasmussen's volume. We will briefly highlight two.

The American Society for Bioethics and Humanities (ASBH) is the largest professional organization for CECs in the United States.[54] At the time of Rasmussen's anthology, the ASBH had seemingly rejected the idea that CECs are moral experts. In the first edition of their *Core Competencies for Health Care Ethics Consultation* they expressly cautioned against the idea that CECs "have special standing in ethical decision making" for fear of "displacing providers and patients as the primary moral decision makers at the bedside" (1998: 31). The organization was especially concerned to prevent abuses in clinical ethics practice, such as CECs' imposing their own values on clinicians, patients, or families.

In 2011, the ASBH updated the *Core Competencies*, which now expresses an openness to ethicists' "sharing expertise": "…a consultant may be asked to share his or her ethics knowledge and expertise as it relates to a broad ethics topic, such as terminal sedation as a palliative intervention at the institution." This version also states that CECs can be warranted in making singular recommendations regarding whether a plan of care is unethical or whether only one course of action is ethically justified (8).

Though this language admits of a great deal of latitude in interpretation, it nevertheless represents a sea change in the way CECs view themselves and present themselves to professional medicine (at least in the United States). Assessing whether such a change is warranted, and if so, how the language here is plausibly interpreted are issues addressed in this volume.

This volume also includes perspectives from those working outside the U.S. context, where ethics consultation has evolved differently, though with many of the same aims. In the U.S., consultation models tend to take one of three forms: a multidisciplinary committee, a team of one or two consultants, or an individual consultant model. Members of these services typically include physicians, nurses, social workers, case managers, risk managers, legal counsel, spiritual care, and ethicists. Members may or may not have formal training in ethics, though are regularly pushes from within the ASBH to formalize a credentialing process for consultants. Regardless of the type of model, the consultation service tends to be requested while there are active questions regarding a patient's care, that is, while a patient is admitted and care decisions are being made.[55] And in all three models, an opinion regarding those decisions is offered, whether in the form of a recommendation or set of morally reasonable options, and that opinion is entered into the patient's medical record.

---

[54] There are some organizations and professional groups for clinical ethics consultants in other geographic regions. For example, the Canadian Bioethics Society holds an annual conference and offers some resources for ethics consultants.

[55] There are exceptions to this. For instance, many committees also engage in reviews of previous cases for purposes of education and quality improvement.

In Europe, in contrast, the common approach is for a multi-disciplinary ethics committee to review the ethical questions that had to be made in past cases for purposes of education and reflection. In the Netherlands, for example, ethicists facilitate Moral Case Deliberations, which employ a conversation method either to arrive at a well-considered decision or to enhance the moral competencies of professionals. The most notable divergence from typical U.S. models is that during moral case deliberations, ethics consultants do not offer advice or attempt to morally justify any conclusion.[56] This reflects an understanding of moral expertise and its scope that is very different from what is suggested by the ASBH.[57]

A second important development is the rise in concerns about disagreement, especially peer disagreement, in moral judgments.[58] While at least some forms of moral expertise are more widely accepted, ethicists disagree (sometimes strongly) over common questions in clinical ethics, for example, whether abortion should be recommended as a treatment option, whether and how to employ the doctrine of double effect, and whether unilateral DNR (physician-ordered) is ever morally preferable.

This problem of disagreement is based on the assumption that experts should generally agree about claims in their fields. If physicists were widely divided over the notion of time or gravity or the speed of light, we would not have strong reasons for accepting their testimony. And even worse, if we have equally good independent reasons to trust two experts, one of whom says *p is true* and the other of whom says *p is false*, we have no reason to trust either of them regarding *p*.[59]

In addition to being a problem in its own right, the argument from disagreement exacerbates what was is known as the *credentials problem* for moral expertise (see LaBarge 2005 and Cholbi 2007 for a fuller discussion). If moral experts are to be of any use, those who need them must be able to identify them. On one hand, other experts presumably know who is an expert in their field—they were likely trained by them or by studying their work. On the other hand, those who are not experts, by virtue of their lack of competence in the field, are in no position to evaluate the reliability of a putative expert's testimony. That is, non-experts cannot rely on the content of a putative expert's testimony alone to evaluate their expertise; they are not competent to evaluate that content for themselves. Instead, they must rely on other markers, whether those be educational credentials, professional credentials, rhetorical competency, track record, etc. While this is not particularly controversial for professions like attorney or tax accountant, when it comes to morality, it is unclear what other markers might be relevant. Degrees in moral philosophy are aimed at academic moral expertise, not practical moral expertise. Rhetorical competency too

[56] Cf. Stolper, et al. (2010).

[57] Cf. Widdershoven and Molewijk (2010); Herrmann (2010).

[58] For a brief explanation of how disagreement can affect beliefs about morality generally, see Jonathan Matheson (2015: 4–5).

[59] Ben Cross (2016) defends this strong version of the argument from disagreement, concluding that the fact that reputable moral philosophers disagree about certain moral claims implies that we should place no degree of trust in either of them regarding those claims.

easily admits of charlatans (even in the medical field), and there is currently a great deal of controversy surrounding the credentialing of CECs. And the notion of a tracking a CEC's outcomes is controversial because, as Scott LaBarge puts it, "the ultimate goal of moral deliberation and action is itself one of the very issues that is at the heart of ethical disputes" (25).

This is where the argument from disagreement comes looming in. If CECs disagree about what is morally preferable, even what a morally preferable outcome is, then it would seem that non-experts have little hope of effectively identifying moral experts, and thus, have little reason to rely on anyone claiming to be one.

1.5 Moving the Debate Forward: The Structure of This Volume

These developments reveal the importance of ongoing scholarship on moral expertise, both for theoretical bioethics and the growing field of clinical ethics consultation. The authors in this volume address a number these debates and more, discussing questions such as: Which type of expertise is most plausible for CECs, if any? What is the content of moral expertise? What degree of authority does moral expertise confer? How should we address problem of disagreement? And can the credentials problem be solved? This collection attempts to move the debate forward by addressing these and related questions. In part I, contributors further the theoretical dimensions of these debates.

Dennis Arjo opens the collection by framing some possible approaches to expertise and then using these to develop a rich concept of moral expertise. We might think of expertise in terms of "knowledge that," which means an expert can draw on a greater fund of true claims about a subject matter than others. Or we might think of expertise in terms of "knowledge how," which means an expert can perform a certain range of tasks better than others. Orthogonally, expertise might involve largely individual pursuits, that do not depend on one's relationship to a community, or it might involve highly social pursuits, that involve enmeshing aspects of one's self in a community of practice. Arjo argues that moral knowledge, given its rich, social complexities, involves both knowledge that and knowledge how of "the norms, including the moral norms, embedded in or structuring the various practices that make for communal life." Arjo elucidates this view of moral knowledge through the Confucian virtue *li*, or ritual propriety, which is also a deeply culturally embedded virtue. If Arjo is right about moral knowledge, then the moral expert is someone who can employ this rich understanding of cultural norms to do or give advice that help societies and those within them flourish.

In Chap. 3, Christopher Meyers argues that moral expertise is not only possible but that there are good reasons to believe that attaining it is fairly straightforward. While some scholars express skepticism, Meyers attributes this largely to a misunderstanding of the sorts of moral advice needed by non-experts. While we have a fairly good grasp of basic moral truths, the difficulties come in determining how they should be applied in complex situations, and how to give advice to others, who may hold different values than we do. Meyers defends four key traits for developing ethics expertise, emphasizing humility as the core character trait and informal reasoning as the core skill.

In Chap. 4, Michael Cholbi argues that, regardless of whether moral expertise exists, and regardless of who might have it, both having moral expertise and dispensing expert moral advice requires knowledge of the correct or best moral theory. Cholbi argues that the need for expertise is based on an asymmetry in knowledge in a subject matter. Those who know less turn to those who know more. Knowing more in ethics means one has true moral beliefs in a non-accidental way. And this is how theory informs moral expertise. But how thoroughly must one grasp moral theory to be an expert? Cholbi considers Dien Ho's (2016) argument that no theoretical *knowledge* is necessary. Ho draws and analogy between science and morality, arguing that, even when scientists cannot resolve fundamental disagreements, they make progress through processes of confirmation and disconfirmation. Scientific theories are tweaked to fit with observable data. Cholbi objects that this is not the case with moral theories. The observable data with which we are working in ethics are our first-order moral judgments, which themselves need to be explained and justified by the theory. Scientific theories gain legitimacy from predictive power, but ethicists need more than that. They need more than simply an agreement between first-order judgments and theories to legitimate knowledge claims and advisory authority; they need a good reason to believe the theory is true.

Even if moral expertise is possible, even plausible, a central problem facing those who might aspire to it is the widespread disagreement among those who specialize in ethics generally, and in sub-fields such as clinical ethics, specifically. While every field has its share of disagreement, ethics seems notorious. In Chap. 5, Jonathan Matheson, Scott McElreath, and Nathan Nobis defend a set of conditions under which we should defer to others on questions of morality. They consider some standard arguments against deferring in cases of morality and conclude that the most promising is the idea that a centrally valuable aspect of moral knowledge is understanding for oneself. If one simply accepts another's testimony about morality, one doesn't acquire the benefit of knowing for oneself. Nevertheless, there good epistemic reasons for thinking that, in some cases, we are justified in relying solely on moral testimony, and thus not understanding for ourselves, because in those cases, we are still more likely to form a true moral belief than if we didn't defer. This discussion then suggests a framework for evaluating disagreement among moral experts, even when we cannot understand the moral complexities at stake.

A further concern related to disagreement is precisely how non-moral-experts might identify moral experts if there are any, explained above as the credentials problem. In Chap. 6, Eric Vogelstein attempts to solve the credentials problem, arguing that we can draw an analogy between moral experts and experts in other abstract fields, such as mathematics. To know whether someone is an expert in calculus, we don't have to understand calculus. We can recognize that someone who studies calculus has a grasp of valid reasoning strategies—truth-conducive methods—that result in true mathematical beliefs. Similarly, we know how to recognize good moral reasoning, even if we aren't sure what to do in every morally complex case. And thus, if someone demonstrates facility with good moral reasoning, then we have a reason to trust their judgment in morally complex cases.

Central to identifying moral experts, should there be any, is identifying relevant skills and traits. Vogelstein has already shown us that, if someone is not adept at moral reasoning, we have a reason to think they are not a moral expert. Similarly, in Chap. 7, Marcela Herdova contends that if someone demonstrates the character trait of hypocrisy, we have a good reason to doubt their moral expertise. She argues, first, if one is a moral authority, then one is a trustworthy source of reliable moral advice. And second, if one is a trustworthy source of reliable moral advice, then one is not a hypocrite. She concludes that being a moral authority requires not only knowing relevant moral facts, but also acting morally, so that, if one is actually a moral authority, then one rarely acts against one's own good advice to others.

The final two chapters in Part I address questions about what sort of advice moral experts might give if there were any we could identify. Through the lens of classic Chinese philosophy, Ai YUAN considers different ways in which a moral expert can and should provide advice in cases of perceived or actual dilemma. YUAN argues that moral experts have to weigh reasons in favor of one course over another, considering the agent's own moral perspective in the process. But beyond this, YUAN suggests that moral experts should also help agents develop the right sorts of moral commitments and equanimity in attitude, taking their expertise beyond isolated decision-points to the agent's moral life as a whole.

In the final chapter of this section, Nancy Potter provides a new, distinctive challenge to the notion of moral expertise. Implementing feminist and decolonialist frameworks, Potter argues against standard views of knowers and knowledge that tend to underlie claims to moral expertise. She argues that ethics and epistemology are co-constitutive of each other, and she suggests that bioethicists can make important contributions in the clinic by making it possible for disempowered persons to testify to their interests and needs.

In part II, contributors enhance discussions of moral expertise at it applies directly to clinical ethics consultation. Johan Bester starts this section by identifying the sorts of skills and traits jointly sufficient for a CEC. He argues that our concept of clinical ethics expertise should be an extension of the moral foundations of medical practice, which is steeped in relationships with vulnerable populations. Whatever skills and traits CECs develop should be cultivated in light of that understanding. And one of the primary difficulties with bringing moral theory directly to the bedside is the prevalence of what appear to us as moral dilemmas—decisions with no good outcome. Drawing on Rosalind Hursthouse, demonstrates a distinction between tragic and non-tragic moral dilemmas using two clinical cases. A central function of ethics expertise is to be able to help others recognize such distinctions and then respond appropriately. Bester develops a model of training clinical ethics consultants that has two central components: analyzing the moral feature of a decision, and weighing those features to arrive at a justifiable response. While moral theory is ineliminable from this process, it plays a narrower role than has traditionally been claimed for moral expertise.

Geert Craenen and Jeffrey Byrnes argue in Chap. 11 that, typically, the primary focus of training in clinical ethics consultation is on theoretical and practical skills. While they agree that these are necessary for expertise in ethics consulting, they contend they are not sufficient. Ethics consultation is no different from consulting services provided by other disciplines in health care, which have longstanding professional practice standards and methods for evaluating care. In addition to the skills identified by the ASBH, they contend that clinical ethics consultants would benefit from a personalist approach to professionalism, which encourages the development of character traits intrinsic to the profession of clinical ethics consulting. They argue that this could be achieved in clinical ethics through apprenticeships modeled after other medical professions. Immersion in clinical life, they argue, can meet this essential criterion for becoming an expert ethics consultant.

In Chap. 12, David Adams highlights a fundamental tension in discussions of moral expertise in clinical ethics consultation. While CECs are often instructed not to impose their opinions on others, they are also expected to provide expert advice. He contends that the type of moral expertise described in the much of the literature is too limiting for CECs and does not really constitute *ethics* expertise. Echoing Meyers and Vogelstein, Adams defends a performative notion of expertise, specifically with respect to the type of reasoning that CECs can offer in complex cases. Rather than identifying ethics experts with respect to their grasp of conventional norms, as the ASBH sometimes suggests, and rather than focusing on true moral beliefs, as some philosophers suggest, ethics expertise should be regarded primarily as a type of reasoning expertise.

Building on the discussions of Craenen, Byrnes, and Adams, Matthew Butkus argues that medical ethics expertise cannot be gained from the armchair. He makes the case that this type of expertise requires on-the-ground and on-going clinical experience. He provides three comparative case analyses to show that merely studying theories or reading about cases or thought experiments will prove inadequate for the sort of expertise that clinical ethics consultants need.

In Chap. 14, Autumn Fiester argues that, while clinical ethics consultants may legitimately claim a very particular and relatively narrow form of ethics expertise, it is not expertise that would justify them in settling normative conflicts or ambiguous values. Part of training as a CEC involves learning strategies for guarding against imposing their own values when consulting in the clinic. And given the importance of values pluralism in American medicine, part of the expertise of the ethics consultant should include helping the medical team avoid values imposition, as well.

Following Fiester, Jennifer Flynn analyzes the accusation of moralism that many clinical ethics consultants face, inside and outside the hospital. Although moralism is today generally viewed negatively, Flynn suggests that moralistic teaching can be appropriate for the properly self-aware and virtuous person. She argues that some perceptions of moralism might be inevitable for CECs, but these perceptions can be mitigated if the CEC is appropriately knowledgeable and skilled in their role. These areas of knowledge and skill are essential to the ethics expertise of CECs, which means that demonstrating their expertise can both lead to accusations of moralism but also mitigate those accusations. Flynn further argues that some degree of moralism is actually required by the ethics facilitation model in CEC, and this is not, in itself, problematic.

In the penultimate chapter, Salla Saxén takes a more relational approach to clinical ethics expertise. The clinical context in a pluralistic society creates a tension for the ethics consultant. On one hand, the CEC must respect the diversity of values held and lived in a society; on the other hand, the CEC is called on to give advice or make recommendation a manner that implies a certain authority with moral concepts. But, of course, value pluralism and moral authority are mutually exclusive concepts. Nevertheless, recognizing this tension suggests an opportunity for CECs. Rather than attempting to escape or resolve the conflict as if it were contentious, the CEC might see it as a paradox that she can use to create a space for genuine moral dialogue. One way of attempting to escape the paradox, for example, is to model ethics consulting as "consensus-building." But Saxén notes that this only resolves the conflict in favor of one side:

> "the consensus-rhetoric obliterates the notion of pluralism, as it silences *the struggle* of values that can be argued to be the very condition of pluralism." The expertise of a CEC, then, can be seen as a competence at balancing power structures to create a space where the complexities of a decision can be expressed and shared protected from pressures to resolve them.

Rounding out the collection is an essay from experienced clinical ethics consultant Evan DeRenzo. DeRenzo has served in the field of clinical ethics for twenty-nine years, her first six as a Senior Staff Fellow in Bioethics at the NIH (National Institutes of Health), and afterward as a clinical ethics consultant in what is now the John J. Lynch, M.D. Center for Ethics at MedStar Washington Hospital Center, a 925-bed acute care hospital. DeRenzo draws an analogy between training as a clinical ethicist and training in other medical specialties, defending the practice of shadowing as an ineliminable part of developing the virtues necessary for clinical ethics expertise. She argues that shadowing experienced clinical ethicists on rounds, committees, and consultations allows the novice to mimic the character traits necessary for informed and empathic moral insight.

References and Further Reading

Agich, George J. 1995. Authority in ethics consultation. *Journal of Law, Medicine & Ethics* 23: 273–283.

American Society for Bioethics and Humanities. 1998. *Core competences for health care ethics consultation*. 1st ed. IL: Glenview.

———. 2011. *Core competences for health care ethics consultation*. 2nd ed. IL: Glenview.

———. 2014. *Code of ethics and professional responsibilities for healthcare ethics consultants*. IL: Glenview http://www.asbh.org/publications/content/asbh-reader.html.

Anscombe, Elizabeth. 1981. Authority in morals. In *Faith in a hard ground*, ed. Mary Geach and Luke Gormally, 92–100. Charlottesville: Imprint Academic, 2008.

Archard, David. 2011. Why moral philosophers are not and should not be considered moral experts. *Bioethics* 25 (3): 119–127.

Arendt, Hannah. 1961. What is authority? In *Between past and future: Six exercises in political thought*, 91–142. New York: Viking Press.

Aulisio, Mark P. 2003. Meeting the need: Ethics consultation in health care today. In *Ethics consultation: From theory to practice*, ed. Mark P. Aulisio, Robert M. Arnold, and Stuart J. Younger, 3–22. Baltimore: The Johns Hopkins University Press.

Ayer, Alfred J. 1954. *Philosophical essays*. London: Macmillan.

Beachamp, Tom, and James Childress. 2013. *Principles of biomedical ethics*. 7th ed. New York: Oxford University Press.

Bergman, Edward J., and Autumn Fiester. 2014. The future of clinical ethics education: Value pluralism, Communication, and mediation. In *The future of bioethics, International dialogues*, ed. A. Akabayaski, 703–711. Oxford University Press.

Broad, C.D. 1952. *Ethics and the history of philosophy*. London: Routledge.

Brummett, Abram, and Christopher J. Ostertag. 2017. *Two troubling trends in the coversation over whether clinical ethics consultants have ethics expertise, HEC Forum*. Springer. https://doi.org/10.1007/s10730-017-9321-8.

Burch, Robert W. 1974. Are there moral experts? *The Monist* 58 (4): 646–658.

Caplan, Arthur. 1989. Moral experts and moral expertise: Do either exist? In *Clinical ethics: Theory and practice*, ed. Barry Hoffmaster, Benjamin Freedman, and Gwen Fraser, 59–87. Clifton: Humana Press.

Cholbi, Michael. 2007. Moral expertise and the credentials problem. *Ethical Theory and Moral Practice* 10: 323–334.

Civan, Andrea, and Wanda Pratt. 2007. Threading together patient expertise. *AMIA Symposium Proceedings*: 140–144.

Coady, David. 2012. *What to believe now: Applying epistemology to contemporary issues*. Malden: Wiley-Blackwell.

Collins, Harry, and Robert Evans. 2007. *Rethinking expertise*. Chicago: University of Chicago Press.

Cowley, Christopher. 2005. A new rejection of moral expertise. *Medicine, Health Care and Philosophy* 8: 273–279.

———. 2012. Expertise, Wisdom, and moral philosophers: A response to Gesang. *Bioethics* 26 (6): 337–342.

Cross, Ben. 2016. Moral philosophy, Moral expertise, and the argument from disagreement. *Bioethics* 30 (3): 188–194.

Crosthwaite, Jan. 1995. Moral expertise: A problem in the professional ethics of professional ethicists. *Bioethics* 9 (5): 361–379.

DeRenzo, Evan. 1994. Providing clinical ethics consultation. *HEC Forum* 6 (6): 384–389.

Döries, A., and K. Hespe-Jungesblut. 2007. Bundesweite Umfrage zur Implementierung Klinischer Ethikberatung in Krankenhäusern. *Ethik in der Medizin* 19: 148–156.

Dreyfus, Hubert and Stuart Dreyfus. 1980. *A five-stage model of the mental activities involved in directed skill acquisition*. Operations research center report.

———. 1986. *Mind over machine: The power of human intuitive expertise in the era of the computer*. New York: Free Press.

Driver, Julia. 2006. Autonomy and the asymmetry problem for moral expertise. *Philosophical Studies* 128 (3): 619–644.

———. 2013. Moral expertise: Judgment, Practice, and analysis. *Social Philosophy and Policy* 30 (1–2): 280–296.

Dubler, Nancy Neveloff, and Carol B. Liebman. 2011. *Bioethics mediation: A guide to shaping shared solutions, revised and expanded*. Nashville: Vanderbilt University Press.

Fiester, Autumn. 2015. Teaching Nonauthoritarian clinical ethics: Using an inventory of bioethical positions. *Hastings Center Report* 45 (2): 20–26.

Fox, E., S. Myers, and R.A. Pearlman. 2007. Ethics consultation in United States Hospitals: A National Survey. *American Journal of Bioethics* 7 (2): 13–25.

Frey, R.G. 1978. Moral experts. *Personalist* 59: 47–52.

Fricker, Elizabeth. 2006. Testimony and epistemic autonomy. In *The epistemology of testimony*, ed. Jennifer Lackey and Ernest Sosa, 225–250. Oxford.

Gesang, Bernward. 2010. Are moral philosophers moral experts? *Bioethics* 24 (4): 153–159.

Goldman, Alvin I. 2001. Experts: Which ones should you trust? *Philosophy and Phenomenological Research* LXIII (1): 85–110.

Gordon, John-Stewart. 2014. Moral philosophers are moral experts! A reply to David Archard. *Bioethics* 28 (4): 203–206.

Hartzler, Andrea, and Wanda Pratt. 2011. Managing the personal side of health: How patient expertise differs from the expertise of clinicians. *Journal of Medical Internet Research* 13 (3). https://doi.org/10.2196/jmir.1728.

Heldal, Frode, and Aksel Tjora. 2009. Making sense of patient expertise. *Social Theory and Health* 7: 1–19.

Hendel, Charles. 1958. *The absurdity of Christianity, and other essays*. New York: Liberal Arts Press.

Herrmann, Beate. 2010. What does the ethical expertise of a moral philosopher involve in clinical ethics consultancy. In *Clinical ethics consultation: Theories and methods, Implementation, Evaluation*, ed. Jan Schildmann, John-Steward Gordon, and Jochen Vollmann, 107–117. Abingdon: Ashgate.

Hester, Micah D., and Toby Shonfeld. 2012. *Guidance for healthcare ethics*. Cambridge: Cambridge University Press.

Ho, Dien. 2016. Keeping it ethically real. *Journal of Medicine and Philosophy* 41 (4): 369–383.

Hoffman, Martin. 2012. How to identify moral experts? An application of Goldman's criteria for expert identification to the domain of morality. *Analyse & Kritik* 34 (2): 299–313.

Hooker, Brad. 1998. Moral expertise. In *Routledge encyclopedia of philosophy*, ed. E. Craig, 509–511. London: Routledge.

Hopkins, Robert. 2007. What is wrong with moral testimony. *Philosophy and Phenomenological Research* LXXIV (3): 611–634.

Hulsey, Timothy L., and Peter J. Hamson. 2014. Moral expertise. *New Ideas in Psychology.* 34 (1): 1–11.

Iltis, Ana S., and Lisa Rasmussen. 2016. The 'Ethics' expertise in clinical ethics consultation. *Journal of Medicine and Philosophy* 41 (4): 363–368.

Iltis, Ana S., and Mark Sheehan. 2016. Expertise, ethics expertise, and clinical ethics consultation: Achieving terminological clarity. *Journal of Medicine and Philosophy* 41 (4): 416–433.

Jones, Karen. 1999. Second-hand moral knowledge. *The Journal of Philosophy* XCVI (2): 55–78.

Jones, Karen, and François Schroeter. 2012. Moral Expertise. *Analyse & Kritik* 34 (2): 217–230.

Jonsen, Albert R., Mark Siegler, and William J. Winslade. 2010. *Clinical ethics: A practical approach to ethical decisions in clinical medicine*. 7th ed. New York: McGraw Hill.

Kahn, Carrie-Ann Biondi. 2005. Aristotle's moral expert: The *Phronimos*. In *Ethics expertise: History contemporary perspectives, and applications*, ed. Lisa Rasmussen, 39–53. Dordrecht: Springer.

LaBarge, Scott. 2005. Socrates and moral expertise. In *Ethics expertise: History, Contemporary perspectives, and applications*, ed. Lisa Rasmussen, 15–38. Dordrecht: Springer.

Lackey, Jennifer. 2018. Experts and peer disagreement. In *Knowledge, Belief, and God: New insights in religious epistemology*, ed. Mathew Benton, John Hawthorne, and Dani Rabinowitz, 228–245.

Licon, Jimmy Alfonso. 2012. Skeptical thoughts on philosophical expertise. *Logos & Episteme* III (3): 449–458.

Locke, John. 1979. *An essay concerning human understanding*. Oxford: Oxford University Press.

MacIntyre, Alisdair. 1988. *Whose justice? Which rationality?* Notre Dame: University of Notre Dame Press.

Matheson, Jonathan. 2015. *The epistemology of disagreement*. New York: Palgrave.

McGrath, Sarah. 2009. The puzzle of pure moral deference. *Philosophical Perspectives* 23 (1): 321–344.

———. 2011. Skepticism about moral expertise as a puzzle for moral realism. *Journal of Philosophy* 108 (3): 111–137.

Meyers, Christopher. 2007. *A practical guide to clinical ethics consulting*. Lanham: Rowman & Littlefield.

Mill, J.S. 2002. On liberty. In *The basic writings of John Stuart Mill*, ed. J.B. Schneewind, 1–119. New York: The Modern Library.

Miller, Dale. 2005. Moral expertise: A Millian perspective. In *Ethics expertise: History, Contemporary perspectives, and applications*, ed. Lisa Rasmussen, 73–87. Dordrecht: Springer.

Mlodinow, Leonard. 2008. *The drunkard's walk: How randomness rules our lives*. New York: Random House.

Noble, Cheryl. 1982. Ethics and experts. *Hastings Center Report* 12 (3): 7–9.

Plato. 1997a. Meno. In *Plato: Complete works* (John M. Cooper, ed., Trans. G.M.A. Grube, 870–897). Indianapolis: Hackett.

———. 1997b. Protagoras. In *Plato: Complete works* (John M. Cooper, ed., Trans. Stanley Lombardo and Karen Bell, 746–790) Indianapolis: Hackett.

Priaulx, Nicky, Martin Weinel, and Anthony Wrigley. 2014. Rethinking moral expertise. *Health Care Analysis* 22 (3): 1–14.

Puma, La, David Schiedermayer John, and Mark Siegler. 1995. How ethics consultation can help resolve Dilemmas about dying patients. *Western Journal of Medicine* 163 (03): 263–267.

Quast, Christian. 2018. Expertise: A practical explication. *Topoi* 37 (1): 11–27.

Rasmussen, Lisa, ed. 2005. *Ethics expertise: History, Contemporary perspectives, and applications*. Dordrecht: Springer.

———. 2011. An ethics expertise for clinical ethics consultation. *Journal of Law Medicine and Ethics* 39 (4): 649–661.

———. 2016. Clinical ethics eonsultants are not 'ethics' experts—But they do have expertise. *The Journal of Medicine and Philosophy: A Forum for Bioethics and Philosophy of Medicine* 41 (4): 384–400.

Raz, Joseph. 1986. *The morality of freedom*. Oxford: Clarendon Press.

Ryle, Gilbert. 1958. On forgetting the difference between right and wrong. In *Essays in moral philosophy*, ed. A.I. Melden. University of Washington Press.

Schmitt, Frederick F. 2006. Testimonial justification and transindividual reasons. In *The epistemology of testimony*, ed. Jennifer Lackey and Ernest Sosa, 193–224. Oxford: UK.

Scofield, Giles R. 1993. Ethics consultation: The least dangerous profession? *Cambridge Quarterly of Healthcare Ethics* 2: 417–448.

———. 1994. Is the medical ethicist an 'Expert'? *Bioethics Bulletin* 3 (1): 1–2 9–10, 28.

———. 2008. What *is* medical ethics consultation? *Journal of Law, Medicine, and Ethics* 36 (1): 95–118.

Shuster, Evelyne. 2014. *The VA crisis is fundamentally an ethics crisis*. Bioethics Forum at The HastingsCenter.org. http://www.thehastingscenter.org/Bioethicsforum/Post.aspx?id=6993&blogid=140. Accessed 02 Aug 2015.

Singer, Peter. 1972. Moral experts. *Analysis* 32: 115–117.

Stolper, Margreet, Sandra van der Dam, Guy Widdershoven, and Bert Molewijk. 2010. Clinical ethics in the Netherlands: Moral case deliberation in health care organizations. In *Clinical ethics consultation: Theories and methods, Implementation, Evaluation*, ed. Jan Schildmann, John-Steward Gordon, and Jochen Vollmann, 149–160. Abingdon: Ashgate.

The Joint Commission. 1992. Accreditation Guide for Hospitals.

UK Clinical Ethics Network. 2011. *Member contact information*. https://www.webarchive.org.uk/wayback/archive/20110510220238/http://www.ethics-network.org.uk/committees/contact-details.

Vogelstein, Eric. 2015. The nature and value of bioethics expertise. *Bioethics* 29 (5): 324–333.

Watson, Jamie Carlin. 2018. The shoulders of giants: A case for non-veritism about expert authority. *Topoi*. https://doi.org/10.1007/s11245-016-9421-0.

Wear, Stephen. 2005. Ethical expertise in the clinical setting. In *Ethics expertise: History contemporary perspectives, and applications*, ed. Lisa Rasmussen, 243–258. Dordrecht: Springer.

Weinstein, Bruce. 1994. The possibility of ethical expertise. *Theoretical Medicine* 15 (1): 61–75.
Widdershoven, Guy, and Bert Molewijk. 2010. Philosophical foundations of clinical ethics: A hermeneutic perspective. In *Clinical ethics consultation: Theories and methods, Implementation, Evaluation*, ed. Jan Schildmann, John-Steward Gordon, and Jochen Vollmann, 37–52. Abingdon: Ashgate.
Williams, Bernard. 1993. Who needs ethical knowlede? *Royal Institute of Philosophy Supplement* 35: 213–222.
———. 1995. Truth in ethics. *Ratio* 8 (3): 227–236.
Wolff, Robert Paul. 1970. *In Defense of Anarchism*. New York: Harper and Row.
Yoder, Scott D. 1998. The Nature of Ethical Expertise. *Hastings Center Report* 28: 12–13.
Zagzebski, Linda Trinkaus. 2012. *Epistemic Authority: A Theory of Trust, Authority, and Autonomy in Belief*. Oxford: New York.
Zaner, Richard M. 1988. *Ethics and the Clinical Encounter*. Englewood Cliffs: Prentice Hall.

Chapter 2
Moral Expertise: A Comparative Philosophical Approach

Dennis Arjo

Questions about moral expertise point to questions about moral knowledge. Typically these questions are asked about the putative domain of such knowledge—What does someone knowledgeable about morality know? Special kinds of facts? Principles? Theories? Some proper relating of these? These questions, and so this way of framing questions about moral expertise, assume a certain picture of knowledge itself. Specifically, the standard approaches to questions about moral expertise assume moral knowledge is propositional, or knowledge of facts—*knowledge that*. To a first approximation, then, moral knowledge is commonly taken to be knowledge of propositions which range across a theoretical—and perhaps partly empirical—domain. A moral expert, then, is someone who knows a lot about that kind of stuff and who is able to leverage her knowledge into sound judgments when it comes to cases.[1]

We are familiar with other ways of thinking about knowledge. Since Gilbert Ryle's *The Concept of Mind*, it has been customary to distinguish *knowing that* and *knowing how*, the latter being the kind of knowledge we have when we are able to do something with some skill, such as ride a bike or ice skate.[2] With some important exceptions the prospects of developing an account of moral knowledge, and so of moral expertise, based on an understanding of *knowledge how* has largely been passed over. This paper looks to contribute to such an account, or at least something akin to it. Specifically I will suggest an account of moral expertise that turns more

[1] See for example, Peter Singer (1972), pp. 115–117, for an argument that philosophers can be moral experts by virtue of their expertise in moral philosophy. For an argument against Singer that accepts the same terms of debate, see David Archard (2011).

[2] See Gilbert Ryle (1949). It should be noted that Sellers was anticipated in drawing this distinction by John Dewey. See Hubert L. Dreyfus for discussion of Dewey's early contribution to the idea that moral knowledge is a kind of *knowledge-how*.

D. Arjo (✉)
Johnson County Community College, Overland Park, KS, USA
e-mail: darjo@jccc.edu

J. C. Watson, L. K. Guidry-Grimes (eds.), *Moral Expertise*, Philosophy and Medicine 129, https://doi.org/10.1007/978-3-319-92759-6_2

on a person's ability to successfully navigate the social world than her knowledge of distinctive kinds of facts, propositions, principles or theories.

2.1 Moral Knowledge as We Find It

Our experience of morality is of a complex of behavioral constraints. Manifested variously as a subjective sense of 'right' and 'wrong,' tacit awareness of external rules and accompanying sanctions for violations, and expectations shared among our fellows, morality is a constant reminder that we are not completely free to do as we please, or to pursue whatever goals may strike our fancy by whatever means. To be a moral agent is to limit, with varying degrees of self-awareness, one's actions and goals to what is within the pale of the permissible.

As is often the case with implicit awareness of the forces that structure our actions, we become most conscious of moral constraints when we are faced with doubts or uncertainties and so are forced to ask "What should I do?" As phenomenologists often point out, philosophers have a tendency to see occasions where we are forced to deliberate about our actions and choices as typical or definitive. In fact, however, the more spontaneous and less self-aware aware movements within ethical space are more basic and representative of moral competence. Consequently, it is there that we should look first for a workable account of moral knowledge. An example will help.

Consider a situation which is not the sort that most readily captures the attention of moral philosophers looking to work through complex and vexing moral dilemmas, but which is in fact fraught with significant moral challenges—consider a parent faced with some particularly challenging behavior on the part of her young child. In characterizing this as a situation posing distinctly moral challenges I would point to several things. First, in sensing she is faced with objectionable behavior demanding a response, our parent is relying on normative standards. She must have some sense of what kinds of behaviors on the part of children are to be celebrated, tolerated, or discouraged. This in turn must be tempered with a sensitivity to confounding factors—perhaps this behavior is "bad," but is it a result of the child's being unusually tired, upset about other matters, understandably frustrated? Complicating things further, our parent must consider not only her child's behavior, but also the feelings it invokes in her as a frustrated adult who may be unduly distressed by what is actually ordinary childish mischief—perhaps it's late evening after a particular trying day at work, perhaps she's just finished dealing with a younger sibling's equally challenging antics. To what extent is her own anger a reasonable response to her child's misbehavior as opposed to a disproportionate rage triggered by other factors? Lastly, when it comes to responding to her child's misbehavior, our parent must navigate a range of possibilities constrained by moral limits—her options reflect potentially competing values, and some seemingly tempting options may well be beyond a generally recognized moral pale. Should she respond with sympathy? Ignore the defiance? Punish the child? If so with what kind of punishment?

The first point I want to make is that it is in this kind of situation, as mundane as it may be, that we find instances of something we can meaningfully call "moral knowledge." What I mean is that a parent's ability to respond well to this kind of misbehavior—her ability to successfully navigate the myriad of morally relevant questions and challenges it poses in order to respond well—is where we should look to gauge her understanding of this small part of the moral world. I will take a moment to substantiate and explain this claim, and then use the example as an occasion for outlining just what kind of knowledge our parent has when she is indeed able to respond well.

I have made the case that in a simple episode of a parent having to respond to her child's misbehavior we can find a number of questions that are properly ethical. There is, to put the point differently, an importantly normative dimension to disciplining children—a lot is at stake, and we have a least an implicit sense that it is the sort of thing that can be done in better or worse ways, where "better" and "worse" range over more than matters of efficiency. It is a short step from here to characterizing successful parenting as requiring moral knowledge. It does not strain ordinary language to say some parents "know" how to handle children who are being difficult while other remain lamentably clueless. Indeed, the sizable industry of child rearing advice offered in books, website, magazines, seminars and so on points to a general sense that "effective discipline" is something that can be taught and learned. While it is perhaps under-appreciated how much of this involves normative questions rather than straightforward empirical claims about what "works," it should not be too controversial to see that one thing parents must learn to do in learning to parent well is to learn to parent morally.

The deeper question then is what kind of knowledge this is. I think it is a good start to assimilate good parenting with *knowledge how* rather than *knowledge that*. The kind of intuitive or not fully reflective understanding of her child, the current situation, the conditions leading up to the troublesome behavior, what would constitute better and worse responses—all this produces effective *action* within the bounds of moral constraints our parent may not be able to clearly articulate. The language that comes naturally in this kind of case points more to skill and skillful behavior than propositional knowledge. Indeed, it is easy to imagine a hapless parent who has digested and memorized the advice of so many parenting experts while remaining paralyzed with uncertainty when it comes to actually doing something with a particular wayward child. This is not to deny that *knowledge that* may be implicated in successful *knowledge how*, or that the latter is not be served by gains in the former.[3] Still, the value of parenting advice, however much informed by theoretical knowledge gleaned from developmental psychology and the like, comes when it translates into effective day to day practice.

[3] For a particularly rich discussion of the ways in which *knowledge that* and *knowledge how* can be woven together see Christopher Winch (2010), Chap. 2.

2.2 Intimacy or Integrity

So far I've argued that effective parental discipline requires moral knowledge, and that to a first approximation the kind of moral knowledge it requires is more readily associated with *knowledge how* rather that *knowledge that.* There is clearly nothing special about the example, so I would also argue that a lot of moral knowledge is to be found embedded in practical contexts where we find ourselves having to navigate with skill situations structured by moral values and commitments. Parenting, working as an employee or managing employees, going to school or teaching students, governing an organization, working in a given trade—all of these and many more are contexts which implicate both skills and awareness of and ability to respond to normative constraints. I want now to develop this account further by incorporating some insights from a work of comparative philosophy that will take us past the familiar *knowledge that/knowledge how* distinction in important ways. For this we turn to Thomas P. Kasulis' idea of "intimacy." Kasulis' work, I think, provides an epistemological framework that will allow us to think of the concept of expertise, and so moral expertise, in some less familiar but very fruitful ways.

In his book *Intimacy or Integrity: Philosophy and Cultural Difference,* Kasulis introduces two distinct ways in which cultures might answer "a fundamental question: how are things related?"—whether the relationship in question is that between knower and known, basic constituents of reality, individuals in relationships, or patterns and colors in a painting.[4] In what Kasulis calls the Integrity Orientation, elements are seen as self-contained and distinct, and relations between them are external. For example, from within an Integrity Orientation, a relationship between two persons is between two fully distinct individuals whose identity exists prior to their coming into the relationship and who maintain that identity in the context of the relationship. From this perspective, individuals have their identity intrinsically, and go on to become, variously, a friend, spouse, colleague, and so on. In each of these relationships, each individual remains, in their core being, the same person. Should any of these relations end, each individual remains person she began as: "[w]ith the termination of an external relationship…, the [individuals] maintain their integrity and exist as unbroken, unviolated, wholes."[5]

By contrast, in an Intimacy Orientation, an individual's identity is seen as a function of her relationships, so that her identity is determined by her connections with her spouse, friends, colleagues, and so on. Here there is no sense that she would be the person she is absent those relations with those people. Since these relations are internal, ruptures—a divorce, the death of a parent, estrangement from a once cherished friend—leave the person forever altered or diminished, no longer the person she once was: "[t]erminating an internal relationship…results in both relatents

[4] Thomas Kasulis (2002), p. 133.

[5] Kasulis, (2002), pp. 59–60.

losing a part of their identity: [each] become[s] less themselves or a least less of what they had been."[6]

Kasulis makes a number of claims about the Integrity/Intimacy distinction. He is particularly interested in using the distinction to capture salient differences between the dominant tendencies of different cultures—the U.S., for example, is marked by a strong preference for Integrity, as evidenced by the American stress on individuality, autonomy, and individual rights and freedoms. By contrast many Asian cultures—Kasulis focuses mostly on Japan—are marked by a preference for Intimacy, stressing family ties, social obligations, and communal values.[7] Kasulis also argues that there are deep connections between, for example, the Japanese preference for understanding personal relationships in the manner of Intimacy and its cultural embrace of Buddhist metaphysical doctrines stressing the inter-relatedness of all things. Both, in Kasulis' picture, are manifestations of an Intimacy Orientation that dominates the culture. I will put these stronger claims aside here and focus on how Kasulis uses the distinction to characterize two very different accounts of knowledge. Exploring the distinctions he introduces here will lead us to the account of expertise I wish to introduce.

2.3 Intimacy and Integrity in Epistemology

According to Kasulis, the dominant epistemologies in Western thought reflect a general fondness for the Integrity orientation. What this means is that despite the differences between various prominent theories of knowledge familiar in Western philosophy, we can identify certain common features.[8] Most notably these are a tendency to think that knowledge consists of true beliefs about an objective world, that anything worthy of being called "knowledge" is at least in principle shared or public, and that knowledge is best gained from a position of dispassionate impartiality. On this picture, the world is as it is, and our epistemic job is to come to understand this world as it is, to conform our beliefs to what is actually the case. Moreover, the world that is known is shared—two people who know that the world is round and that it is 5 billion years old know the same things about the same world and in the same way. This stress on publicity is crucial, as it suggests the possibility of definitive procedures for discovering facts or obtaining knowledge that are readily accessible, at least in principle, to all competent knowers. This is part of what it is

[6] Kasulis (2002), pg. 60.

[7] These cultural contrasts are heavily qualified—Kasulis stresses that no culture can be exclusively characterized using either of the orientations and that there can be subcultures that are dominated by the orientation that is less prominent among the mainstream.

[8] It is worth stressing that these remarks refer to broad characteristics of the more dominant theories of knowledge—traditions that arose in part in opposition to mainstream epistemology such as pragmatism and phenomenology arguably display greater affinities with an Intimacy orientation. Not surprisingly, many see points of convergence between these theories and classical Asian thought.

for the Known to be "objective," and it points to the irrelevance of subjective facts about the Knower. How she feels about the world's shape or age is of no relevance. As Kasulis puts it:

> the integrity-based model of knowledge assumes a publicly verifiable objectivity. Through objective distance the knower is an observer of the real...Given the knower's independence from the known, any other potential knower will...be able to attain the same knowledge of reality.[9]

The paradigm of such knowledge is of course scientific knowledge where both what is known—empirical reality—and the means by which it is known—the scientific method—are held to be objective and immune to the vagaries of the biographical details of individual Knowers.

The contrasts with knowledge as seen from an Intimacy Orientation are quite sharp. Here the paradigm is not the public and dispassionate investigation of self-standing facts perfected in science, but rather it is the ability of an expert to understand a specialized zone of activity in ways that require deep and practiced familiarity. Kasulis uses the example of judges in sporting competitions like gymnastics. Gymnasts are evaluated on their performance in various activities that all judges witness, and there is a common scale each judge must use to reach her conclusion as to which of the gymnasts has performed the best. All this may suggest a familiar kind of publicity. Nonetheless, judging a gymnastic performance in a knowledgeable way is a very different process than that by which someone deploys a scientific procedure in order to come to a conclusion about an empirical matter. The competence to judge gymnastics comes by way of a competence that comes only with spending years learning the sport and learning to see the many subtle nuances that distinguish good from great performances. Importantly, this kind of knowledge does more than relate a Knower and a Known.

What is known in Intimacy becomes a part of the Knower, in the sense that someone committed to a sport like gymnastics comes to orient her life around it—being in that world becomes part of who she is. As Kasulis points out, such deep familiarity typically requires an emotional attachment rather than cool, rational detachment. Only someone who truly loves the sport will spend the time and energy needed to master it. Kasulis characterizes this kind of knowledge as "dark" in that it is not readily accessible to outsiders, and often the exact reason for a judgment remains unarticulated. Rather, such knowledge is available only to the initiated, and is often implicit or tacit. As Kasulis puts it:

> the Olympic judges' perception of subtle differences in the quality of style is not a knowledge that can be tested by just anyone. It derives from the judges' expertise developed over years of participation in the sport...[T]he evaluation of style occurs in the overlap between the judge and the sport. Those who have never taken part in the sport or thought of the sport as part of themselves have no objective basis for such expert judgments.[10]

---

[9] Kasulis (2002), p. 72.

[10] Kasulis (2002), p. 78.

Importantly, none of this should be taken to imply that there is no objectivity in knowledge as Intimacy. That a range of gymnastic judges are seeing the same things and evaluating them according to genuine epistemic standards rather than subjective hunches or feelings is demonstrated by the impressive convergence of scores—we generally find near agreement as to which gymnast has outperformed which, and judges whose scores are regularly outliers are objects of suspicion. That there is something worth calling "knowledge" here is also demonstrated by the possibility of codifying the performance standards to a certain degree. As anyone who has listened to the commentary of an accomplished gymnast during a broadcast of the Olympics comes to appreciate, it is possible in a general way to spell out what "the judges are looking for" when evaluating a given routine. It is certainly not random or a matter of personal preference. That said, the ability to apply these standards is not easily transferred. For someone like me, even a detailed checklist is going to be fairly worthless, as I simply lack the intimate familiarity with the sport needed to see even mildly subtle differences between two performances in real time. Lacking this competence, I will miss most mistakes smaller than a gymnast landing on her backside, even if in the abstract I know what I am supposed to be looking for.

2.4 Moral Knowledge as Intimacy

To return now to our unhappy parent, hopefully it is clear how whatever skills she may bring to bear on her challenge, her knowledge can be well understood from the standpoint of Kasulis' Intimacy Orientation. The responsiveness to her daughter's personality, current behavior, emotional needs, her own mood, the ability to judge the appropriateness of each of the various disciplinary options available to her as they may impact this particular child—all these seem to point to a "dark" understanding born of the deep familiarity characteristic of Intimacy. Insofar as the scenario envisioned involves moral questions, as I argued it does, it seems reasonable to conclude that part of what our parent must know is the nuances of the distinct ways in which her moral values and commitments—whatever they may be (more on this presently)—are realized in the particulars of this situation. Greater knowledge in such contexts is born of greater familiarity and sensitivity and is manifested in the ability to notice small but relevant factors in evaluating and responding to daily challenges. Here too then is a case of knowledge as Intimacy, in this case moral knowledge. To the extent our example is just that, a stand in for a range of morally charged everyday encounters, there is a case to be made that moral knowledge is at least often amenable to such an analysis. To put the point generally, one kind of moral knowledge, at least, is knowledge gained through familiarity with a morally normed practice.

As indicated earlier, there is a clear sense in which such an analysis of moral knowledge would assimilate it most readily with standard accounts of *knowledge how*, and certainly a component of how Kasulis understands knowledge as Intimacy involves what is readily understood as skills. There are, however, additional resources in Kasulis' analysis, as there is nothing that precludes elements of *knowledge that* as being among the components of what is related in Intimacy. As we saw with the example of gymnastics, a competent judge must certainly possess a range of straightforward propositional knowledge in the form of rules of competition and the workings of the scoring system she must employ. What points to Intimacy is less the kind of knowledge identifiable according to the *knowledge that/knowledge how* division than how it is incorporated into a practice.[11]

2.5 Sources of Moral Content

To take things further we need to look more closely at just what provides the content of the moral component of the Intimacy we find in the contexts we are looking at. What I mean by this can be gleaned if we return to gymnastics. Gymnastics is a highly structured sport whose competitions are organized by local, national and international organizations. These groups enjoy the authority to determine the rules of competition and standards of performance—they decide, for example, that stepping out of bounds in a floor routine results in a .1 point reduction. Gymnastics is also a sport with a history of well-defined techniques that are progressively mastered by gymnasts as they learn their craft. To this, we can add a shared sense of values guiding the construction of full routines according to the difficulty of different moves and the aesthetic qualities of their various combinations. Lastly, it should be noted that here, too, there are norms of fairness, sportsmanship, work ethic, and the like that are both codified by the governing bodies and left more implicit. These, too, are part of what budding gymnasts must learn. All of this is the *content* that must be mastered by a competent judge—all of this is what she must be intimate with if she is to gain the ability to judge well.

What we are looking for now, then, is a general statement of what provides the content for the *moral* knowledge I have argued emerges as part of Intimacy in the various contexts we've considered. Going back to our initial example, what would be comparable to what a gymnastics judge must learn in the case of our parent disciplining her child? It seems clear the answer has to be the various sources of cultural norms regarding parenting. That is, parents learn, with varying degrees of success, how to care for children by way of learning their culture's ways of caring for and raising children. As even a cursory glance at the anthropological literature on child rearing across cultures reveals, the ways in which children are fed, clothed, educated, socialized, and so on varies a great deal, and indeed it can vary a great deal within a single "culture" according to differences between subcultures or

[11] Indeed, insofar as *knowledge that*—scientific knowledge, for example—is generated through practices requiring skills born of experience and deepening familiarity, it too reflects elements of Intimacy. One learns how to be a scientist, or how to *do* science. The process by which scientific results are actually generated are indeed "dark" to the uninitiated, even if those results can be presented in ways that abstract away from their origins in the labs of practitioners.

different communities defined along, for example, ethnic, religious, or socio-economic lines. My claim, then, is that the ethical norms of parenting vary with these cultural divisions as well, and so a given parent's sense of propriety as to how to handle a misbehaving child, among the any other challenges of parenting, will stem from her familiarity with and membership in a specific cultural milieu that provides the moral norms. Note such knowledge may include, in certain cultural settings, familiarity with the advice offered by scholars and researchers well-versed in child psychology and developmental psychology. Importantly, however, on the model being sketched here, such knowledge is of interest only to the extent it is incorporated into actual child rearing practices, and as it is understood by parents raising children. In general then, I am arguing that moral knowledge is simply a piece of knowledge of the norms, including the moral norms, embedded in or structuring the various practices that make for communal life.

2.6 The Relativism Worry

To philosophically trained ears this analysis is surely worrisome, as it seems to collapse the critical distinction between a description of what people may *believe* is morally acceptable when it comes to something like raising children and a properly normative account of what is *actually* acceptable. Such a distinction is critical if we are to have a place from which to evaluate existing practices, whether in our own cultural setting or in others. When it comes to morally charged topics, we want to allow for morally better and worse responses and allow for the possibility that practices that may have long been widely embraced would be actually better abandoned. In short, the account on offer seems to be veering dangerously in the direction of moral relativism, and so away from any proper account of "moral *knowledge*" at all.

To counter this impression I want to do two things. One is to present an account of a moral tradition that explicitly ties moral knowledge with mastery of specific cultural practices. Here I will turn to classical Confucianism and the idea of *li*, or "ritual propriety," as well as the moral ideal embodied by the *junzi*, or "exemplary person." My suggestion will be that the junzi, as presented in Confucianism, is a moral expert precisely by way of her exceptional ability to navigate the social world in which she finds herself. I will then rehearse an argument that addresses worries that, as a moral tradition, Confucianism tends towards relativism. Though there is, I will concede, a certain ineliminable element of relativism in a moral tradition like Confucianism, it is not, I will argue, a pernicious sort, and there is enough "moral realism" to be found within it to blunt the concerns noted in the previous paragraph. This is because Confucianism can be read as appealing to an objective conception of human flourishing that can be used to judge the moral adequacy of particular practices.

2.7 Confucianism: A Brief Primer

Classical Confucianism presents a distinctive "view of a life we should like," a life, that is, that we ought to find satisfying, or worth pursuing given the basic realities of human concerns.[12] At the core of this picture is an account of moral development and cultivation—canonical texts such as the *Analects*, *Mencius*, and *Xunzi* offer an account of the direction and process of moral growth and refinement. The optimal outcome of moral development is a kind of person called the *junzi* (君子), or exemplary person, the person we should strive to become if we are to live well.[13] Throughout the canonical texts the *junzi's* character is revealed through an array of virtues, most notably *li* (禮: propriety), *yi* (義: appropriateness), *zhi* (智 or 知: wisdom), *zhong* (忠: loyalty), *shu* (恕: reciprocity), *xin* (信: trustworthiness), and especially *ren* (仁: humaneness or benevolence).[14] While there is disagreement among scholars and commentators about exactly what these virtues amount to, all commentators agree the moral qualities of the *junzi* are fundamentally social and relational. Among these, *ren* is the most critical, and it is often treated as a meta-virtue, a quality of character that comes with the acquisition of other morally desirable traits.

Commonly translated as "benevolence," the character *ren* names a complex and multifaceted characteristic of a morally accomplished person and her behavior. As "benevolence" suggest, a person who is *ren* is someone whose concerns include the welfare of others: "'[the *junzi*] establishes others in seeking to establish themselves and promote others in seeking to get there themselves."[15] The *junzi* is typically contrasted with the *xiao ren*, or "small" or "petty" person who is concerned only with "profit" or self-interest. But there is considerably more to it. Developing and manifesting the sentiments underlying *ren* implicate both the affective and cognitive dimensions of human psychology. Rather like Aristotle's virtuous person whose perfected character and intellectual sophistication renders her practically wise, the *junzi* distinguishes herself through an effective combination of wisdom and affective discipline that enables her to act well in the world.[16] Additionally, *ren* is a

[12]Amy Olberding (2012), pp. 57. Olberding introduces the phrase "view of a life we should like" in contrast to an explicit and rigorous theory of human flourishing, something she concedes we do not find in early Confucian texts. The initial, intuitive idea such a theory is meant to elucidate and defend—that of a better, more satisfying and complete way of life—is what she means to capture in the idea of life we should like. A rich and compelling image of this, she argues, is what we find in text such as the *Analects*. Below I will also appeal to "flourishing", but such appeals should also be understood in this pre-theoretical sense.

[13]The classical Confucian moral imagination recognizes a higher type, the *sheng ren*, or sage, but such people are exceedingly rare, and this not a status we can realistically aspire to. Though later recognized as such, Confucius denied that he was sage and claims to have never met such a person—the sages he recognizes lived long ago.

[14]As explained in the next paragraph, *ren* is translated in a variety of ways—I here defer to the more popular renderings, but there is much to be said for the alternatives introduced there.

[15]Rogers T. Ames and Henry Rosemont Jr. (1999), 6.30.

[16]Hence the link between *ren* and the virtues listed earlier. The cognitive dimension of *ren* points to the importance of knowledge and wisdom, or *zhi*, while the practical dimension points to the ability to translate such wisdom into effective action, or *yi*. All of this requires, in turn, a refinement of our moral sentiments: hence *shu, zhong*, and *xin*.

relational virtue that is seen as lying at the very core of our humanity. As translations such as "humaneness" or "consummate person" suggest, *ren* connotes a sense of having fulfilled one's potential as a human being by expanding and perfecting one's relationship with others. Accordingly, the *junzi*, a person whose character and behavior who is *ren*, is someone who knows how to act well in social, communal and relational contexts and who has an affective profile that enables her to do so naturally or effortlessly. Only such a person, the tradition suggests, is able to live a truly human life well.

There is much here to suggest that Confucianism resonates well with more familiar forms of Western virtue ethics, such as that of Aristotle.[17] There are two distinctly Confucian ideas, however, that point to a heightened appreciation of the moral importance of defined social roles and contextually prescribed behavior that is not emphasized in Western theories. One of these ideas—*xiao*, or filial piety—points to Confucianism's unapologetic embrace of social hierarchy. The second, and our main concern going forward, is *li*, or ritual propriety. Together these concepts highlight the extent to which Confucianism embeds ethical questions and concerns in the concrete practices of everyday life.

The character *li*—often rendered as "rites" or "rituals" and in some contexts "propriety"—is another character denoting a web of concepts and practices that eludes a simple translation. As "rites" or "ritual" suggest, the character refers to religious practices, specifically rituals of divination and sacrifice. However, more broadly the character points to any carefully scripted or choreographed social interaction. As "propriety" suggests, it is also used to denote affective dispositions that incline a person to defer to the standards governing such interactions—in the tradition this is tied especially to a honed sense of shame. Our present purposes invite us to understand *li* in these more expansive senses. Accordingly, I will follow Roger Ames and Henry Rosemont Jr. when they define *li* as:

> those meaning-invested roles, relationships, and institutions which facilitate communication and which foster a sense of community. The compass is broad: all formal conduct from table manners to patterns of greeting and leave taking, to graduations, weddings, funerals, from gestures of deference to ancestral sacrifices—all of these, and more are li.[18]

This set of rituals, rites, mores, standards of etiquette, and the like provides the behavioral repertoire that structures social interactions—*li* is what tells how to behave at funerals and wedding and dinner parties, in classroom, boardrooms and churches. But as this passage suggests there is much more to *li*.

As Steve Angle has noted, Confucian claims about *li* blur the lines between what might more readily be described as etiquette and morality, a distinction Western philosophy typically keeps sharp.[19] *Li* is where the various facets of *ren* are practically realized—the *junzi* demonstrate her benevolence only according to the dictates of *li*. This requires a measure of knowledge, as *li* must be learned.

[17] See for example, May Sim (2007) and Jiyuan Yu (2007).

[18] Roger T. Ames and Henry Rosemont Jr. (1999) pp. 51.

[19] Stephen C. Angle (2012), Chap. 6. See also Amy Olberding (2016).

Accordingly, knowledge of *li* is a component of the moral wisdom or *zhi*: one will learn how to express one's feelings of respect for one's parents, or to interact with one's teacher, by learning those aspects of *li* relevant to precisely that. Moreover, learning *li* is transformative: it is in the performance of *li* that our moral sentiments—the basis of virtues like *zhong*, *xin*, and *shu*—are developed, honed, and nurtured. In other words, it is in the performance of *li* that one both expresses and develops the affective dimension of virtue, and it is the learning of *li* that one develops the wisdom that makes the appropriate and corresponding behaviors (*yi*) possible.

It is also clear that the kind of knowledge embodied in knowledge of *li* is well captured by Kasulis' notion of Intimacy. It is knowledge earned through growing familiarity with normed governed practices into which one is initiated—in this case, practices we begin to master as children as we are socialized to be functioning members of our families and communities. It is also knowledge of matters we would be hard pressed to separate from ourselves given the depth to which such processes of socialization help define who we are. As Kasulis frequently notes, the "darkness" of culturally embedded knowledge—its inaccessibility to the uninitiated—accounts for the frequency and peculiarly frustrating nature of cultural misunderstandings. What can seem so clearly right and unremarkable to those who are at home in a given cultural setting can be alien and disconcerting to outsiders. Lastly, the kind of knowledge embodied in *li* resists full codification, and even where its codification is possible, simply following the rules does no count as genuine propriety. This last point is worth emphasizing.

In the Confucian moral imagination the *junzi* and *xiao ren* are joined by a third person who fall somewhere between the two. This is the *xiangyuan* (鄉原), or "village worthy." This is a person who has at least largely mastered the outward expression of *li*—she reliably displays appropriate behavior in the typical affairs of the day. Nonetheless, the village worthy is an object of moral disapproval in Confucian texts, largely because her behavior is not an expression of *ren*, or true virtue. So part of the problem is that there is something fake about the village worthy—her politeness, her ritualistic expressions of concern or respect for authority do not stem from the appropriate sentiments that she pretends to be conveying. Her motives, we might say, remain that of the petty person, even if her outward behavior resembles that of the *junzi* as she seeks to reap the rewards of being thought virtuous without undergoing the hard work of becoming truly virtuous. As Confucius pointedly puts it, the village worthy is a "thief of virtue," gaining the esteem and praise of others by faking the behavior of the *junzi*.

Importantly, the village worthy's inauthenticity is not her only failing. Though largely convincing—enough so that having a village worthy as a neighbor would not be the end of the world—her behavior will eventually betray her. In subtle ways the village worthy's lack of true understanding of *li* will emerge as her flawed character leads to inappropriate action. Most importantly, it will be in those situations where familiar scripted patterns of behavior suddenly become unreliable. Where the *li* one learned must be applied in unusual and uncommonly difficult circumstances, the *junzi*, but not the village worthy, will know what to do. As Stephen Angle argues, uncritical deference to established rules of behavior will not tell us what to do when "there is no relevant ritual rule at all, or when rituals conflict, or when…rituals conflict with the humane response to an exigent situation."[20] The *junzi*, but not the village worthy, is an expert.

In sum, *li* teaches us how to address others according to their relation to us, and how to express feelings such as affection, respect, disapproval, appreciation, grief, and sympathy, according to shared expectations so that they are more effectively communicated. Going the other way, *li* also produces the behavior that hones the sentiments being expressed. *Li* is what structures interactions between teachers and students *as* teachers and students, or between parents and children *as* parents and children. This is critical to instilling a willingness to defer to appropriate authority on the one hand and the restraint needed to exercise authority responsibly on the other. For this reason, learning *li* is a critical part of a person's moral education, a central means by which her emotions and their expression are disciplined according to shared standards enabling her to take her place as a valued member of the community. An essential component of this process is the acquisition of the moral knowledge, I have argued, that is typically implicit in social competence. As Amy Olberding puts it:

> [*li*] aims to provide a form of moral training that can render learners equal to the moral work of ordinary life, inculcating appropriate cognitive-emotional dispositions toward others, as well as honing social perception and bodily expressions. It does this most basically by encouraging adherence to well-established rules for human conduct and communication but aims ultimately at a moral competence that renders as behavioral and bodily instinct conduct that can significantly enhance moral experience and community.[21]

In the first parts of this paper, I argued that a familiar, if underemphasized, kind of moral knowledge is achieved in the mastery of culturally embedded practices such a parenting. The insights of Confucianism, I think, elucidate this further by giving us a vivid sense of the moral import of the shared expectations that shape our behavior in ordinary social contexts. As we have seen, Confucianism allows us to move past a more limited conception of ethics that would relegate the more mundane affairs of everyday life—and the norms the govern them—to lesser realms of etiquette, manners, or mores. Rather, we would do well to see such a realm—the place where primary relationships are lived, children reared and educated, the business of communities conducted—as the place where a sizable portion of what deserves to be called moral knowledge is found. As we have also seen, the kind of

---

[20] Angle (2012) pg. 97. Angle argues this points to the need for a higher value—*ren*—that is distinct from and served by *li*. Angle concedes this is a somewhat controversial reading of Confucianism as some will argue *li* itself is the source of moral value in Confucianism. For reasons that will emerge in the next section, I side with Angle on this question. Angle also argues that Confucius is a bit unfair to the village worthy, as surely it is some kind of a mark of moral progress that one is willing and able to defer to moral standards even if it is for less that fully noble reasons. Indeed, a willingness to do so may be a necessary first step.

[21] Olberding (2016) pp. 242.

knowledge at issue is well captured by Kusulis' account of Intimacy. Returning to the *junzi*, we can now say such a person well represents one model of what we might mean by a moral expert. What the moral expert is expert at is knowing how to be a good spouse, parent, neighbor employee or employer, and so on as "good" is understood in her community. In the Confucian picture, such an expert is simply someone who is culturally accomplished, someone who in their very person has learned and been shaped by the *li* of her thriving culture so that she is able, as Confucius eloquently puts it, "give [her] heart-and-mind free rein without overstepping the bounds."[22] Unlike the less accomplished, such a person need to bring to mind explicit rules of behavior or standards—hers is the Intimate knowledge of the thoroughly initiated.

2.8 Relativist Worries Revisited

By now, this account of moral knowledge and expertise will have done more to increase the worries about relativism raised above than alleviate them. Clearly, the norms and practices captured by the concept of *li* will be relative to specific cultures, and what is considered an appropriate from of greeting or interacting at the dinner table or dressing for a funeral in one cultural setting will be found to be deeply offensive in another. If, as I have suggested, Confucianism sees *li* as the vehicle by which we come to have moral knowledge, surely it is indeed committed to a brand of moral relativism—what constitutes sound moral judgment in one cultural setting will be moral foolishness in another. If so, it is hard to see in what sense social competence or expertise can count as moral competence or expertise, properly considered, if we are inclined to reject relativism and insist that some culturally entrenched practices are morally repugnant.

The first part of this inference is certainly sound—Confucians are fully aware of the differences in the practices and values of different cultural periods in Chinese history. For example, they make much of the difference between their own way of life and those of foreign lands. They also acknowledge the constant evolution of the *li* within the culture in which they themselves live, recognizing that what is considered *yi* or appropriate now may well have been inappropriate in another time even within the middle kingdom. So yes, what we learn is right or appropriate or even polite will turn on the social context in which we grow up, and it may be odds with what others learn in different times and places.

What does not follow is the conclusion that the Confucian picture leaves us with no resources for critically evaluating either our learned practices or those of other cultures. It is this critical edge that will enable us to speak meaningfully of moral expertise. This is because the moral authority of the *li* ultimately derives from its

[22] Ames and Rosemont Jr. (1999) 2.4. Lest we think it is easy to become a *junzi*, Confucius notes that was not until he was 70 that he felt he had arrived. Hence the claim that it constitutes a kind of expertise.

ability to structure a community in productive and successful ways. Put differently, *li* is in service to higher values—one becomes good by way of mastery of *li*, but what it is to be good can be identified independently.[23] If *li* shapes our moral psychology, on the Confucian picture it can do this in better or worse ways as judged by both the kinds of individuals it produces and by the kind of community whose practices it defines. *Li* that elevates few from the ranks of the self-centered petty person, or that produces a society rent by conflict and bereft of social cohesion, cannot be judge to be *successful li.*

On a deeper level, Confucianism presents a picture of human flourishing that ties it in essential ways to ideas of community—human beings are fundamentally social beings in Confucianism. But this does not mean that any kind of community will do. Confucian texts point to a robust moral psychology that makes specific and substantive demands on how human societies must be organized in order to prosper and to allow those living in them to prosper. While it is consistent with Confucianism that a variety of social organizations might create conditions sufficient for human flourishing, this is distinct from the claim that any and every social organization will do.[24] The conditions for human flourishing are constrained by the psychological salience of certain human relations, beginning with familial relations, and the practical realities of what is needed for them to develop well. Returning then to the matter of moral expertise we can now present a fuller Confucian account. The kind of knowledge possessed by the *junzi*—the intimate knowledge of the performance of *li*—counts as *moral* expertise, rather than as mere cultural competence, insofar as the culture in question more broadly creates the conditions necessary for human flourishing.

2.9 The *Junzi* as Moral Expert: Some Lessons for Bioethics

By way of conclusion, I would like now to sketch an application of the model of moral expertise developed here to the question of moral expertise in medical ethics, the context in which questions about moral expertise often arise. To find purchase, this model will look for practices in which we find established norms that are Intimately known by accomplished practitioners. That there are such practices is

[23] If this sounds a little paradoxical, consider an example such a playing piano. Knowing how to play piano is embodied and displayed in somebody's playing piano, but this is distinct from the musical qualities that are exemplified in accomplished performances. It is in reference to these that we judge this person's piano playing or the value of different techniques, practice regimes, and so on. Expertise in playing piano is embodied in the playing but it is, as I have put it here, in service to the aesthetic properties on display in good playing.

[24] Classical Confucianism itself can sometimes be criticized on these grounds as there is truth to the charge that its fondness for the social and cultural practices of the early Zhou Dynasty seems to rest on little more than its familiarity. On the other hand, the willingness to criticize practices current in the later Zhou, and the willingness of Confucius and others to embrace changes enacted since the early Zhou, suggest a recognition that that the *li* was not self legitimating.

clear enough—we now recognize those who job it is to analyze, evaluate and offer advice about ethically charged medical situations typically involving end of life decisions or difficult choices about burdensome treatment. Such choices are made not in the abstract, but as part of the very real practices of medical care by those chosen to serve on ethics committees and the like. Those who are accomplished at this, having been initiated into the various facets of applied medical ethics, would count as moral experts within this realm. The moral knowledge they possess is of the ethical norms and values informing medical practice at its best. Such knowledge would be continuous with or, better, realized as a cultural knowledge that allows practitioners to respond to the concrete and highly particular needs of patients, their families, and the institutions in which medicine is practice. As critics of current medical ethics are quick to point out, in practice, members of ethics committees are trained in and typically aver to existing legal standards, widely recognized "best practices," and the values and beliefs reflective of contemporary liberal democracies.[25] All of this is codified in various ways, including definitive documents such as *Core Competence for Health Care Ethics Consultation* published by the American Society for Bioethics and Humanities.[26] The test, however, is how well all this translates to accomplished performance in the clinical setting.

Of course, all this is properly tendentious. On the account on offer, whether such an analysis really allows us to conclude that professionally proficient medical ethicists are *moral* experts depends on more than their being good at what they do. It must also be true that what they do is good—the practices at which they are accomplished must serve the higher values constitutive of human flourishing, whatever those actually are. As Edmund P. Pellegrino has argued in a number of works, medicine itself is a "moral enterprise" and those working within it have good grounds for believing that at its best the practice of medicine promotes genuine human goods.[27] The persistence of deep controversy about both the aims and methods of medicine in, for example, end of life decisions, points the difficulties in defining the conditions of human flourishing. The classical Confucians themselves worked in a social setting marked by disagreement about such matters.[28] The range of beliefs about what constitutes a good life in current liberal democracies is enormously more vast, and indeed, profound disagreement about fundamental questions about what constitutes a good life is now taken to be a defining characteristic of contemporary liberal democracies. How to accommodate such diversity while maintaining some degree of political and cultural coherence is a matter that occupies much current political philosophy.

[25] See for example H.T. Engelhardt (2012). Similar points are made by a number of the essays collected in this volume.

[26] American Society for Bioethics and Humanities (2011).

[27] See for example, Edmund D. Pellegrino (1995). See also Edmund D. Pellegrino and David C. Thomasma (1993).

[28] For an extended argument that Confucian ethical teaching was not the most compelling available in its own historical period, see Chad Hansen (2000).

Critics of contemporary medical ethics as a practice make much of these unresolved debates about foundational moral values. H.T. Engelhardt complains, for example, that critical moral and epistemic questions are being begged in the standard appeal to prevailing legal norms and liberal values in medical ethics, given the moral diversity typical of modern liberal democracy. Uncritical acceptance of individual autonomy as an inviolable good, for example, cuts against more traditional views that insist "individuals' wills be subordinated to higher principles or communal values." As Engelhardt puts it, "in the face of moral pluralism there will be different standards of evidence and of sound rational argument. What is one to say about all this? Given this disarray, what legitimacy can bioethics, especially clinical ethics consultation, possibly possess?"[29]

There is a valid point in this, but it is not quite the one Engelhardt is hoping to make. As Stephen Wear notes in defending current practice, those who make clinical consultations or serve on established ethics committees do not have the luxury of waiting for the resolution of the foundational disagreements marking contemporary liberal thought. Decisions must be made, and in practice there is little to be done except to navigate the difficult matters of medical decisions with the parties involved according to accepted standards of practice and generally accepted values such as respect for patient autonomy, informed consent, and non-maleficence. But, Wear points out, it is not as though the values incorporated in contemporary medical ethics are arbitrary or arrived at in mysterious ways. Even if unsettled and often challenged, they reflect the lived tradition within which contemporary medical ethics has emerged, that of liberalism itself, and these decisions are made in better and worse ways as judged by liberalism's standards.[30] They are, to appeal again to Confucianism, part of what informs the *li* of contemporary liberal democracies—this is the established practice of which practitioners must have Intimate knowledge. Such knowledge may point to a kind of expertise, but whether we should consider it *moral* expertise depends finally on our judgments about liberalism itself.

References

American Society for Bioethics and Humanities. 2011. *Core competencies for health care ethics consultation: The report of the American Society for Bioethics and Humanities*. 2nd ed. Glenview: American Society for Bioethics and Humanities.

Ames, Roger T., and Rosemont Henry Jr. 1999. Introduction. *In The analects of confucius: A philosophical translation*. New York: Ballantine Books.

Angle, Stephen C. 2012. *Contemporary Confucian political philosophy*. Cambridge: Polity Press.

Archard, David. 2011. Why moral philosophers are not and should not be moral experts. *Bioethics* 25: 119–127.

[29] Engelhardt (2012).

[30] See Stephen Wear (2005). See also Lisa S. Parker (2005). Parker offers an account of ethical expertise that draws on Care Ethics in a way that I think resonates with the account offered here.

Engelhardt, H.T. 2012. A skeptical reassessment of ethics. In *Bioethics critically reconsidered: Having second thoughts*, ed. H.T. Englehardt. London: Springer.

Hansen, Chad. 2000. *A Daoist theory of Chinese thought*. Oxford: Oxford University Press.

Kasulis, Thomas. 2002. *Intimacy or integrity: Philosophy and cultural differences*. Honolulu: University of Hawai'i Press.

Olberding, Amy. 2012. *Moral exemplars in the analects: The good person is that*. New York: Routledge.

———. 2016. Etiquette: A confucian contribution to moral philosophy. *Ethics* 126: 422–446.

Parker, Lisa S. 2005. Ethical expertise, maternal thinking and the work of clinical ethicists. In *Ethics expertise: History, contemporary perspectives, and applications*, ed. Lisa Rasmussen. Dordrecht: Springer.

Pellegrino, Edmund D. 1995. Towards a virtue-theory normative ethics for health professionals. *Kennedy Institute of Ethics Journal* 5: 254–277.

Pellegrino, Edmund D., and David C. Thomasma. 1993. *The virtues in medical practice*. Oxford: Oxford University Press.

Ryle, Gilbert. 1949. *The concept of mind*. Chicago: University of Chicago Press.

Sim, May. 2007. *Remastering morals with Aristotle and Confucius*. Cambridge: Cambridge University Press.

Singer, Peter. 1972. Moral experts. *Analysis* 32: 115–117.

Wear, Stephen. 2005. Ethical expertise in the clinical setting. In *Ethics expertise: History, contemporary perspectives, and applications*, ed. Lisa Rasmussen. Dordrecht: Springer.

Winch, Christopher. 2010. *Dimensions of expertise: A conceptual exploration of vocational knowledge*. London: Continuum.

Yu, Jiyuan. 2007. *The ethics of Confucius and Aristotle: Mirrors of Virtue*. Oxford: Routledge.

Chapter 3
Ethics Expertise: What It Is, How to Get It, and What to Do with It

Christopher Meyers

Ethics experts abound and are in all walks of life. Consider: The wise mentor, whose experience, observation skills, and judgment allow her to sagely advise; the judge, who finds ways to work within the law to produce solutions that enhance lives; and the ethicist, who has devoted a life to studying millennia of moral theory and concepts and has combined that with often years' worth of challenging in-setting experience. In each of these examples – just a few of the many available – one can reasonably assume such persons will be more skilled than average at reaching better solutions to tough ethical quandaries.

For some such people, the expertise seems to flow effortlessly: Choices emerge naturally, based on years-long acquisition of wisdom; for others, it comes about only as a result of careful ethics reasoning. Whichever version, an affirmation of ethics expertise does not imply that any given person will be an expert at all times on all moral issues. It does, however, affirm that persons who have acquired the right skill set will be more likely to find or to guide others to better ethical solutions.

Note, however, that in listing those examples, my focus is on the *practice* of ethics: In their actions and in their advice, these experts reveal an ability and a commitment to choose ethically. That focus, one that will dominate most of the discussion in this chapter, is concerned less with *what* one is practicing when one does good ethics, and more with *how* to do it, with the reasoning skills that underlie ethics expertise. Said differently, I'm not doing a metaphysics of ethics. Rather, the tenor of my remarks is consistent, as I'll explain more fully below, with a Rossian/Aristotelian approach: All persons who have sufficient experience and judgment – what Ross called "mental maturity" (Ross 1930/1988: 29) – recognize that certain moral claims (e.g., one should keep one's promises, avoid causing harm, and work

Some of what follows, including specific verbiage, is taken from Meyers, 2003 and 2007a.

C. Meyers (✉)
Emeritus, California State University, Bakersfield, CA, USA
e-mail: cmeyers@csub.edu

J. C. Watson, L. K. Guidry-Grimes (eds.), *Moral Expertise*, Philosophy and Medicine 129, https://doi.org/10.1007/978-3-319-92759-6_3

to improve the world) have moral force. And, assuming some caveats about virtuous commitment, the more knowledge and experience one has, the more likely it is that one will make better ethics choices – including in one's role as a clinical ethicist.

This practice emphasis is not to discount theory; rather, it *assumes* it. As I argue elsewhere (Meyers 2016, 2018a, b), good ethics reasoning builds upon the insights revealed in the history of moral theory, with better choices successfully achieving alignment among principles, results and character. For persons engaged in the daily work of practical ethics – clinical consultants included – this approach is so standard as to border on the obvious: Even when we cannot be confident we have determined the one best option in a tough ethical problem, we can at least narrow the choices to a very few better ones, ones that show respect for persons, produce more good (or at least less harm), and are consistent with honorable character. And some persons are better at finding and guiding others to those better choices; that is, they are ethics experts.

If that conclusion is in fact so obvious, why is expertise even an issue? The controversy can be traced, I think, to five explanations:

1. A theoretical commitment to the normative, to there being right answers to ethical problems and to accepting that some persons are more expert at finding them, encountered a deep and pervasive challenge in the moral skepticism of 1950s positivism (and subsequent subjectivism and emotivism). See, for example, A.J. Ayer's particularly blunt critique: "It is silly, as well as presumptuous, for any one type of philosopher to pose as the champion of virtue. And it is also one reason why many people find moral philosophy an unsatisfying subject. For they mistakenly look to the moral philosopher for guidance."[1] Combine this with ongoing debates about moral realism, and it is no surprise there continues to be strong resistance to claims of ethics expertise, both within the broad philosophical community and in clinical ethics.
2. To say some persons are more expert at ethics *reasoning* is not to say the same persons are also expert at *behavior*. Nor does it affirm they will also be good ethics counselors. Being expert at reasoning through problems, even to the point of determining correct ethical solutions, by no means guarantees one will also be expert at motivating persons –oneself included[2] – to act accordingly.
3. There is a common perception that moral decision-making is different, more personal, than other forms of reasoning. The correct ethical choice is, this perception holds, connected to religion or culture, even subjective to individual tastes. Thus, "Who are you to claim expertise and thereby challenge *my* ethical standards?" For the religious, the correct answer is provided via their sacred documents and doctrines; for the cultural or individual relativist, choices are and only can be reduced to group or individual preferences. For all such persons, it is nonsensical, even intrusive, to suggest correctives. Add to this the often very real arrogance attached to ethics theorizing – what Nick Fotion (2014) calls a "strong"

[1]Quoted in Caplan 1989: 64.

[2]See, however, my comments below. See also Marcela Herdova's essay in this volume (Chap. 7).

approach to theory – and it makes perfect sense that many persons would reject *any* claim of expertise. Expanded on below, a short response here is that ethics reasoning is very difficult, laden with a wide array of theoretical, conceptual, political, epistemological, and emotional factors, but it is not *qualitatively* different than other forms of difficult reasoning.
4. One may be quite expert in certain areas of ethical reasoning and a relative dolt in others. For example, based on experience and research, I am pretty good at the ethical questions present in clinical and journalistic environments, but am usually way out of my depth in technical fields like scientific research and computing. To claim all-inclusive expertise is again to steep oneself in arrogance.
5. Even those who are expert in, say, clinical environments – genuinely skilled at making sense of the ethical, medical, legal, and political realities present in such cases – do not necessarily also make good clinical ethics *consultants*. For that, as I discuss in detail below, one also needs emotional aptitude, experience, people skills, and a rich understanding of the cultural norms of one's particular institution.

In short, ethics expertise has two major components: Expertise at ethics *reasoning*, to the point of reliably being better at evaluating problems and coming up with better answers, with the necessary connection such reasoning has to the history of moral theory; and expertise at ethics *consulting*, with the associated range of skills and attitudes.

In this essay, I address both, being careful to delineate what is and isn't included in each. I start by briefly addressing mainly theoretical objections and conclude that well-trained and committed persons can be experts at ethics reasoning, while also stressing that expertise is not an either/or proposition but rather, like wisdom, admits of degrees.

I then address some of the skeptical arguments from bioethics literature and show that it is wholly possible to be an expert ethics *consultant*. I lay out what I take to be the criteria for such expertise and conclude with a reminder that good ethics reasoning is extraordinarily difficult work. The expert needs highly developed skills, including a sophisticated awareness of the long tradition in moral theory and its associated arguments; a nuanced understanding of core concepts and how they relate to particular problems; a proficiency at logical reasoning; a sufficient grasp of the complex array of facts that underlie any ethical issue; and the people skills necessary to judge affected parties' veracity and emotional engagement.

Acknowledging this difficulty further affirms four related caveats:

1. Expertise is variable, across people, contexts, and skill sets;
2. The difficulty of ethics reasoning means one will make mistakes, often when the most urgent of ethical concerns – persons' lives and well-being – are at stake;
3. Humility, thus, should be a dominant character trait for ethicists; and
4. Given that one cannot be a true expert on all the criteria, the best decisions emerge from team-based approaches, wherein different persons' respective expertise combine for better choices.

3.1 A Defense of Expertise

As the points above suggest, to defend ethical expertise is also to defend some version of ethical truth. In ideal circumstances – almost never present in tough real-world problems – the expert should be able to determine the *one* correct answer, or at least she can narrow the options to a very few better choices, also thereby excluding a wide array of bad ones. This approach, thus, also rejects all but the more sophisticated versions of ethical relativism.[3] In simple terms: It is just plain *wrong* – prima facie – to torture innocent persons for no reason, or, for that matter, to gratuitously break promises, cause unwarranted harm, treat persons unjustly, and so forth.

The sorts of ethical quandaries clinical ethicists get called on are of course far more complex and the associated moral reasoning is often difficult, rife with epistemological deficits and emotional resistances. But then so is reasoning in astrophysics or evolutionary biology (or a whole host of other difficult fields), and we do not hesitate to embrace their potential for truth acquisition. Is there something qualitatively different about ethics reasoning?

I think not, in largest part because I also reject supernatural metaphysics – of any sort, but certainly with respect to ethics reasoning. Rather, (more) correct answers in ethics come from a reasoning model that relies on the key insights from the tradition: We should be virtuously committed to relying upon core moral principles as we attempt to effectuate best outcomes, while also striving for what *works* – naturally works, socially works, and politically works. In short, my approach is an amended version of W.D. Ross's quasi-realism (Ross 1930/1988; Meyers 2018a, b, 2016): Once we attain sufficient experience, we come to recognize, in abstraction, the validity of basic ethical principles. When these inevitably clash in dilemmas, the expert relies on the tools of logic to determine better, maybe even best, choices.

On this very practice-driven approach to ethical problem solving, good reasoning is 'merely' a question of determining, to the extent possible, what all is at stake in a problem and using standard informal logic tools to evaluate inferences among ideas. I say "to the extent possible" because such stakes can be quite expansive and one often does not have sufficient access to all the pertinent facts – including institutional culture and politics – nor clear comprehension of all relevant ethical concepts or principles and how they apply in the case. Neither can one always accurately predict how specific options will play out or how those affected will react to them. But the more effectively one reasons through all this, the more likely it is she will reach the best solution. And it is almost certainly the case that she will reach *better* ones, since she will have successfully narrowed them down to only the most plausible, as that is determined by their alignment with principles, results, and character and by their workability.

[3] Any number of approaches to metaphysics and epistemology (e.g., lifeworld, forms of life, and naturalism) are relativist in that they reject the possibility of *absolute* truth declarations, while also embracing "universal" truth ascriptions that are determined via a claim's internal consistency or reflective equilibrium with other accepted claims within that system (Meyers 2016).

All this is what good informal reasoning does – it gives the reasoner the necessary tools for reaching better, if not always the best, choices. Note, thus, the irony in the number of philosophers who reject truth in ethics but are the first to insist that university students simply *must* take a logic or critical thinking course, the clear purpose of which is to teach one how to better analyze information to reach better, more truthful – even sometimes the best, *truthful* – conclusions. The same holds for ethics reasoning, the purpose of which is evaluate a distinctive element of human – and possibly non-human animal – behavior. It can thus be evaluated accordingly, that is, through an analysis of which choices pragmatically promote principles, results, and character. Such an analysis, done well – with the right information, conceptual clarity, and careful reasoning – can in fact reach better conclusions.

Thinking of ethics reasoning in these terms reveals it to be largely an empirical enterprise, what Kant denigrated as mere anthropology. But good ethics certainly demands good facts and expert consultants must have, as detailed below, a toolkit of social science methods. But good ethics goes beyond most social science approaches in that it embraces *normativity* – some behaviors are better than others and *should* be promoted as such. Ethics reasoning of this sort is also dependent on the kind of conceptual and theoretical analysis that is philosophy's stock in trade. And, as should be clear, the best clinical ethics consultants are highly competent in determining what is at stake in problems, analyzing connections, engaging in conceptual clarity, working out plausible consequences, and accurately estimating how various choices align with individual and institutional character – again all standard components of the philosophical repertoire.

3.2 Reconsidering Ethical Expertise in Bioethics

In addition to resistance in philosophical quarters, much of the bioethics literature shies away from expertise language.[4] Consider the following comments, by Michael Bayles and Rosemary Tong, respectively:

> It is ludicrous for a hospital to have an ethicist on a beeper for call to the bedside to make instant thumbs up or thumbs down decisions. The work of ethics involves careful and detailed analysis and reflection, precisely what is not possible in the practical world. Hence, while ethicists can sometimes handle hard cases better than practitioners, their relative strength is not at the level of action. Rather the role of applied ethics is to reflect upon such situations and help practitioners be clearer about what to look for and how much weight to assign to considerations when they must make decisions. Ethicists are not specialists on a par with perinatologists, tax lawyers, and structural engineers (Bayles 1984: 116).

> Even if all the members of an ethics committee had formal training in ethical reasoning, only their *procedural* skills as professional ethicists—and not also their *substantive* conclusions as moral agents—could be non-problematically offered to other moral agents. Ethics

[4] See, however, the Summer 2016 (Vol. 41) special issue of *The Journal of Medicine & Philosophy*. Dedicated to ethics expertise, all the authors adopt some variation of a pro-expertise position.

> is not a set of conclusions that experts pass on to non-experts; rather, it is a set of decision-making tools that non-experts as well as experts must use if they are to reach their own conclusions about what is morally permitted, required, or forbidden to them.... Morality is a *personal* quest (Tong 1991: 418; third emphasis added).

Note there is at least an ironic tension here, since many of those in Bayles' and Tong's camp also take considerable pride from the impact that philosophically grounded bioethics, combined with the work of patient advocates and legal activists, has had in ethically improving medical practice. In the last 40 years alone, far greater emphasis has been placed, for example, on enhancing informed consent requirements, promoting confidentiality, shifting the medical culture away from paternalism and toward autonomy, creating greater protections for research subjects, enhancing the rights of children to have more of a voice in their medical care, and, maybe most importantly if not always with complete success, reinforcing respect for patients as whole persons, rather than as mere physical specimens. That is, bioethicists accept that, in these respects at least, health care is a morally better environment than it was prior to the bioethics movement.

Such irony aside, arguments against expertise and toward an autonomy-driven version of ethics consultations have dominated the literature, as represented in the highly influential *Core Competencies for Healthcare Ethics Consultation* (Multiple authors 2013). Mark Aulisio and Robert Arnold, among the *Core Competencies'* principal authors, voice those views in a commentary to an earlier version of this argument:

> Approximating ethical truth ... should not be the goal of ethics consultation. There is a deep sense in which, on our view, 'ethical truth' is irrelevant for ethics consultation. This is largely due to the fact that issues which arise in the clinic emerge in a context in which individuals retain their political rights to live according to their own ethical views, even if those views turn out to be false from some other particular ethical point of view—even, per hypothesis, the "correct ethical view." We suggest that there is a deep sense in which clinical ethics consultation is more in the domain of the political than the moral, and more in the domain of practice than theory. Given its context, the appropriate question for ethics consultation is most often "Who should be allowed to decide?" rather than "Which view most approximates ethical truth?" (Aulisio and Arnold 2003: 279)[5]

On the surface, this may sound like an endorsement of morally neutral tolerance, to the point of at least bordering on relativism. But I think that is an incorrect reading: Their view, again representative of much of the literature, has a deeply embedded and normative endorsement of the right to autonomy, as understood within a political environment, namely, "to live according to their own moral views, even if those views turn out to be false even from, ... per hypothesis, the 'correct moral view.'".

As I will reinforce below, there is much to be said for being more attuned to the politics of consultations, given their often highly contentious nature and the professional and social contexts in which they reside; managing such politics is part of the process for determining what options will pragmatically work. But note their

[5] Similar arguments are given by Casarett et al. (1998) and Dubler and Liebman (2013).

emphasis on autonomy is also a clear endorsement of ethical truth, that is, to the truth of autonomy being a key, even *the* key, ethical principle.

Now, one can certainly make the case that autonomy should be the default, if also defeasible, principle. While I don't embrace that view,[6] the relevant point here is that such a position is not morally neutral; Aulisio and Arnold's procedural aim is to promote autonomy, since, evidently, it is the only moral principle that one can non-controversially endorse. But this is of course still a clearly normative position, and one made all the more problematic when one realizes just how hard it is to actually achieve autonomous decision-making in clinical environments, particularly in end-of-life choices (Ackerman 1982; Fortunato et al. 2017). Achieving autonomous decision-making is elusive, at best, because of illness, fear, pain, patients' socialization, and because power too often prevails, even when the discussion procedure is conscientiously designed to promote individual autonomy and genuinely democratic processes. Furthermore, it is surely preferable to combine both, that is, to have a process that respects and encourages at least quasi-autonomous participation *and* that gets as close as possible to the right answer, even if that sometimes means overriding patient or surrogate autonomy – think of demands for medically ineffective treatment.[7]

I worry, in fact, that placing so much emphasis on autonomy can actually be a way of *avoiding* moral commitment: If one is unwilling to declare one's moral point of view, one can *claim* a morally neutral tolerance – even if, as Aulisio and Arnold admit, one genuinely believes another moral point of view is correct. Better, they say, to assume that autonomy should prevail, since – per above – who is the ethicist to challenge another's standards? Well, the ethicist is, or should be, someone who is skilled at the reasoning tools that do, in fact, make her an expert. I will elaborate below on what it means to be an expert *consultant*, but for now, realizing one has higher expertise at ethics reasoning does not mean one should act as a moralist, running around the clinical setting, waving her ethics wand, insistently imposing her will on everyone. No, for the political, cultural, professional, and institutional reasons to which Aulisio and Arnold point, she must also be a gifted diplomat, pushing here, backing off there, carefully picking her battles. And, critically, when she sees decisions going contrary to her reasoned judgment, she also has a clear duty to try to mitigate any associated harms by working carefully with those who will enact the choice, paying particular attention to any morally problematic consequences.

To summarize: Ethics reasoning is not qualitatively different than other areas of difficult human inquiry, which also means that ethical truth-seeking is not only a coherent enterprise, it is, or should be, the goal of all ethics reasoners – certainly including clinical ethics consultants. Let me thus turn to a discussion of expertise in ethics consultations. I start this section by continuing the discussion of supposed

[6] The "pragmatic alignment" described above does not give prima facie priority to any principle, but rather judges in every context which one(s) ought to prevail.

[7] This problem, increasingly common in clinical ethics, can plausibly be traced to the terrific job we ethicists have done over the last decades to teach patients and families that they have the right to dictate their medical choices; think of it as the flip side of the autonomy movement.

neutrality; namely, I argue that ethical neutrality – even absent the appeal to normatively–rooted autonomy – is likely impossible and that the explicit acknowledgment of one's normative advocacy brings with it a clear duty to acquire the intellectual, theoretical, emotional, and people skills necessary to a level of ethics reasoning that would justify such advocacy.

3.3 The Myth of Objective Neutrality

Recall Tong's conclusion that the expertise philosophers bring to ethics consultation is their *procedural* skills; philosophers are, she says, particularly well-trained at problem analysis, at impersonally or objectively figuring out what is at stake in ethical problems – who will be affected, what values (or principles or rules) are involved, what are likely consequences. They can then use these to help practitioners reach their own "personal" conclusions regarding the morally best choice.

Tong is certainly right that well-trained philosophers are especially skilled at such analysis and thus can provide real assistance in getting at the heart of ethics problems. As noted above, however, these skills emerge from the standard philosophical methodology of deduction, induction, intuition, and empirical examination. And this method is of course not arbitrarily chosen – philosophers rely on it precisely because they believe reasoned scrutiny gives us better, more accurate, answers. Put another way, philosophers rely upon rational analysis for good reason – because they believe it can get them at least closer to the truth of the matter in these difficult ethical dilemmas.

But note that such skills are anything but objective and normatively neutral. Any analysis of a problem comes with a normative underpinning. The very determination of the considerations that warrant analysis is an explicitly normative enterprise, since it assigns value to some concerns, and to some people, rather than to others. That such normative assumptions are often not transparent, even to the person holding them, obviously does not mean they are not present. Indeed, beliefs without an ethically informed foundation would be far worse, little better than ad hoc reactions to empirical information.

To make matters worse, the very perceptual tools persons rely on for analysis are likely normatively embedded; that is, it may not be possible to perceive even simple facts without filtering them through a conceptual scheme that includes background normative judgments. The full argument is too complex for the limited space here, but a short version is that the professional, social, and institutional cultures in which persons reside help frame how they make sense of the world. Patricia Werhane explains it as follows:

> We all perceive, frame, and interact with the world through a conceptual scheme modified by a set of perspectives or mental models. Putting the point metaphorically, we each run our "camera" of the world through certain selective mechanisms: intentions, interests, desires, points of view, or biases, all of which work as selective and restrictive filters. We each have

> what I call our own metaphysical movies of the world, because they entail projections of one's perspective on the given data of experience (Werhane 1999: 49).

In which case objective analyses of the sort recommended by Tong and others is a myth; they will always be informed via the normatively-laden conceptual scheme the analyst brings to the table. In short, it is both more honest and more rational to acknowledge a point of view and to give reasoned arguments in support of corresponding recommendations. When done correctly, following a good model of ethical reasoning and consultative engagement, those recommendations will imbue a corresponding level of expertise.

The "more honest" point needs emphasis: In any case consultation, the ethicist has almost assuredly reached an opinion and, knowingly or otherwise, will subtly communicate that view in her case analysis. Better, therefore, to be up front and own one's ethical stance. And if she has acquired and implements good ethical reasoning, she should embrace and communicate *through* her expertise (with, of course, the caveats noted above and below about how best to express that expertise). Happily, clinicians are explicitly taught to reflect critically upon expert advice and to use the process as an opportunity to learn, steps they should of course also take with any ethics advice. But imagine a resident's distress at calling for a cardiology consultation, only to have the specialist arrive, engage in sophisticated problem analysis, and then depart without having given any advice as to how to help cure the ailment.

Seeing the ethicist as an analog to the medical specialist, while not precise,[8] is in fact more consistent with the medical ethos, with its ever-narrowing specializations and concomitant attachment of expertise. This is especially true in the teaching hospital environment, where medical students and residents are explicitly trained to seek the help of expert specialists. Such practitioners demand more than mere problem analysis, and rightly so. To provide a rich, complex analysis of a problem – determining what is at stake and then pretending to ethical neutrality – comes across to those working in the trenches as at best awkward and incomplete, and at worst as cowardly. Indeed, it often makes the clinicians' eventual decision-making all the more difficult: The ethicist takes a problem with which clinicians are already struggling and then, through her analysis, makes it all the tougher. In doing this, she also, per the above, directly impacts how the participants ethically view the problem. The least she can do, thus, is own her normative involvement and (carefully – see below) make recommendations.

[8] One view is that medicine, being more scientific, has clear objective facts, while ethics is made up of essentially and irresolvably contested concepts. Both premises are, I believe, mistaken; medicine is not an objective science and ethics is not hopeless relativism. For a nice discussion of this see Yoder (1998).

3.4 The Messiness of Ethics Consultations

"Messy" doesn't actually do justice to the complexity of clinical ethics cases. On any given problem, there is likely to be a room full of rampant and often conflicting egos; power dynamics that supersede even academic conferences; a wide range of (frequently contested) medical, financial, legal, and institutional facts; and, in the midst of it all, a patient or family scared out of their wits. Being an expert on, say, a nuanced take on the categorical imperative is going to be of very limited value in this environment.

But it is equally problematic to jump from there to Bayles's conclusion, quoted above, that "it is ludicrous for a hospital to have an ethicist on a beeper for call to the bedside to make instant thumbs up or thumbs down decisions. The work of ethics involves careful and detailed analysis and reflection, precisely what is not possible in the practical world." Bayles is reacting to a consultant model – that of a solo ethicist, rather than a team – that, while more prevalent in the 1980s, is still frequently an expected part of an ethicist's role: Namely, that one is available 24/7, often only by phone, and for which real advice is being sought.[9] Even granting some hyperbole in his language ("instant thumbs up or thumbs down"), why is it any more "ludicrous" for an ethicist, a person whose professional life has been devoted, one presumes, to the study and practice of ethics reasoning, to make these decisions than for a clinician to do so? Better to leave it to the medical resident in the midst of a thirty-six-hour shift, or to the terribly overworked nurse buried in paperwork, or to the attending physician trying to juggle patient care, resident education, committee meetings, and various other institutional pressures and constraints? Even the moral skeptic must acknowledge that the ethicist can hardly do worse than harried clinicians, and generally will do better, assuming she truly has acquired the skills of ethical reasoning and interpersonal engagement described above and below.

Such skills acquisition clearly reinforces the importance of proper training, of the sort one is unlikely to receive in the all-too-popular two-week intensive bioethics or mediation workshops. As the ongoing discussions about the formal professionalization of clinical ethics consulting attest, it is a distinct intellectual and skills-based field, one that demands extensive training well beyond that achievable in such workshops.

[9] One should be careful not to perceive that all ethics consultants work in large academic settings, with a full team of readily available experts. I am the consultant at four area hospitals for which it is at best hit-or-miss that we can pull together a team. Far more common is that I will get an off-site call on an urgent matter for which they need the best advice I can muster in the circumstances. As uncomfortable as this makes me, I also know that my limited input is better than none at all.

3.5 Expertise and the Clinical Ethicist

If one assumes, as I do and as I think any philosopher must, that some persons are more expert at informal reasoning generally, then, per the arguments in the first part of this essay, there is no reason to suppose that one cannot be an expert at ethics reasoning – in the clinic or elsewhere. To be expert at such work means, again, to be expert at informal logic, interpersonal engagement, empirical investigation, political discernment, and to have at least sufficient competency in medical and legal facts. It also means to have such character traits as perseverance, courage and humility.

Piece of cake, right? In other words, reinforcing the earlier point, it is extremely unlikely any one person will be expert at all of these. Furthermore, I have been using "ethicist" as if that is a monolithic concept and role, when in fact there is no standardized version. Ethicists range from full-time professional staff at high level academic hospitals who are often part of a team of similarly employed consultants, to part-time paid "beeper ethicists" (to borrow Bayles's term), often otherwise employed at a local university, to otherwise full-time members of the hospital staff (nursing, social work, clergy) who have received more training than their hospital counterparts, but who see this work as at most a corollary to their primary responsibilities.

Although I think there are serious problems with the full-time ethicist role (Meyers 2007b), it is unlikely that someone in the last category will have the time and training to attain any real level of expertise as a clinical ethicist. For my purposes here: An ethicist is someone who largely if not wholly defines their professional life in that role, in whatever specific employment capacity they inhabit. Note this means one can be an expert at ethics reasoning but not work in a position for which such expertise is called upon – academic philosophers who specialize in practical ethics but never consult come to mind. In what follows here, however, I focus on the knowledge, skills, and character traits that enhance expertise in the role of clinical ethicist.

One other prelude: The smattering of qualifiers throughout this paper ("sufficiently," "largely," "genuinely," "truly expert") underlines the earlier point that expertise is not an either/or proposition but, like wisdom, admits of degrees; one can be better or worse at each of the criteria or excellent at some and weaker at others. So how much is enough to be deemed an expert? When that person is likely to consistently come up with better solutions to tough ethical problems than her peers. Note, though, that this is a relative characterization, dependent on the environment in which one works and who one's peers are. It also assumes that expertise is nearly always enhanced via a team of consultants, with the associated mix of skills and character traits.

Wherever the ethicist practices, though, she should of course be continually striving to be as expert as possible. Let me, thus, take the criteria in turn:

1. Expert at Ethical Reasoning

The ethicist must have a solid understanding of moral theory and the core insights contained therein. This includes such standards as what it means to promote respect for persons, to universalize moral rules, to promote aggregate qualitative utility, and to use practical wisdom to discern how a virtuous person would choose from among competing moral goods. She must also be cognizant of the meaning[10] of such important moral concepts as "person," "the good," "autonomy," and "honor."

She must further be skilled at informal reasoning, including the ability to differentiate reasons from conclusions and opinions from facts and to understand the logical relationship among ideas – which ideas serve to plausibly imply a given conclusion, versus which, while potentially important in their own right, are logically disconnected from that conclusion. She must also be able to evaluate those logical relations in a mainly dispassionate way, stepping back enough so that any emotional legacy she may bring to the conversation does not unduly influence her reasoning process. And she must be willing to engage all of this *transparently*, openly revealing her reasoning steps and why she drew some inferences rather than others.

2. Expert at Interpersonal Engagement

Persons trained philosophically are more likely to have enhanced expertise in ethical reasoning, whereas they will typically be left in the interpersonal dust by those educated in, for example, social work and nursing. These skills are among the defining features of those professions, and thus of their training programs. Philosophers, by contrast, rarely receive interpersonal training beyond the minimum necessary to manage a classroom or get along with one's colleagues in committee meetings. Hence, we again see the importance of a consultative team, where expertise and responsibilities are shared and mutually reinforced.

The key element of interpersonal engagement is communication. Many ethical dilemmas emerge because one medical service is not clear about another's treatment plan, let alone about what has been communicated to patients or family, or because nursing or social work is reticent, given power differential, to voice concerns in treatment plans. Add to this the ego involvement many physicians bring, along with the emotional turmoil felt by patients, family, and clinical staff, and the ethicist must be particularly skilled at enhancing and mediating effective dialogue. This means getting the right people in the room, creating conditions that give them the safety to voice their concerns, and carefully listening – not just to the words, but to tone and silence. These skills, as I discuss below, overlap with political discernment, since they include successfully reducing the power asymmetry routinely embedded in such conversations.

[10] I do not mean to suggest there are universally accepted, uncontested definitions of these concepts. The expert, in fact, will be versed in those debates and have reached at least tentative conclusions as to which versions are the most plausible.

Finally, someone needs to be skilled at communicating the results of these conversations with patients and families. Given the kind of training ethicists have historically received, and given the level of trust patients and families place upon the white coat, the ethicist should generally not be the primary communicator. Better to leave that to the trusted physician, sometimes with reinforcement from the ethicist or other health care professional.

3. Expert at Empirical Investigation

Good ethics is dependent on good facts and on common agreement over those facts. Many perceived ethical dilemmas come down to (often unknowing) disagreement over the facts of the matter, rather than to genuine value differences, and once everyone gets on the same page, the problem resolves. The expert ethics consultant, thus, cannot be merely a passive recipient of others' descriptions; she must actively seek out information from multiple sources. It is all too easy, for example, to accept a resident's case presentation as accurate and complete, only for the attending physician to come along later with a significantly divergent take. Compound this with the different frames through which different types of health care professionals (and patients and families) view cases, people, and circumstances, and it becomes all the more critical for the ethicist to actively engage all to try to get at the most accurate factual interpretation (Carrese et al. 2012).

Such active investigation extends beyond medical facts to the legal, financial, and religious. While the (non-JD) ethicist cannot practice law, she certainly needs to be cognizant of major bioethics court rulings and legislation and to be able to cite them in relevant circumstances. This is especially true during times of major legal transition, such as California's recent legalization of physician aid-in-dying. As I write this, it has been legal for just over a year and the level of continuing confusion over the most basic of legal facts is striking. It is also striking how often important financial implications – for example, the impact an out-of-town nursing home placement will have on a family – are left out of case discussions. It may not be the ethicist's job to determine the costs and budgetary impact, but, where relevant, she should make it an element of group consultations.

Religious tenets are even more perilous territory, in part because of the general reluctance of medical professionals to question a family's stated beliefs. When, however, those beliefs produce choices that conflict with best medical judgment – especially when the decision-maker is choosing on behalf of the patient – it is wholly legitimate to delve into whether they are giving an accurate representation of their faith's tenets, via input from an appropriate religious authority (Meyers and Eskew 2009).

Add psychological considerations (family dynamics, denial, professional distrust) and the impact of institutional or professional culture (Werhane 1999), and one can see how important it is to engage in active empirical investigations. As is the importance of a team approach, for no single individual will have the time or resources to manage all of it. Part of consultative expertise, thus, is putting together a team where each member is a relative area expert.

4. Expert at Political Discernment

Medical organizations are hierarchical institutions, with established pecking orders and ongoing jockeying for power and place. As with all organizational settings, power impacts who will talk with whom and with what degree of respect. The ethics consultation is not immune from this – those with power dominate, while others are reticent to participate, let alone challenge, often with the result that vital information is excluded from the conversation.

One method for at least reducing the impact of power is to strive for consensus among all participants through a facilitated conversation. Better versions of this rely on models of deliberative democracy (Gutmann and Thompson 2002), in which decisions are communal and based on rational engagement. To enhance the likelihood of consensus, conversations can be structured to ensure "that everyone who would be affected by a decision is able to participate" (Casarett, Daskal and Lantos: 8).

Imagine, thus, a committee-based ethics consultation in which all those relevantly affected by any resultant choices are equal participants in a dialogue that seeks to produce consensus through reasoned debate. Does this idealized version exist in real-world case consultations? In my thirty-plus years of experience, it is exceedingly rare. Most committees have not adopted the formal mechanisms (e.g., allotted time for speaking, prohibitions against interruptions, a moderator who creates safe space for hesitant participants) that would encourage, let alone require, that type of discussion, in part because the very thing the mechanisms are trying to obviate – power asymmetry – is the major obstacle: Those with power have to agree to establish the mechanisms in the first place and then to follow them in practice. That is, they must be willing to give up power, hardly a common phenomenon (Young 1997; Fraser 1989).[11]

The impact of power is less pronounced in (typically academic) institutions that have well-established ethics programs that are seen as an integral part of the hospital's effective provision of quality health care. Less pronounced, though, is not absent, especially when gender is added to the mix, i.e., when the traditional holders of power – physicians – are typically men and other committee members – social workers, case management, and nursing – are women. For the many institutions in which ethics is largely a tertiary function, where ethics consultations are mainly ad hoc with floating committee members and ill-defined processes, persons with pre-established positions of power – physicians and administrators – dominate the conversation; they control the tone, the content, and often even whether the conversation takes place at all. In this context, it is unreasonable, *unfair*, to ask a vulnerable nurse or social worker – let alone a patient or family member – to assertively engage an attending physician, especially one who is at the meeting at best begrudgingly.

The ethicist who embraces her role as a power equalizer (Meyers 2007b, 35) can do much to reduce the asymmetry, but only if she has enough institutional authority that she feels sufficiently empowered to speak on behalf of the vulnerable. But to do

[11] As Young and Frazier note, in addition to role status, power is also largely attached to gender, race, and age, namely, power still largely resides with older white men.

that well, she must also be politically savvy, knowledgeable of hierarchies – both those common to the profession (for example, neurosurgeons generally have more power than pediatricians) and historically generated within her particular institution – and able to discern how power impacts specific consultations. She can then use her role to attempt to mediate the associated asymmetries, to give power to those who lack it. This can be as easy as asking, say, a unit nurse for her opinion on the case, or as hard as directly challenging a power player's facts or conclusions. I would note, however, that any such power mediation requires that the ethicist be willing to take risks, that is, to be truly independent of institutional power structures (Meyers 2007b).

3.6 Training

It should be clear that I consider the ethicist's role to be highly demanding. Ethical reasoning alone is extraordinarily difficult, given the biases, fears, hopes, and dreams we all bring to such work, and for the expert ethicist, reasoning is just one piece of the puzzle. For those who take it on as their primary professional identity, considerable training is essential and should include at least the following:

Extensive, post-baccalaureate training in ethics theory and practice, of the sort characteristically taught in philosophy MA or PhD programs;
Extended experience in clinical settings with sufficient exposure to medical terminology and disease processes;
Some academic exposure to the psychology and dynamics of organizational culture;
Some academic exposure to bioethics legal history; and
Some training in mediation techniques.

Fortunately, there are now any number of excellent degree programs that meet these criteria[12]; I would just emphasize that the many short-term, usually certificate-based, programs might be excellent as a refresher or to give secondary members of a consultative team a deeper understanding of ethics work, but they alone cannot substitute for the far more extensive training needed for genuine expertise.

3.7 Conclusion

Note that nowhere in these discussions have I said anything about the expert ethicist also being an exemplar of ethical *behavior*. It is possible to be a relative expert at all the listed skills and still be a sleazeball in one's personal life. The very limited

[12] The Hastings Center keeps a running list of post-graduate programs in bioethics. See: www.thehastingscenter.org/publications-resources/bioethics-careers-education/graduate-programs-2/

empirical research suggests ethicists are no more likely than other professors to behave ethically; they might even be worse (Schwitzgebel 2014). A common joke points to one explanation: Expertise in ethical reasoning simply makes one more corrupt because one is better able to use sophistry – clever rationalization – to come up with a justification for most anything. Schwitzgebel's answer is more sympathetic: There is no reason to think there should be a causal connection between the study of, maybe even the professional practice of, ethics and individual behavior (Schwitzgebel 2013).

I think this is partly correct.[13] For those who merely study, write about, and teach moral theory, meta-ethics, and problems in practical ethics, that is, those who do not engage in regular in-setting consultations, the "rationalization" assessment makes sense, given the distancing typically attached to such purely intellectual exercise. In Tong's language, "abstract thought, detached from faces and real-life situations, depersonalizes all involved.... [Members of] ethics committees must be careful lest their attempt to be impartial makes them treat people as equal, faceless, interchangeable atoms in the universe" (Tong 1991: 419). Combine the competitive argumentation that is so common to philosophy with the recognition that classroom discussions and academic writing typically means keeping the subject at an intellectual and merely abstract arms-length, and it is easy enough to think of ethics problems as mere puzzles to be solved.

Consider, though, a central, and I think correct, tenet of virtue theory: With time, experience, good mentorship, and a commitment to self-improvement, one becomes better at ethical understanding and behavior. Consider also the kind of experience that comes with the professional practice of clinical ethics, where the 'puzzles' are as hard as anything a text or classroom discussion can dish up *and* they are real – real people's lives, real emotions, real pain and joy. One would have to be callous in the extreme not to be profoundly affected by, changed by, *improved* by, regular work with such cases.

This brings us back full circle to what it means to have ethics expertise. On my argument, it is mainly about the acquisition of skills. At a general level those are the skills of ethics reasoning, that is, of using the tools of informal reasoning to make it more likely that one will be better than the average person at analyzing and providing solutions to tough ethical problems. The expert consultant then combines those reasoning skills with ones of interpersonal engagement, empirical investigation, and political discernment, as acquired through proper training and extensive experience and informed by the wisdom-enhancing reality of making choices that profoundly impact others' lives.

To *behave* like an ethics expert entails all those skills *plus* the right character traits, including benevolence, generosity, courage, perseverance, and humility. Again, setting this as the standard for overall ethics expertise – of judgment *and* action – places it quite high, as voiced by Sidney Callahan:

[13] Herdova, in this volume (Chap. 7), takes this argument even further.

> Excellent ethical judgment, or better yet, ethical discernment, demands personal moral wisdom and an ardent commitment to seek the true and the good, [all present in a] good person,.. [one who possesses] emotional literacy, mature self-discipline, cultivated interpersonal sympathies, and a steadfast personal commitment to high moral standards of worth (Callahan 1994: 25).

But of course, the standard *should* be quite high, because the work of ethics consulting is both hard and involved with the most important of considerations: Consultants are called upon only when the cases are particular vexing and resulting decisions are almost always about life or death choices.

In case all this sounds just too daunting, recall that ethics expertise it is not an either/or status: One can be more or less skilled and have adopted the needed traits to greater or lesser extent. That is, one can be more or less of an ethics expert. But one cannot be tepid about one's concerted *commitment* to getting it right, to taking on the challenges – of skill and character – to attain higher levels of expertise. Furthermore, whatever level one is at, that expertise will almost certainly be enhanced by an effective consultative team, with each member bringing her own skills and traits of character, resulting in the whole exceeding the sum of its parts.

References

Ackerman, Terrence. 1982. Why doctors should intervene. *Hastings Center Report* 4: 14–17.

Aulisio, Mark P., and Robert M. Arnold. 2003. Ethics consultation: In the service of practice. *The Journal of Clinical Ethics* 14: 276–282.

Bayles, Michael. 1984. Moral theory and application. *Social Theory and Practice* 10: 97–120.

Callahan, Sidney. 1994. Ethical expertise and personal character. *Hastings Center Report* 14: 24–25.

Caplan, Art. 1989. Moral experts and moral expertise: Do either exist? In *Clinical ethics: Theory and practice*, ed. Barry Hoffmaster, Benjamin Freedman, and Gwen Fraser Clifton, 59–87. New York: Humana Press.

Carrese, J. A et al. 2012. The members of the American Society for Bioethics and Humanities Clinical Ethics Consultation Affairs Standing Committee. Undated. HCEC Pearls and Pitfalls: Suggested Do's and Don't's for Healthcare Ethics Consultants. Available at http://asbh.org/uploads/publications/Pearls_and_Pitfalls.pdf. Accessed 18 Oct 2017.

Casarett, David J., Frona Daskal, and John Lantos. 1998. The authority of the clinical ethicist. *Hastings Center Report* 28: 6–11.

Dubler, Nancy N., and Carol B. Liebman, eds. 2013. *Bioethics mediation: A guide to shaping shared solutions*. New York: The United Hospital Fund.

Fortunato, John T., Jason Adam Wasserman, and Daniel Londyn Menkes. 2017. When respecting autonomy is harmful: A clinically useful approach to the nocebo effect. *The American Journal of Bioethics* 17: 36–42.

Fotion, Nick. 2014. *Theory vs. Anti-theory in ethics: A misconceived conflict*. New York: Oxford University Press.

Fraser, Nancy. 1989. *Unruly practices: Power, discourse, and gender in contemporary social theory*. Minneapolis: University of Minnesota Press.

Gutmann, Amy, Dennis Thompson, and Dennis. 2002. *Why deliberative democracy?* Princeton NJ: Princeton University Press.

Meyers, Christopher. 2003. A defense of the philosopher-ethicist as moral expert. *Journal of Clinical Ethics* 14: 259–269.

———. 2007a. *A practical guide for ethics consultants: expertise, ethos and power*. New York: Rowman and Littlefield.

———. 2007b. Clinical ethics consulting and conflicts of interest: structurally intertwined. *Hastings Center Report* 37: 32–40.

———. 2016. Universals without absolutes: a theory of media ethics. *Journal of Media Ethics* 31: 198–214.

———. 2018a. Ethics theory and ethics practice. In *Ethics across the curriculum: Pedagogical perspectives*, ed. Elaine Englehardt and Michael Pritchard . New York: Springer (forthcoming). pp. 131–145

———. 2018b. *The professional ethics toolkit*. Oxford: John Wiley and Sons (Forthcoming).

Meyers, Christopher, and Stewart Eskew. 2009. Religious beliefs and surrogate medical decision-making. *Journal of Clinical Ethics* 20: 192–200.

Multiple authors. 2013. Core competencies for healthcare ethics consultation: Report of the American Society for Bioethics and Humanities. Glenville, IL: ASBH.

Ross, William David. 1930/1988. *The right and the good*. Indianapolis: Hackett Publishing.

Schwitzgebel, Eric. 2013. The moral behavior of ethics professors and the role of the philosopher. The Splintered Mind: Reflections in Philosophy of Psychology, Broadly Construed. Available at: http://schwitzsplinters.blogspot.com/2013/09/the-moral-behavior-of-ethics-professors.html.

———. 2014. The moral behavior of ethics professors: Relationships among self-reported behavior, expressed normative attitude, and directly observed behavior. *Philosophical Psychology* 27: 293–327.

Tong, Rosemary. 1991. The epistemology and ethics of consensus: Uses and misuses of 'Ethical' expertise. *Journal of Medicine and Philosophy* 16: 409–426.

Werhane, Patricia. 1999. *Moral imagination and management decision making*. New York: Oxford University Press.

Yoder, Scott D. 1998. The nature of ethical expertise. *Hastings Center Report* 28: 12–13.

Young, Iris Marion. 1997. Communication and the other: Beyond deliberative democracy. In *Iris Marion Young, Intersecting voices: Dilemmas of gender, Political philosophy, and policy*, 60–74. Princeton: Princeton University Press.

Chapter 4
Why Moral Expertise Needs Moral Theory

Michael Cholbi

Philosophical and ethical literature concerning whether moral expertise exists and who (if anyone) might possess it has proliferated in recent years. My purpose here is not to address either of these questions, at least not directly. Rather, my concern is to investigate a matter that has been somewhat neglected in this literature, namely, the relationship between moral expertise and moral theory. With a few exceptions, debates about the nature and distribution of moral expertise have proceeded with relatively little attention to moral theory. No doubt part of the explanation for this inattention is that many of the contributors to these debates are ultimately concerned with whether a particular class of individuals, namely clinical bioethicists, are (or should be treated as) moral experts. (See among many Rasmussen 2011; Priaulx 2013) A central theme of my discussion will be that the neglect of moral theory in philosophical debates about moral expertise is unfortunate inasmuch as moral expertise is much more entangled with moral theory than contributors to these debates typically acknowledge. In particular, I shall attempt to show that moral expertise is theory-dependent. By this, I mean, first, that moral expertise consists, at least in part, in knowledge of the correct or best moral theory, and second, that knowledge of moral theory is essential to moral experts dispensing expert counsel to non-experts. Roughly then, if utilitarianism is correct, then a moral expert must embrace utilitarianism and invoke it in support of her moral testimony; if Kantianism is correct, then a moral expert must embrace Kantianism and invoke it in support of her moral testimony; etc. Hence, moral experts would not be moral experts absent knowledge of moral theory, nor could they play the role we would expect them to play in moral inquiry and deliberation absent such knowledge.

My plan is as follows. In Section 4.1, I suggest that debates about moral expertise are better served not by efforts to define the notion but by attempting to identify those features of moral expertise that are responsible for scholarly controversies

M. Cholbi (✉)
California State Polytechnic University, Pomona, CA, USA
e-mail: mjcholbi@cpp.edu

J. C. Watson, L. K. Guidry-Grimes (eds.), *Moral Expertise*, Philosophy and Medicine 129, https://doi.org/10.1007/978-3-319-92759-6_4

about its nature and distribution. Section 4.2 proposes that there are two such features, captured in what I call the epistemic condition and the testimonial condition for moral expertise. Sections 4.3 and 4.4 seek to demonstrate that neither of these conditions is theory-independent; that is, in order for a putative moral expert to satisfy these two conditions, she must possess and make use of knowledge of the correct or best moral theory.

There is one sense in which moral expertise might depend on moral theory that I do not address here. Some philosophers have thought that irrespective of whether there are individuals with the moral knowledge necessary to be moral experts, there is itself something morally questionable about our treating others as moral experts, perhaps because obeying the moral dictates of others would amount to foregoing or denying our own autonomy or agency. (Wolff 1970; see also D'Agostino 1998 and Driver 2006) A moral theory might, in other words, condemn moral agents' deferring to moral experts. If so, then a moral theory rules out not expertise itself but the practice of treating others as experts. This 'deontic' dependence is not the sort of dependence of moral expertise with which I am concerned here. My concerns are with whether moral expertise is theory-dependent metaphysically, in that knowledge of the correct or best moral theory constitutes moral expertise, or epistemically, in that such knowledge has a part to play in the experts' moral testimony being rightfully treated by others as expert testimony.

In investigating the relationships between moral expertise and moral theory, I do not aim to hash out the merits of rival moral theories in an effort to identify which of these theories is correct or most defensible. Whether the fact that a moral theory renders moral expertise intelligible is a mark for or against its plausibility depends crucially on the degree to which rendering moral expertise intelligible is an important desideratum on moral theories generally. Sentiments vary on that matter, I expect. Some moral philosophers may insist that a moral theory needs to make sense of moral expertise, whereas others, upon discovering that a particular theory problematizes, or even precludes, moral expertise, would respond to this finding with indifference or even enthusiasm. I do not weigh in on this metaphilosophical dispute here. I merely hope to clarify the relationship between moral theory and moral expertise and leave it to others to draw out the implications my conclusions may have in appraising rival theories.

4.1 The Stakes of the Moral Expertise Debate

Precisely what moral expertise consists in is far from obvious. Julia Driver (2013) has recently suggested that debates about moral expertise come to loggerheads when different senses of moral expertise are conflated. She notes that moral experts may be expert moral judges, especially adept at arriving at correct moral judgments; expert moral practitioners, those who act morally well more than others might; or expert moral analysts, those who have greater insight into the nature of morality. Driver suggests that this diversity of forms of moral expertise is unsurprising given that expertise in a given domain may take on different guises. Compare linguistic expertise. Those able to speak a given language with great fluency, those with excellent knowledge of that language's grammar, and those capable of crafting compelling poetry in that language can all plausibly be called experts in that language despite their respective aptitudes being only contingently correlated to one another.

Driver's observations underscore that the project of *defining* moral expertise is likely to be misguided. 'Moral expertise' is not an everyday term from which we can extract different candidate definitions that can then be compared to ordinary usage, and it is therefore likely that any attempt to define moral expertise will strike partisans in debates about its nature and distribution as attempts to shape such debates by definitional fiat. Hence, I make no pretense of defining moral expertise here. Rather, I follow what I believe is a more fruitful method, namely, one that begins by asking why moral expertise is hotly debated in the first place and what is concretely at stake in such debates. What practical difference does it make if there are moral experts — or put differently, what must moral experts be if their existence makes a practical difference?

In this connection, that the question of whether bioethicists (or moral philosophers) are moral experts has been so prominent in these debates is telling. Technologically advanced liberal democratic societies put moral expertise in a precarious social position. On the one hand, such societies tend to suppose that consultation with experts is essential to crafting wise decisions or policies. Respectful of science, such societies usually have large numbers of officials, scholars, etc., who, thanks to their expertise, wield disproportionate influence over public choice and action. At the same time though, these same societies tend to embody the Rawlsian picture of liberal society, with their members endorsing a diversity of reasonable conceptions of the good while aiming to respect value pluralism and individual autonomy. (D'agostino 1998; Kuczewski 2007; Kovács 2010) These two features of such societies generate competing demands that render *moral* expertise problematic in ways that other forms of expertise seemingly are not: The authority of experts should be respected, but the judgments of would-be moral experts are not and should not (thanks to value pluralism, autonomy, and so on) be invested with the same authority as the judgments of other expert authorities. This problematic is in evidence in debates about whether clinical ethicists are moral experts. In an institutional context in which patient autonomy is given normative priority and paternalism is frowned upon, what place is there for an individual whose expertise is not scientific or medical but ethical, and in what sense ought patients defer to these experts?

Debates about moral expertise, I propose, therefore acquire their practical stakes from worries about whether, in liberal democratic societies in particular, there ought to be a class of individuals acknowledged and treated as having disproportionate moral authority. They are thus debates are whether anyone should be understood as suitable for playing the *social role* of moral expert.

My larger concern here is what place knowledge of moral theory would need to have for moral experts to be suited to play this social role. But the more immediate point is that we need not settle every detail about the nature of moral expertise in order to home in on what moral expertise would have to consist in so that there could be individuals who play the social role about which controversy has arisen. Let us now consider in greater depth the social role of experts and what features moral experts must have in order to account for the controversies surrounding their existence and distribution.

4.2 Two Features of Moral Experts

One platitude about moral expertise is that it is a species of expertise. I contend that the most succinct way to capture the notion of expertise is to say that experts are those whose testimony in their respective areas of expertise ought to be trusted. That expertise involves experts being trusted illustrates that expertise gains conceptual traction in those areas of human endeavor in which the knowledge relevant to that endeavor is not equally distributed, and as a consequence, some individuals must defer to the judgments of others if the latter are to make correct choices within that endeavor. While a scientific layperson ought to trust the testimony of a quantum physicist regarding how subatomic particles behave, two expert quantum physicists typically have no need to *trust* one another's testimony regarding quantum physics. This is not because disagreement among experts in some area is impossible. Rather, experts are in possession of area-specific knowledge that enables them to arrive at their own considered judgments in that area, so that when disagreement among them may arise, individual experts draw upon that area-specific knowledge so as to appraise or verify other experts' testimony. Experts thus have no need to trust other experts, for to trust another's testimony is to invest confidence in its veracity despite being unable to wholly verify or certify the truth or justifiability of that testimony in the way that experts can. Being peers, experts do not appraise each other's judgments in their exact area of expertise with the deference that characterizes non-experts' stance toward experts.

There might appear to be instances wherein experts in a given field ought to defer to other experts in that same field. For instance, experts may rightfully defer to other experts when they have reason to believe that their expertise would be compromised by self-interest or a lack of partiality. The adage "a lawyer who represents himself has a fool for a client" underscores this possibility. An expert lawyer does not lose her expertise when she represents herself, but her capacity to offer expert counsel may be weaker precisely because of her personal proximity to the matters calling for expert judgment. But note that this example is not best described as an expert deferring to another expert. Rather, a would-be expert recognizes that because her expertise may not have its usual level of reliability, she is not properly treated as an expert in this particular instance. Hence, she is in effect a non-expert deferring to an expert.

That experts are those whose testimony in their respective areas of expertise ought to be trusted also accounts for why expertise has little traction in those areas where knowledge is widely distributed (we have little need for experts in the addition or subtraction of single digit numbers) or where there is good reason for skepticism that there is knowledge in any robust or objective sense (experts in unicorn biology, say.) (McConnell 1984).

Expertise thus has a set of background social conditions that elicit it and make it salient, namely, when

(a) individuals vary in their levels of knowledge relevant to some endeavor,[1]
(b) some individuals acknowledge that they lack the knowledge adequate to make choices within that endeavor, and
(c) individuals in that class seek out experts in that endeavor with the aim of accepting the experts' testimony for the purpose of making choices within that endeavor.

Let us call this set of conditions *the expert context.*

Expertise thus rests on an asymmetry between knowers. Non-experts look to experts in order to judge, choose, or act on the basis of truths they acknowledge an inability to discern or justify adequately on their own. "Recognition of expertise," David Archard (2011: 120) observes, "gives the non-expert a good reason to endorse the judgment of the expert, a judgment she would not otherwise make or have a good reason to make." (See also Watson, forthcoming) Non-experts' deference to a bona fide expert's testimony thus improves the epistemic standing of the non-experts' judgments in the expert's area of knowledge, but not by providing evidence *intrinsically* relevant to the truth of those judgments. In deferring to another's expertise, a non-expert is implicitly disavowing her ability to fully or adequately evaluate the evidence relevant to the judgments about which she defers. The non-expert's evidence for accepting the judgment in question consists in large measure of the fact that an expert attests to the truth of the judgment. The expert's competence in evaluating propositions within her area serves as the non-expert's primary reason for accepting the expert's testimony. The expert is, we might say, an epistemic surrogate for the non-expert.

When the moral expert offers moral testimony, that testimony is correct (*when* it is correct) in a non-accidental way. For she does not issue correct moral testimony simply by chance. Her presumed knowledge of the moral domain not only explains her testimony, i.e., her moral beliefs do not only account for why she makes the moral utterances she does. Her knowledge is also what lends that testimony its credibility. But what exactly does the moral expert have knowledge of?

A moral expert's area of expertise is *practical* and *normative*. (Iltis and Sheehan 2016) Moral expertise is practical in the sense that it is constituted by knowledge of first-order moral propositions attributing deontic status to choices, traits of moral character to individuals, etc. This claim is subject to a careful qualification: A person's moral expertise may be highly domain-specific (expertise in the morality of war or of corporate accounting, say), so that while her advice within that domain

[1] Expertise need not involve only propositional knowledge. A good many crafts, arts, etc., involve expertise that consists largely in "knowledge how." See Dennis Arjo's contribution to this volume (Chap. 2) for how the concept of expertise might include "knowledge how."

can be trusted, her moral advice concerning other domains may be no more reliable than the average person's. Moral expertise is normative rather than descriptive because it includes knowledge of which putative moral truths are in fact true rather than any kind of sociological or psychological knowledge of what moral claims individuals or groups accept or why they accept them. This is not to say that moral expertise has no descriptive or empirical component. For it will often be true that having greater moral knowledge than the average person is explained in part by having greater knowledge of non-moral but morally relevant facts. For example, an expert in end of life ethics needs to have at least rudimentary factual knowledge of which medical interventions tend to sustain or end life. (Conversely, a person may be disqualified from being a moral expert by virtue of ignorance of applicable empirical facts). The moral expert's expertise is moral because and to the extent that her knowledge of first-moral propositions exceeds and cannot be conceptually reduced to her knowledge of some body of non-moral facts. Moral expertise is thus compatible with any metatethical stance that affirms the existence of genuinely normative moral truths.

This characterization of moral expertise leaves some loose ends (for example, how to distinguish moral norms from other norms, such as those of etiquette (Foot 1972)),[2] but it points the way toward two conditions individuals must meet in order for them to play the controversial social role of moral expert.

Expertise presupposes an epistemic asymmetry between experts and others. But how is this asymmetry best understood, i.e., to what degree on in what respect are experts more knowledgeable than others? (Driver 2013, Watson forthcoming) Answering this question seems to require having some idea of how knowledgeable non-experts are about first-order morality. In my estimation, that moral controversies or dilemmas receive so much popular and scholarly attention should not obscure that first-order moral knowledge, particularly with respect to relatively straightforward moral phenomena, is pretty widely distributed among human moral agents. Rare are human moral agents who do not know of the moral presumptions against deception or causing harm or injury or in favor of keeping promises (their behavioral adherence to these presumptions is another matter). Moral experts must therefore have significantly more first-order moral knowledge than this. Let us say that a person whose first-order moral knowledge exceeds this threshold satisfies the *epistemic condition* for playing the social role of moral expert.

A second condition for moral experts to play their social role reflects what I earlier called the expert context. A moral expert does not merely have a disproportionate amount of moral knowledge. Other non-expert individuals must, in order for her to play the social role of expert, see her moral testimony as more reliable than theirs. These non-expert individuals are striving to make moral judgments they understand themselves to be insufficiently competent to make absent expert testimony or counsel. As I once expressed it:

[2] See Dennis Arjo's contribution to this volume (Chap. 2) for a discussion of how social norms can have moral significance.

> Much in the same way that nothing counts as a chair that cannot be reliably sat upon, so too no one can count as an expert in morality who does not satisfy the very expectations of those who wish to utilize said expertise for practical purposes. The main expectation is the provision of reliable moral advice. (Cholbi 2007:329)

Being a practical affair, morality is ultimately about choosing and doing, and so in deeming someone a moral expert, we are not primarily expressing our esteem or wonderment, as we might upon learning that an individual can recite π to a large number of digits or provide an off-the-cuff accounting of the causes of the Great War. We are instead expressing a confidence in their moral counsel. Hence, we would be loath to treat someone as a moral expert absent a justifiable confidence in the comparative reliability of their moral testimony. A moral expert, then, is someone whose high level of moral knowledge justifies her being treated as a reliable purveyor of moral advice because other individuals are always *pro tanto* warranted in trusting that individual's moral testimony. Call this second condition the *testimonial condition* for moral experts' discharging their social role.

It is crucial to recognize that moral knowledge and its transmission or acceptance by others do not march in lockstep. There is certainly nothing inconsistent about S being a moral expert in the sense of having much more first-order moral knowledge (S meets the epistemic condition) while S's moral testimony is nevertheless not worthy of others' trust (S fails to meet the testimonial condition). The converse does not hold, however: Any knower whose moral testimony is trusted must, in order for that trust to be warranted, actually possess first-order moral knowledge. In the absence of such knowledge, the trust of non-experts would be misplaced. For it is this knowledge that renders the moral expert "*deserving* of trust with respect to their moral judgments." (Driver 2006: 625, emphasis added)

4.3 Theoretical Knowledge and the Epistemic Condition

To return to the question at hand: To what extent must individuals have theoretical knowledge of morality in order to fulfill the social role played by ostensible moral experts?

It might seem obvious that a moral expert must know the correct or best moral theory. After all, experts in a given area standardly possess a body of knowledge that can at least loosely be called 'theoretical'. This will be the case with respect to most academic knowledge, but it will also be true in cases of more craft-like knowledge. A skilled woodworker, for instance, may not know the principles of physics that apply to wood, but she will certainly know general principles about how wood is affected by various causal processes and which processes to follow in order to realize desired designs or effects. Why deny, then, that theoretical knowledge is essential to moral expertise?

Dien Ho (2016) believes that the possibility of moral expertise requires neither that there be any correct moral theory nor that experts know which moral theory is correct. Ho's position rests on two arguments. I take up the first argument here, the second argument in the subsequent section.

Ho develops his position regarding moral theory's role in moral expertise in response to a common skeptical argument (Ho 2016: 1–3): Moral expertise is impossible unless there are true moral theories of which we can have knowledge. But because either there are no true moral theories, or we cannot know which moral theories are true, then moral expertise is impossible (Frey 1978; Crosthwaite 1995; McGrath 2008; Cross 2016). Ho responds to this skeptical argument by appealing to observations regarding scientific theories and expertise. There can sometimes be "fundamental disagreements" in science that do not "lead to skepticism towards scientific expertise," Ho observes (2016: 371). Basic disputes in physics, for example, need not be resolved in order for us retain our confidence in the existence of expert knowledge in physics. Quantum mechanics and general relativity are incompatible theories, but their incompatibility does not seem to entail that there are no expert physicists, much the less that physics expertise is impossible. But if fundamental theoretical disagreement in science does not vitiate the prospects for scientific expertise, neither should fundamental theoretical disagreement in ethics vitiate the prospects for moral expertise. In the terms laid out in the previous section, Ho seems to be denying that experts need have knowledge of moral theory in order to satisfy the epistemic condition for individuals to play the social role of moral expert. Experts can issue expert moral judgments, and moral inquiry can proceed on familiar terms, even if experts lack knowledge of the true or correct moral theory. Therefore, expert moral knowledge need not be theoretical knowledge.

Ho does not develop a robust account of why theoretical disagreement in science does not cast doubt on scientific expertise. But here is a conjecture that is at least compatible with Ho's remarks on the matter: Of the various criteria used to evaluate scientific theories, predictive power is most central. A scientific theory that issues predictions that are not borne out by relevant observations should be rejected. But sometimes multiple theories are compatible with relevant observations, in which case, all other things being equal, there does not seem any basis for preferring one such theory over another. We do not know which of these theories is true or correct, and indeed, on some 'instrumentalist' conceptions of science, there is not much more to a theory's validity than its predictive success. (Popper 1962) Hence, partisans of any predictively satisfactory theory can rightfully function as experts despite their claims to theoretical knowledge being contestable. Hence, Ho argues, perhaps we can get along just fine in scientific inquiry, reaching true conclusions, etc., even if there exists uncertainty at the level of theory. And so to the extent that experts contribute to scientific inquiry, their contributions need not presuppose the truth of any theory. Extrapolated to the moral case, moral expertise need not involve theoretical knowledge.

I shall assume *arguendo* that Ho is correct that the lack of theoretical knowledge does not undercut scientific expertise. Yet if this is Ho's intended defense of the dispensability of theoretical knowledge to moral expertise, it does not seem to show that theoretical knowledge is inessential to moral expertise though.

First, Ho's position seems to show that when there are no first-order disputes, theoretical disagreement does not impugn claims of expertise. If several scientific theories are predictively adequate, that some experts defend one theory while others defend another theory does not call into question their playing the epistemic roles of experts. But the expert context in which moral experts play their distinctive roles are contexts in which the moral equivalent of scientific observations — first-order moral judgments or 'intuitions' — are themselves contested. Moral non-experts look to moral experts in part to ascertain what they *ought* to 'observe', not to settle the theoretical significance of uncontroversial or already established moral 'observations.' Thus, the contexts in which theoretical disagreement in the sciences does not impugn scientific expertise (where predictions converge across theories) look to be a special case, largely inapplicable to the very contexts in which moral experts will be called upon for their moral expertise.

We will return to questions about theoretical knowledge and the expert context in the next section. But perhaps this criticism is uncharitable to Ho. It could be that the dispensability of theoretical knowledge to expertise is most *in evidence* in contexts where there is first-order agreement, but it is dispensable in any context. Yet even conceding this, it is not clear that representing the relationship between moral theories and moral judgments as akin to the predictive relationship between scientific theories and scientific observations does justice to the place that moral theories have in moral knowledge.

Suppose that some scientific theory T implies some observation O, and that O is observed in some experimental setting. O thereby confirms T. On its face, moral inquiry may appear to conform to this picture. Suppose that a moral theory M implies (in conjunction with relevant empirical facts) a first-order moral judgment N. Suppose further that N strikes us as antecedently plausible, i.e., we have a 'pre-theoretical' intuition in favor of N. We might say that N thereby 'verifies' M. Conversely, if N is antecedently implausible, then this counts against or 'disconfirms' M. N is thus evidentially relevant to the truth of M, though of course in neither case need we assume that N provides *conclusive* evidence for or against M. More specifically, should N 'disconfirm' M, we might take this as grounds for modifying or qualifying M rather than rejecting it altogether. Following the method of reflective equilibrium, we may undertake multiple iterations of this process with the aim of identifying the moral theory or principles that reflect the maximally coherent relationship between candidate moral principles and our intuitive moral judgments. (Rawls 1971). So far then, it may seem that Ho's analogy between scientific and moral inquiry holds. Moral intuitions 'predicted' by a moral theory vindicate that theory in a way structurally similar to the way observations vindicate a scientific theory.

In the scientific case, the centrality of predictive success is what lends credibility to the theoretical agnosticism to which Ho refers. That T predicts O and O is observed seems adequate to the aims of scientific inquiry. To add 'and furthermore T is *true*' is either to be redundant or to wade into philosophically contentious territory. T ought to be accepted (in part) *because* its predicted observations are borne out. The justificatory relationship here thus runs from the observations to theory.

The same cannot, I think, be said, of moral theory and moral judgments. That scientific and moral inquiry share an apparently similar hypothetico-deductive structure obscures differences between the role that theories play in these domains.

To see why, consider again a process of verifying a moral theory by appeal to intuitions or first-order moral judgments. Suppose that utilitarianism is theory M and that M (in conjunction with relevant empirical facts) implies the intuitively plausible claim N, that (say) it can sometimes be morally permissible to accede to a terminally ill patient's request to actively hasten her death. (Rachels 1975) M implies N, and N's plausibility suggests that this implication relationship holds. So far, so good. It would not, however, be correct to think that the processes of verification uncover the same justificatory structure as in the scientific case. Moral theories make 'predictions' we can test, yes. But in testing them, we take ourselves to be testing an assumed explanatory or justificatory relation between moral theories and moral judgments, a relation stronger than mere predictive efficacy. If utilitarianism implies that it can sometimes be morally permissible to accede to a terminally ill patient's request to actively hasten her death, then the truth of utilitarianism seems to account for the truth of this first-order judgment.[3] The facts that utilitarianism posits as morally relevant (facts about the promotion of welfare, etc.) explain *why it is the case* that it can sometimes be morally permissible to accede to a terminally ill patient's request to actively hasten her death. As Ho notes, the "aim of a discipline determines the antecedent need to address deep metaphysical questions." (2016: 4) In the case of morality, our aims include providing justifications of our first-order moral judgments in terms of theoretical moral conceptions stronger than the logical implication of the former by the latter. An adequate moral theory, unlike (perhaps) an adequate scientific theory, does not just get the extensional relationships among concepts or properties correct. It explains those relationships as grounded in moral facts that are presumptively not local or 'one-off'. (Timmons 2013) As in the scientific case, moral theories are 'tested' against first-order judgments. But the fact that we accept a moral theory because it implies plausible first-order moral claims does not entail that the justificatory relationship runs from the first-order claims to the theory. Rather, the truth of the theory justifies the first-order claims. Moral inquiry aims at something more than predictive success, namely, explanatory grounding of our first-moral claims. It — and those who claim expertise in it — thus need to "address deep metaphysical questions" in ways that science and scientific experts may not need to.

A failure to distinguish discovery from justification can obscure this difference. When using (say) reflective equilibrium, we are aiming at a mutual attunement of our moral theories and our first-order moral judgments. There is a sense, then, in which (for example) that utilitarianism implies that it is sometimes morally permissible to accede to a terminally ill patient's request to actively hasten her death gives us reasons to accept utilitarianism. The biconditional

> Utilitarianism is true if and only if it is sometimes morally permissible to accede to a terminally ill patient's request to actively hasten her death

cannot be true unless both utilitarianism is true and that it is sometimes morally permissible to accede to a terminally ill patient's request to actively hasten her

[3] This is not to preclude that some other theory could also imply this first-order judgment.

death. But the fact that we might 'discover' that the latter is intuitively plausible and thereby confirm the former does not entail that the truth of the latter explains the truth of the former. The explanatory relationship is in fact the reverse.

Our reasons for accepting first-order moral judgments thus cannot be detached from the moral theories that imply them as readily as observations can be detached from the scientific theories that imply them. Theoretical considerations play a role in moral deliberation and moral reasoning that they need not play, and often do not play, in scientific deliberation and scientific reasoning. (Hooker 1998) With respect to moral expertise, moral experts will need to rely upon theories (or theory-like considerations) in a more direct way than will scientific experts. For moral theories are reason-giving in a way that scientific theories are not: Scientific theories can be acceptable even if they only indicate extensional relationships among their concepts. Moral theories, on the other hand, derive their plausibility from the relationships among their concepts being both genuinely explanatory and justificatory. A moral theory gives us largely theory-based reasons to believe a first-order moral judgment, whereas a scientific theory gives us reasons to anticipate observations that themselves bear on whether we have reason to accept the theory. To revert to our earlier example, utilitarianism does not merely imply that acceding to a patient's request to actively hasten her death is morally permissible. It purports to give us a reason that accounts for its moral permissibility.

Ho's analogy between moral theory and inquiry on the one hand and scientific theory and inquiry on the other hand thus proves suspect. For unlike in the scientific case, the evidential or justificatory relationship between moral theories and first-order moral judgments cannot be modeled on or reduced to how theories imply or 'predict' first-order moral judgments. To the extent that moral experts can contribute to moral inquiry, they would seem to need to possess knowledge of the correct moral theory (or theory-like considerations) in order for their moral testimony to be supported by moral reasons. And in the absence of providing reasons in favor of their own moral convictions, it is hard to see how moral experts could contribute to progress in moral inquiry at all.

4.4 Theoretical Knowledge and the Testimonial Condition

Ho offers a second argument for the independence of moral expertise from moral theory, one that suggests that experts can satisfy my testimonial condition without reference to theoretical moral knowledge. Ho argues that because theoretical knowledge is often unnecessary in order for moral disagreement to be rationally resolved, experts need not have theoretical knowledge in order to provide expert counsel regarding moral questions. To think otherwise is to falsely assume that "without a normative framework, one cannot solve any moral problems." But because that assumption is false, individuals can rightfully claim to be experts, and play the typical role of moral experts, even absent knowledge of the true or correct moral theory. (Ho 2016: 381).

Here Ho seems to cast doubt on the place of theoretical moral knowledge in moral expertise by indicating that such knowledge is not essential in order for experts to satisfy the testimonial condition. Experts' moral testimony can and ought be trusted despite experts not invoking moral theory in support of that testimony. Hence, even if (contrary to the considerations adduced in the previous section) theoretical moral knowledge is necessary in order for moral experts to function as experts, such knowledge is not essential to non-experts' investing their trust in the moral testimony of experts.

Ho is certainly correct that moral disputes are often resolved without the invocation of robust theoretical claims. In many ordinary deliberative contexts, individuals may seek to persuade one another, or seek consensus, regarding some moral question without giving so much as a thought to how various answers to that question might be theoretically grounded.

"When we try to determine what we ought to do, we do not take some broad ethical theory, plug in the particulars of the situation, and see what recommendation falls out," Ho observes. "Moral problems, unlike calculus, are usually not solved by filling in the values for the variables" a moral theory designates as morally relevant. (2016: 374)

But Ho's inference that because shared moral deliberation often proceeds in largely atheoretical terms moral experts' testimony can be atheoretical seems unwarranted. Indeed, Ho again overlooks that the expert context is one where theoretical knowledge is likely to be sought out. Echoing Judith Thomson (Thomson 1990), Ho observes that disputes about moral issues are often resolved via "discourse within a narrow context in which we assume some shared moral judgments, and we do not challenge the broad foundation of morality." (2016: 375) But the perceived need for consulting moral experts arises primarily or most acutely in contexts in which the participants in a moral discourse have concluded that their own deliberative capacities are inadequate to the moral question at hand — that with respect to *these* moral phenomena at least, they are *not* sufficiently expert. They may find themselves dumbfounded by novel moral phenomena, beset by competing moral intuitions about those phenomena, unable to render those intuitions about these phenomena consistent with judgments about other seemingly similar phenomena, etc. Any resolution of these moral questions likely to satisfy these non-experts will need to engage with morality's theoretical foundations. In terms familiar from Aristotle, these non-experts' struggles with 'the that' of morality reflect struggles with 'the why' of morality. Their inability to answer the moral questions that challenge them often reflects a lack of theoretical understanding, of which theoretical principles are relevant to those questions and what those principles imply. If an expert is to assist them, that expert must therefore invoke the theoretical knowledge needed to provide that understanding. Put differently, in many contexts of moral inquiry, the contributors to that inquiry operate as competent peers, seeing themselves and others as roughly equally capable to address the moral questions at issue. But as noted earlier, the relationship between moral experts and those who might accept their moral testimony is asymmetric, for the contexts in which moral inquirers look to experts are those in which those inquirers judge their own moral competencies to be *in*adequate, i.e., they are implicitly casting doubt on their shared moral judgments and assumptions. And the guidance they seek is theoretical in nature.

We can get a better handle on the place of moral theory in expert moral testimony by imagining a moral expert who espouses an 'anti-theoretical' stance. Take, for instance, Jonathan Dancy's particularist moral theory (Dancy 2004), which proudly denies that moral knowledge can be systematized or codified. According to his theory, principles (which we would ordinarily expect to play a crucial role in systematizing or codifying our moral knowledge) have an extremely limited role in moral thought or choice. There are not, according to Dancy, defensible moral principles nor is moral deliberation characterized by the invocation of principles to address specific cases. Indeed, Dancy goes so far as to embrace a radically contextualist or holistic picture of practical reasons, denying even that there are true principles concerning the relevance of a given consideration (e.g., that an act would cause harm) to the moral justification of action.

Dancy's particularism represents the skeptical end of a spectrum regarding the prospect of moral knowledge being systematized or codified.[4] But note that this skeptical stance seems to have implications concerning moral expertise: If this form of particularism is true, then individuals whose moral knowledge satisfies the epistemic condition may nevertheless fail the testimonial condition and thereby be disqualified to play the social role of moral experts. Other moral experts, possessed of the same body of moral knowledge as a particularist moral expert, would of course find whatever moral counsel such an expert provides highly reliable (though, of course, being experts themselves, they would presumably have lesser need for or interest in such counsel). But those who sought out the particularist's moral counsel would be understandably uncertain about what level of confidence to have in such counsel. Consistent with the theoretical commitments of particularism, a would-be particularist moral expert could not straightforwardly appeal to principles to explain why her counsel is warranted in any specific case. For the particularist's expertise consists less in knowledge of such principles than in knowledge of which considerations are relevant to the case at hand and how they are relevant. Nor could she readily employ standard methods of moral argumentation in such explanations. For example, using generalizations and counterexamples in an effort to persuade her audience of the reliability of her moral claims would be greatly complicated by the fact (at least it is a fact according to particularism) that any consideration invoked in a generalization supporting a moral claim could play a different justificatory role in another moral context.

My point here is not to argue against particularism of any kind. Rather, I merely wish to stress that there are profound tensions within Dancy-style particularism between the epistemic and testimonial dimensions of expertise, that is, *given* how particularism represents the moral knowledge possession of which is necessary for a would-be expert to satisfy the epistemic condition, any individual who satisfies

[4] Other more moderate particularisms (for instance, Little 2001) may admit the possibility that moral experts can meet the testimonial condition.

this condition is not likely to satisfy the testimonial condition. For particularist moral knowledge is opaque in a manner that makes moral expertise very difficult to judge, particularly for those who are not experts and are seeking the counsel of moral experts. Particularism is therefore likely to face an uphill battle in trying to address Scott LaBarge calls the "credentials problem," i.e., the problem of how non-experts can be rationally warranted in accepting the moral testimony of putative experts. (LaBarge 2005)

Again, I do not take these observations to refute particularism. If there are moral experts, then particularism is cast in doubt, and conversely, if particularism is true, then perhaps there are no moral experts. I take no stance on which of these options is correct. All the same, the case of particularism underscores a critical point regarding how moral theories relate to moral expertise: There are, or at least can be, gaps between the warrant a moral expert has for her moral beliefs, including her theoretical beliefs, and the warrant that others may have for trusting those experts' moral testimony. Particularism, I just argued, appears to have an unusually wide such gap.

The case of particularism illustrates that a moral expert would be seriously hampered in playing the role dictated by the expert context if either she did not (or seemingly could not, in the case of the particularist moral expert) invoke theoretical claims. In this regard, the considerations show that theoretical knowledge is needed in order for experts to satisfy the testimonial condition parallel those that show that theoretical knowledge is needed for experts to satisfy the epistemic condition. The practice of shared moral inquiry and deliberation is oriented around reason giving. As soon as moral inquirers acknowledge a need for expert aid, they are compelled to assess a putative moral expert's testimony based on the reasons that support such testimony. This exercise mirrors first-personal moral reasoning, but occurs when moral agents acknowledge the shortcomings of their own first-personal moral reasoning. And in looking to experts, their trust in the experts' testimony should hinge in part on the experts' invoking theoretical claims (or making theoretical argumentative moves) in support of their testimony.

It may be thought that my defense of the centrality of theoretical knowledge to moral expertise ends up insisting that non-experts be experts. For I have argued that in order for non-experts to invest their trust in the moral testimony of putative experts, those experts must give evidence of theoretical knowledge. But it might then be inferred that non-experts are evaluating the experts' testimony as if they (the non-experts) were experts themselves, evaluating the experts' theoretical commitments as if they possessed the very expert moral knowledge they seek from the moral expert.

Invariably, non-experts will have to make judgments about the credibility of expert testimony, a challenge complicated by their own lack of expertise. But they need not be experts themselves in order to reach intelligible judgments about moral experts' moral testimony. For one, they may appraise that testimony by reference to what might be thought of as the experts' theoretical virtues. An expert who gives (seemingly) inconsistent testimony, who cannot provide a basis for morally differentiating similar phenomena, or who cannot answer rudimentary objections is not one that non-experts ought to trust. Such theoretical virtues likely underdetermine truth in moral theory inasmuch as partisans of rival theories may possess them to the same degree. All the same, non-experts have reasons to hold such an expert's moral testimony in suspicion. My invocation of these theoretical virtues is not meant to offer a complete account of the considerations that non-experts ought to take into account in deciding whether a putative moral expert's moral testimony deserves to be trusted. I merely mean to highlight that non-experts have rational bases on which to appraise the credibility of such testimony that do not presuppose that the non-experts have expert knowledge. More importantly, an expert's inability to give any *theoretical accounting* of her testimony regarding first-order moral questions would, I propose, be a 'red flag' alerting non-experts that the expert's claim to reliable to first-order moral knowledge should be second guessed.

In sum then, by neglecting the peculiarities of the expert context, Ho wrongly extrapolates from the fact that typical discursive contexts do not invoke moral theory that the atypical contexts in which experts come on the scene need not invoke moral theory. We should agree with Ho that it is foolish to think that a theoretical framework is necessary to "solve any moral problems." But we should be skeptical that *experts* can satisfactorily solve moral problems absent reference to any theoretical framework. Ho thus convincingly demonstrates that theory is not needed when experts are not needed. But there is good reason to think that theory is needed just when experts (apparently) are.

4.5 Conclusion

I have provided no reason to think there are experts in moral theory, nor any guidance regarding the true or correct moral theory knowledge of which would be necessary for an individual to be a moral expert. I have attempted to defend the centrality of theoretical knowledge to moral expertise against those who deny that moral experts must have theoretical knowledge or that experts' moral testimony can be rightfully trusted by others if that testimony is atheoretical. In order for individuals to have moral expertise, their moral testimony must be ultimately rooted in theoretical knowledge.. Theory has a reason-giving function in moral discourse that it lacks in other discourses, including scientific discourse. Moral experts must therefore be theoretical experts.

References

Archard, David. 2011. Why moral philosophers are not and should not be moral experts. *Bioethics* 25: 119–127.

Cholbi, Michael. 2007. Moral expertise and the credentials problem. *Ethical Theory and Moral Practice* 10: 323–334.

Cross, Ben. 2016. Moral philosophy, moral expertise, and the argument from disagreement. *Bioethics* 30: 188–194.

Crosthwaite, Jan. 1995. Moral expertise: A problem in the professional ethics of professional. ethicists. *Bioethics* 9: 361–379.
D'Agostino, Fred. 1998. Expertise, democracy, and applied ethics. *Journal of Applied Philosophy* 15: 49–55.
Dancy, Jonathan. 2004. *Ethics without principles*. Oxford: Clarendon Press.
Driver, Julia. 2006. Autonomy and the asymmetry problem for moral expertise. *Philosophical Studies* 128: 619–644.
———. 2013. Moral expertise: Judgment, practice, and analysis. *Social Philosophy and Policy* 30: 280–296.
Foot, Philippa. 1972. Morality as a system of hypothetical imperatives. *Philosophical Review* 81: 305–316.
Frey, R.G. 1978. Moral experts. *Personalist* 59: 47–52.
Ho, Dien. 2016. Keeping it ethically real. *Journal of Medicine and Philosophy* 41: 369–383.
Hooker, Brad. 1998. Moral expertise. In *Routledge encyclopedia of philosophy*, ed. E. Craig, 509–511. London: Routledge.
Iltis, Ana S., and Mark Sheehan. 2016. Expertise, ethics expertise, and clinical ethics consultation: Achieving terminological clarity. *Journal of Medicine and Philosophy* 41: 416–433.
Kovács, József. 2010. The transformation of (bio)ethics expertise in a world of ethical pluralism. *Journal of Medical Ethics* 36: 767–770.
Kuczewski, Mark G. 2007. Democratic ideals and bioethics commissions: The problem of expertise in an egalitarian society. In *The ethics of bioethics: Mapping the moral landscape*, ed. Lisa A. Eckenwiler and Felicia Cohn, 83–94. Baltimore: Johns Hopkins University Press.
LaBarge, Scott. 2005. Socrates and moral expertise. In *Ethics expertise: History, contemporary perspectives, and applications*, ed. L. Rasmussen, 15–38. Berlin: Springer.
Little, Margaret Olivia. 2001. On knowing the 'why': Particularism and moral theory. *Hastings Center Report* 31: 32–40.
McConnell, Terrance. 1984. Objectivity and moral expertise. *Canadian Journal of Philosophy* 14: 193–216.
McGrath, Sarah. 2008. Moral disagreement and moral expertise. In *Oxford studies in metaethics*, ed. R. Shafer-Landau, vol. 3, 87–108. New York: Oxford University Press.
Popper, Karl. 1962. *Conjectures and refutations*. New York: Harper Torchbooks.
Priaulx, Nicky. 2013. The troubled identity of the bioethicist. *Health Care Analysis* 21: 6–19.
Rachels, James. 1975. Active and passive euthanasia. *New England Journal of Medicine* 292: 78–80.
Rasmussen, Lisa M. 2011. An ethics expertise for clinical ethics consultation. *Journal of Law, Medicine, and Ethics* 39: 649–661.
Rawls, John. 1971. A theory of justice, 2 1999. Cambridge, MA: Harvard University Press.
Thomson, Judith. 1990. *The realm of rights*. Cambridge, MA: Harvard University Press.
Timmons, Mark. 2013. *Moral theory: An introduction*. 2nd ed. Rowman and Littlefield.
Watson, Jamie Carlin. Forthcoming. The shoulders of giants: A case for non-veritism about expert authority. *Topoi* 37: 39. https://doi.org/10.1007/s11245-016-9421-0.
Wolff, Robert Paul. 1970. *In defense of anarchism*. New York: Harper and Row.

Chapter 5
Moral Experts, Deference & Disagreement

Jonathan Matheson, Scott McElreath, and Nathan Nobis

5.1 Introduction

We sometimes seek expert guidance when we don't know what to think or do about a problem. In challenging cases concerning medical ethics, we may seek a clinical ethics consultation for guidance. The assumption is that the bioethicist, as an expert on ethical issues, has knowledge and skills that can help us better think about the problem and improve our understanding of what to do regarding the issue.

The widespread practice of ethics consultations raises these questions and more:

- What would it take to be a moral expert?
- Is anyone a moral expert, and if so, how could a non-expert identify one?
- Is it in any way problematic to accept and follow the advice of a moral expert as opposed to coming to moral conclusions on your own?
- What should we think and do when moral experts disagree about a practical ethical issue?

In what follows, we address these theoretical and practical questions about moral expertise.

J. Matheson (✉)
University of North Florida, Jacksonville, FL, USA
e-mail: j.matheson@unf.edu

S. McElreath
William Peace University, Raleigh, NC, USA
e-mail: scott.mcelreath@peace.edu

N. Nobis
Morehouse College, Atlanta, GA, USA
e-mail: nathan.nobis@morehouse.edu

J. C. Watson, L. K. Guidry-Grimes (eds.), *Moral Expertise*, Philosophy and Medicine 129, https://doi.org/10.1007/978-3-319-92759-6_5

5.2 Expertise

In an increasingly specialized world, the need for expertise continues to grow. Most people could not tell you how their computer works. Most people could not fix their car if it broke down. Most people cannot determine why they are sick or how they can get better. Put bluntly, we need experts. We need people who we can rely on for both their information and their abilities. But what exactly is expertise?

The first thing to note is that expertise is relative to a domain. No one is just a flat-out expert. Rather, one could perhaps be an expert car mechanic, an expert pianist, or an expert physicist. Expertise can also be more or less fine-grained. One might be an expert on Honda automobiles, but not an expert on vehicles more generally, an expert on a certain type of jazz but not a musical expert more generally, and so on. Expertise is relative to a time as well. Expertise can come and go. It can come through the acquisition of the relevant knowledge and skills, and it can go through the loss of such knowledge and skills. Experts in a domain need to stay 'up to date' in their domain to retain their status as experts. This will require keeping up with the state of knowledge and keeping a certain set of skills finely honed.

Following Goldman (2001) we can distinguish several factors relevant to expertise. First, expertise is composed of both a cognitive component and a skill component. To be an expert you need to have a significant amount of knowledge within the domain of your expertise and you must be able to apply that knowledge to action. Different domains of expertise will place a differing emphasis on these two components. Some domains of expertise are more skill, or know-how, oriented. Being an expert mechanic or pianist, for instance, is primarily about possessing a certain set of skills even though some propositional knowledge is required as well. In contrast, other domains of expertise are more intellectually oriented. Being an expert physicist or epistemologist, for instance, is primarily about possessing a certain body of knowledge, even though some skills are required here as well. Our focus in this paper will be on intellectual expertise in general and moral expertise in particular.

While intellectual expertise is primarily concerned with having a fund of knowledge, intellectual experts must also be able to deploy and to apply that knowledge. Intellectual expertise requires the skills of applying the expert's knowledge to new cases and discovering answers to new questions in their domain of expertise. An expert isn't merely a "scholar" of existing knowledge; an expert must be able to apply and extend that knowledge to new questions and problems. Intellectual skills include being able to construct a valid argument, recognize and respond to objections, revise an argument in light of counter-evidence, and correctly weigh reasons. These skills also include reflecting on how an argument, or principle, applies to other issues, being able to find inconsistencies between views, and being able to make revisions to arrive at a consistent set of views.

Secondly, expertise is comprised of both comparative and non-comparative elements. To be an expert in some domain you must possess significantly *more* knowledge in that domain than most. This is the comparative element. Experts are unusual. If everyone were an expert in some domain, then no one would be an expert in that domain. Experts also have significantly greater knowledge and/or skills than most people. That said, expertise is not a wholly comparative matter. Someone could have the most knowledge about some issue, and yet only know very little. It takes more to be an expert than to be the best of a bad lot. So, experts in a domain must possess a significant amount of knowledge and skill in the domain of their expertise. A non-comparative threshold must be passed to have expertise.

Finally, expertise is relative to a population. This point relates to the comparative aspect of expertise. Experts need to have an unusual amount of both knowledge and skill, but such a comparison only takes place against a reference class. Someone may be an expert with respect to one population group but not with respect to another. For instance, a philosophy professor may be an expert on Kant with respect to the general population, but not with respect to the other members of her department.[1] This all leads to a final account of expertise:

> Someone S is an expert in domain D at time T with respect to population P just in case S possesses an unusually extensive body of knowledge in D at T and S has unusually extensive skills to apply that knowledge at T to new questions and problems compared to others in P.[2]

5.3 Moral Expertise

With this account of expertise in hand, we can begin to examine moral expertise. Moral expertise is a form of expertise that is relative to a moral domain. It is doubtful that someone is an expert on all of morality.[3] There are simply too many subfields and issues within morality as a whole to master all of it. For instance, within the domain of the moral are the fields of normative ethics, applied ethics and metaethics. Moral expertise is more likely to obtain within more limited moral domains. The smaller the domain, the more likely expertise is to obtain since the scope of relevant knowledge (if not also abilities) will also shrink.

For instance, we can think of a moral domain centering on a moral question.[4] There are moral domains, for example, which are focused on bioethical questions such as these, and many more:

---

[1] For instance, consider an atypical population in which all humans have been killed except for a group of neuroscientists. A question arises as to whether the members of this group are still expert neuroscientists since their knowledge and skills are no long unusual. On our view, they are still experts at neuroscience relative to some counterfactual populations, but not relevant to the population of existing humans. Thanks to Jamie Watson for bringing such an example to our discussion. For an alternative account, see Collins and Evans' (2007) account of "ubiquitous expertise" according to which large segments of the population can be experts on (say) the English language.

[2] This closely follows Goldman (2001: 92).

[3] Compare with Coady (2012) who says similar things about the possibility of a science expert.

[4] This is not to say that moral expertise only obtains regarding such questions. Plausibly there are also moral experts in broader domains such as research ethics, clinical ethics, animal welfare, etc.

- Is physician-assisted suicide ever morally permissible? If so, when and why?
- Are physicians always required to tell their patients the truth? If so, when and why?
- Is abortion ever morally permissible? If so, when and why?

Experts in any of these domains must possess an unusually extensive amount of knowledge relevant to that domain. Such knowledge will consist of both descriptive and normative knowledge. That is, an expert in the domain of physician-assisted suicide must have extensive descriptive knowledge about the practice (medical knowledge, social knowledge, etc.) as well as descriptive knowledge about the state of the debate (hospital policies relevant to these issues, state and federal laws, knowledge of the options for final moral views on the issue, the moral arguments that have been made, the objections given to those arguments and replies, etc.) Such a moral expert must have significant normative knowledge (knowledge of what potential moral considerations there are, the kinds of moral reasons there are, the comparative weight of these reasons, etc.)

In addition to this set of knowledge, the moral expert must also be able to apply this knowledge to solve problems and questions. This is why a moral expert might serve as a consultant, to help address an ethical problem, or as an advisor or educator, to help plan for future scenarios. That is, such a moral expert would need to be able to apply his or her general knowledge of the topic to a particular case (particularly when it is a novel case), being able to factor in the particular details of that case. Moral experts need both moral know-how and knowledge of moral propositions.[5]

5.4 Non-Experts Identifying Moral Experts

Given the above rough characterization of moral expertise, is anyone a moral expert? We think the answer is, "yes." We know that, for established ethical issues, there are people who possess a deep understanding of the relevant facts, issues, and arguments – indeed the entire body of major scholarly literature surrounding a topic – and are able to use that understanding to engage new problems and questions about the topic. Further, some of these people also have the personal and communication skills to competently serve as advisors to families in need of navigating an ethical challenge in an informed way. Given our account of expertise, such individuals would qualify as experts in the domain of the relevant moral issue.

However, the mere existence of moral experts is of little help to those seeking moral guidance. To make use of someone's expertise we must be able to identify

[5] Our discussion presumes some kind of realism about morality, that is, that there are moral truths or facts that are sometimes justifiably believed or known. Whether moral expertise and the issues of this chapter would be understood differently on the variety of non-realisms is an issue for another occasion.

them as an expert in the relevant domain. Expert identification can be particularly difficult for someone who is not themselves an expert in that domain. After all, if you lack the relevant knowledge and skills, how can you determine who has them? If you don't know anything about physics, how could you identify an expert physicist? Applied to our concern, how might a non-moral expert identify a moral expert?

Let us begin by describing three ways that someone can be a non-moral expert about some domain, D. First, she may have no background information about the knowledge or skills needed to be an expert in D. Maybe Carol never learned about euthanasia or critical thinking in college. Presumably, a critical thinking course would have helped her identify the intellectual skills needed to be a moral expert. Second, a non-moral expert may have some background information, but it is old or scant. Connor learned about euthanasia and critical thinking in college, but that was a long time ago and he has not thought about it since then. Third, a non-moral expert may have background information about the knowledge needed to be an expert in D, but not the skills needed, or the other way around. Ed, an epistemologist, may be able to identify the intellectual skills needed to be an expert in euthanasia, but he knows very little about euthanasia. Or Danny, a doctor who specializes in end-of-life care, may be familiar with euthanasia policies, theories, and concepts, but not familiar with how to apply critical thinking concepts.

Besides seeking further education – something that not many people have the time to do – how might these individuals who lack moral expertise acquire more information that might help them decide if a given person is an expert in some moral domain? While this question deserves more attention than we can give it here, we hope to show that the worry is not as overwhelming as it may initially seem. To this end, we will briefly outline some places to start.

(1) **Observe the behavior of the potential expert to see if she acts differently, more effectively, or better than others in her field.** Suppose Cindy has been nominated to serve as a euthanasia therapist. She currently provides end-of-life care to patients in the hospital. A non-moral expert may shadow Cindy and compare her work to others who have a similar job. A non-moral expert may observe that Cindy asks different questions to her patients and with a more caring tone, that Cindy needs less time than others to work with her patient, and that she extraordinarily combines professionalism with sympathy. A non-moral expert has some evidence that Cindy is an expert at euthanasia therapy.[6]

(2) **Observe the behavior of the people who are directly affected by the potential expert to see if they are improved more than they are by others in the potential expert's field.** Because of Cindy's unique methods, her patients may make more reasonable, well-informed, and timely choices for themselves. Again, we here have some evidence that Cindy is an expert at euthanasia therapy.

[6] To say that a subject, S, has "some evidence" for a proposition, P, leaves it open whether that evidence is sufficient for S to reasonably believe that P.

(3) **Observe how other observers evaluate the potential expert.** Ideally, a non-moral expert would ask experts what they think about a potential expert such as Cindy, but that's not likely given that a non-expert is not sure how to identify experts. So, the more observer observations, or testimonies, that a non-expert can get, the better. The following are questions to ask an observer:

a) Do you agree with the potential expert?
b) Do you respect the potential expert?
c) Does the potential expert have appropriate accreditation, degrees, or work experience?
d) Do you put your trust in the potential expert's knowledge and skills?
e) Do you praise the potential expert's past actions?
f) Do you see the potential expert as unbiased?
g) Does the potential expert positively engage with people who disagree with her?

The more answers to these questions a non-moral expert might get, the more evidence she might have for or against the potential expert's expertise (Goldman 2001: 92–93).

We do not know when to tell a non-moral expert to stop seeking more information about whether a potential moral expert is an expert. Much depends on the quantity and quality of the information she receives. But we do know that taking the advice in this section is necessary to making a reasonable judgment when identifying a moral expert. Importantly, we see no good reason to think that in principle non-experts cannot reasonably identify experts.[7]

5.5 Moral Deference

Suppose a non-expert on a moral issue has successfully identified a moral expert. Is it appropriate for the non-expert to believe what the expert says simply because of her expertise? That is, is there anything problematic with deferring to a person's moral expertise? This question is the focus of this section.

Much of what we believe is believed on the basis of testimony. For example, our numerous beliefs about temporally and geographically distant events are mostly believed on the basis of someone else's say-so. Among our testimonially-based beliefs are beliefs where we have simply *deferred* to another:

> S1 defers to S2 about p when S1 comes to believe p merely because S1 discovers that S2 believes p.

[7] See Anderson (2011); Collins and Weinel (2011); Gelfert (2011); Matheson (2005); and Miller (2013) for more detail about how non-experts might identify experts.

Clearly not just anyone should be deferred to on just any matter, but in general deferring to an expert on a matter of their expertise is seen as appropriate.[8] For instance:

- It is appropriate for me to defer to a civil war historian as to how many soldiers in the Union army were immigrants.
- It is appropriate for me to defer to a chemist as to the molecular structure of caffeine.
- It is appropriate for me to defer to an entomologist as to what kind of insects are in the attic, etc.

However, many find there to be something problematic about other kinds of deference. For instance, moral deference in particular has come under a great deal of scrutiny. That is, many have questioned whether:

- It is appropriate for me to defer to a moral expert as to whether it is morally wrong to eat meat.
- It is appropriate for me to defer to a moral expert as to whether the use of attack drones is morally permissible.
- It is appropriate for me to defer to a moral expert as to whether it is morally obligatory for me to give more to charity, etc.

Is there an important difference between moral deference and other kinds of deference? If so, what grounds this difference?

First, let's examine a case of moral deference.

Meatless Melanie

Melanie meets Maggie at an academic party. Melanie is a historian and she comes to find out that Maggie is an applied ethicist who has spent most of her career examining the moral case for vegetarianism. In their conversation, Melanie doesn't ask Maggie to present the main arguments for or against vegetarianism, but simply asks Maggie what her position is on the issue. Maggie says that she believes that it is morally wrong to eat meat. Considering Maggie's expertise on the matter, Melanie defers and also adopts the belief that it is morally wrong to eat meat.

Intuitively, something is amiss with Melanie's deferring to Maggie. Minimally, something about Melanie's deference is worse than the non-moral cases of deference given above. In the case, it is clear that Melanie is justified in believing that Maggie is in a better epistemic position than she is on the matter, so the problem with the deference is not that Melanie defers to someone who is not sufficiently informed on the matter. Further, things seem even worse if we add to the case that Melanie is aware of all of the relevant non-moral facts that Maggie is aware of. That is, we can imagine that while Maggie has thought about the morality of eating meat much more than Melanie has, there is no difference in their knowledge of the relevant non-moral descriptive facts (they are both equally aware of the psychology of

[8] For more detailed discussion on the nature of deference, see Zagzebski (2012) and Lackey (forthcoming).

animals, how animals are treated, etc.). So, the case of Meatless Melanie gives some strong intuitive support to the claim that there is something amiss with moral deference.[9] Following McGrath (2011), let's call this claim 'DATUM':

> **Datum** There is something amiss with moral deference that is not present in ordinary cases of deference.

Given that something seems amiss with moral deference, what could it be? What accounts for the difference between the non-problematic cases of deference explored above and moral deference? DATUM cries out for an explanation given that in general we don't find any problem with deferring to those who are in a better epistemic position on the matter. So, the problem with moral deference is not simply that it is a case of deference; deference in different circumstances is unproblematic. Further, the problem with moral deference is not simply that it is the formation of a new moral belief. We do not find anything problematic in coming to adopt a new moral belief by being led through reasoning to this conclusion. Moral deference is problematic in a way that moral persuasion is not. That is, the problem with forming a moral belief by deference is not shared by other ways of coming to form a moral belief. So, there appears to be something uniquely amiss with forming moral beliefs by deference. Perhaps some questions simply should not be outsourced.

Several explanations of DATUM have been offered in the literature. In what follows, we will examine these explanations to determine whether any of them provide a reason to not defer to a moral expert on a moral matter.[10]

5.6 Normativity

One proposed explanation of DATUM is that moral deference is a kind of normative deference, and the problem lies more generally with normative deference. On this view, the impropriety of moral deference comes from the fact that in doing so the subject defers on a matter of values and right conduct.

While initially tempting, this suggestion should be rejected for several reasons. First, some cases of normative deference do not seem to be at all problematic. Consider the following:

- It is appropriate for me to defer to Emily Post on matters of etiquette.

---

[9] This is not to say that any case of moral deference shares this same sense of impropriety. In cases where there is a great deal at stake and little to no time to act, moral deference may not seem problematic at all. In fact, it may even seem to be required. Thanks to Jamie Watson to bringing up such a scenario to us.

[10] For more on this debate, see Crisp (2014); Decker and Groll (2014); Driver (2006); Enoch (2014); Hazlett (2015); Hills (2009, 2010, 2013); Hopkins (2007); Howell (2014); Jones (1999); Laurence (1993); McGrath (2009); Mogensen (2015); Nickel (2001); Sliwa (2012); and Vavova (2014).

- It is appropriate for me to defer to the sommelier as to what wine to pair with dinner.
- It is appropriate for me to defer to my lawyer as to which defense I ought to pursue.
- It is appropriate for a student to defer to his logic teacher as to what he may and may not infer from his premises.

In each of these cases, the deference is reasonable and lacks the intuitive problems observed in cases of moral deference. Further, there are some cases of non-normative deference that appear to be amiss in a way similar to cases of moral deference. Consider the following:

Freshman Freddy
Freddy is a first-year university student taking introduction to philosophy. On the first day, he hears from his professor about what the course will cover. In particular, Freddy is excited to hear that they will be discussing whether humans have free will. This is something he hasn't really thought about before. After class, Freddy quickly approaches his professor and asks here whether we have free will. Freddy's professor tells him that she does believe that humans have free will. Freddy defers and comes to believe that humans have free will.

Gill & God
Gill has started thinking about whether God exists. He has found the issue confusing and hard to engage. He soon learns that his university is hosting a lecture on the issue from a renowned scholar in the philosophy of religion. After the talk, most of which Gill couldn't follow, he asks the scholar about whether he believes that God exists. The scholar says that he does. Gill defers and comes to believe that God exists.

In both of these cases, the subject defers about a non-normative proposition. In neither case is the proposition about what is good/bad, right/wrong, or how things should/shouldn't be. Nevertheless, something appears to be amiss in these cases of deference in the same way that something appears to be amiss in cases of moral deference. So, some normative deference appears entirely appropriate and some non-normative deference appears to be as problematic as moral deference. Given all of this, that moral deference entails deferring about a normative claim is not a good explanation of what is amiss in moral deference.

5.7 Accessibility

A second proposed explanation for DATUM is that moral deference is inappropriate in ways that other kinds of deference are not since moral facts are equally accessible to everyone. If moral facts are equally accessible to everyone, then deferring about a moral claim would be like deferring to someone who is not in any better of an epistemic position than you on the matter. Something would clearly be amiss if one colleague deferred to his or her epistemic peer on the matter. Such deference is puzzling if not entirely inappropriate. For instance, if you and I are standing outside in the rain, and I believe it is raining solely on the basis of deferring to you, then something has gone wrong. Similarly, if the moral truth was equally available to us, my forming a moral belief by deferring to you on the matter would be bizarre if not mistaken. So, the equal accessibility of morality may be able to explain why cases of moral deference are amiss – they parallel other cases where deference is inappropriate.

As McGrath (2011) has pointed out, the claim that moral claims are equally accessible to all can be understood in two ways. First, it could be that moral truths are equally clear to everyone – that we are all equally accessing moral facts. However, as McGrath notes, such a claim is implausible given the vast amounts of moral disagreement of which we are all too aware. Moral disagreements exist despite the fact that there are open-minded, sincere inquirers on both sides of the issue. If moral truths were equally accessible in this sense, the extent of moral disagreement wouldn't be nearly as vast as it is. Second, it could be that in principle everyone is equally *able* to access moral truths – that moral truth is equally available to everyone in principle. It is not clear that this claim is true either, but even granting its truth, there are good reasons to doubt that it is able to sufficiently explain DATUM. Even if everyone in principle is able to access moral truth equally well, it doesn't follow that there is anything amiss with moral deference. For instance, it is reasonable for me to defer to my doctor even if, in principle, I could go to medical school myself and learn everything that she knows. Even if I can in principle access some fact, if I have not yet done so, it is reasonable for me to defer on the matter.

5.8 Value Differences

A final explanation of DATUM is that moral deference precludes something that is morally more important than a justified moral belief. One such candidate is moral understanding. Moral understanding is clearly valuable. However, if I simply defer to someone on a moral matter, this will perhaps preclude me from understanding why that moral truth is a truth. If I come to believe p by deferring to a moral expert, then I fail to understand why p is true. To understand why a proposition is true, one needs to possess and understand the reasons that show that it is true. In moral deference, the speaker's reasons are not transferred to the hearer. The reason the hearer comes to believe the asserted claim in a case of deference is simply because the speaker said so. Given this, moral deference precludes moral understanding.

Another such candidate is moral virtue. Moral virtue is clearly valuable. And if I simply defer to someone on a moral matter this will not allow me to act virtuously since I will not thereby acquire the right reasons, emotions, motivations, and dispositions to accompany my action. Moral deference thus prevents the kind of integration integral to possessing a virtue. So, on this explanation, deferring on a moral matter is at odds with the acquisition of greater moral goods – it falls short of the ideal and demonstrates a defect in the deferrer. Given this, such an account can offer an explanation of what is amiss with moral deference.

While such an explanation provides an account of what is amiss with moral deference, it does not show that moral deference is inappropriate. On this account, moral deference is not suitable for giving us all that we should want. We should want moral understanding, and we should want to develop and possess moral virtues, and moral deference does not provide either of these goods. However, while moral deference does not give us everything that we should want, it does give us something that we want – a justified moral belief. A justified moral belief is not everything, but it is certainly better than the available alternatives (an unjustified belief or no belief at all).

There is a parallel here with moral action. Doing the right action isn't everything. Someone who just does the right thing doesn't do as well as he could. Such an individual lacks the relevant moral virtues and it is good to have moral virtues. Such an agent can properly be criticized for not acting virtuously. However, while there are important things lacking from our agent's action it has something really important going for it – it was the right thing to do. It is surely better to do the right action while lacking the relevant virtues than either alternative open to the agent in such a situation (to do the wrong action or to do no action at all). So, while a mere right action does not give us everything that we want, given the situation that the agent is in (lacking the relevant moral virtues) it is the best we can get, at least for now.

Similarly, while a justified moral belief is not all that we should want for a subject, given that developing moral virtues or having moral understanding on the issue are not available options for her right now, a justified belief is the best we can get, at least for now. The 'for now' is important here as well, since presumably a justified moral belief needn't be the end of the matter and is plausibly even the best next step in attaining moral understanding or developing moral virtue. Acquiring these other moral goods are only further hindered by lacking a justified moral belief. In fact, the kind of moral education required to develop moral understanding and moral virtue must start with moral deference. So, while moral deference cannot give someone everything they should want, it does give them something important and can help them attain those greater goods.

Given all of this, we have an explanation for what is amiss with moral deference, but this explanation does not have it that we should not defer on moral matters. At most, this account has it that we shouldn't stop our inquiry and moral development at the point of deference.

5.9 Disagreements and Moral Expertise

While there may be no problem in principle with moral deference, a different kind of challenge comes from considering disagreements among moral experts. Moral disagreements are ubiquitous, and this phenomenon holds even among moral experts. Given that moral experts often disagree about moral matters, the prescription 'defer to the experts' may not be so easy to follow. On many moral matters, there is no consensus among the moral experts, and this raises a challenge for moral deference to the experts. In what follows we will examine what one should believe and what one should do when there is no consensus amongst the experts.

It will be helpful to distinguish several different scenarios that could obtain when there are multiple experts within a domain:

FULL CONSENSUS: Every relevant expert agrees about p.

PARTIAL CONSENSUS: Full consensus is not achieved, but there is a clear dominant view amongst the relevant experts regarding p.

DISARRAY: The opinions of the relevant experts are sufficiently dispersed so as to prevent either full or partial consensus regarding p.

What one is reasonable in believing about a moral matter will depend upon what one is reasonable in believing about which of the above scenarios obtain. If one is reasonable in believing that there is full consensus regarding the moral proposition, then that person is reasonable in adopting the same attitude as the experts toward that proposition. For instance, if Stan is reasonable in believing that all the relevant moral experts believe that torturing innocent children for pleasure is morally wrong, then Stan is also reasonable in himself believing that torturing innocent children for pleasure is morally wrong.

If one is reasonable in believing that while there is not a full consensus on a moral matter amongst the relevant moral experts, that nevertheless a clear majority adopt the same belief about some moral proposition, then that person is reasonable in adopting that same belief about that proposition. For instance, if Sue is reasonable in believing that while not all of the relevant moral experts believe that it is impermissible to harvest the organs of a healthy individual without consent to save five others, a vast majority of the relevant moral experts do believe this, then Sue is reasonable in believing this proposition as well. This parallels what it is reasonable to believe in cases of partial consensus in other domains. For instance, if Sam is reasonable in believing that 98% of the relevant experts believe that human activities are contributing to climate change, then Sam will be reasonable in believing this as well. Similarly, if Shawn is reasonable in believing that 9 out of 10 dentists believing flossing improves dental health, then Shawn is reasonable in believing this proposition as well.

Why think so? Recall that an expert about p is much more likely than a non-expert about p to have a true belief about p. So, given our epistemic ends of believing truths and not believing falsehoods, we will do better at satisfying these twin goals by going with what the expert believes. In cases of partial consensus there is no one doxastic attitude that all the relevant experts adopt toward the target proposition, but there nevertheless is an overwhelming consensus. The best explanation of such a distribution of opinions among the relevant experts in such a scenario is that the majority opinion is correct. For instance, if 10 equally qualified math experts were given the same problem and nine arrived at the same answer, it would be rational to believe that the 9 are correct. While it is possible that the 9 are all incorrect, this possibility is far more unlikely given what we know about the situation.

The third possible scenario is disarray. If one is reasonable in believing that disarray holds, then one is reasonable in believing that there is not even a partial consensus amongst the relevant experts. In such a scenario, there is not a dominant view amongst the relevant experts, but rather opinions are fairly evenly divided. If one is reasonable in believing that this scenario obtains regarding some moral proposition, then one is reasonable in suspending judgment regarding that proposition. Such a scenario parallels a political race that is 'too close to call' in that even if at some point one position has a lead, the lead is too unstable and does not support believing things will hold.[11] For example, if one is rational in believing that regarding the proposition that you are morally required to give all of your disposable income to help those in poverty, the relevant moral experts are in a state of disarray, then that person is reasonable in suspending judgment regarding that claim. This too parallels what is true in other non-moral domains. If Sarah is reasonable in believing that amongst the relevant experts that slightly more of them believe that a Republican will win the next election, she should nevertheless suspend judgment about this claim.

Why think so? In a case of disarray, matters are too unsettled amongst the relevant experts to make rational a belief on the matter. Even if one is reasonable in believing that one 'side' of the moral matter enjoys slightly more support than the others this does not suffice to make it rational to adopt that side. After all, many factors go into weighing expert opinions. So far, we have seen that expert opinions are to be trusted above those of lay people, but not all expert opinions are equal, either. Even among a group of experts there will be differences in their respective epistemic positions[12] regarding a claim within their domain of expertise. While all experts are epistemically well-positioned, not all experts are equally well-positioned. Since what the layperson is reasonable in believing will depend upon how those weights are distributed, in a state of disarray, those differences can shift which position is best supported by the expert opinions.

Further, it is quite difficult to determine these matters, particularly for a lay person. Even having cleared the hurdle of identifying the relevant experts, making this kind of much more precise judgment is typically beyond the abilities of a layperson – they are often not reasonable in making such judgments. But, if the lay person is not reasonable in making such judgments, and such judgments can alter which position is best supported by the expert opinion, the layperson is not reasonable in determining which position is best supported by the expert opinion in scenarios of disarray.

To make matters worse, such judgments of relative epistemic positions amongst the experts are not the only factor that can tip the balance of expert opinion in a state of disarray. Another factor worth mentioning is the independence of the relevant opinions. Independently formed opinions that agree carry more weight that agree-

[11] This analogy is given by Carey and Matheson (2013).

[12] We can think about someone's epistemic position on matter as corresponding to how likely they are to have a true belief on the matter. What precise details matter for one's epistemic position is contested, but there is much agreement that their evidence, intelligence, and intellectual virtues are relevant.

ing opinions that were not formed independently.[13] So, another factor that is relevant in determining the overall balance of expert opinion is the relative independence of the various opinions. We are quite familiar with all of the ways that non-epistemic factors can create agreement. This is seen in the frequency of shared political beliefs in numerous regions of the country, shared religious beliefs in numerous countries, and shared philosophical beliefs amongst graduates of the same schools. So, the independently formed shared opinions carry greater epistemic weight.

This is not to say that two or more people with a shared history who formed the same opinion don't both lend weight to that opinion, rather, such agreement lends less weight than that same number of people who independently formed that same opinion. Given this, the independence of the opinions of the experts in the relevant domain will also be important in determining where the balance of expert opinion lies. This is yet one more factor relevant to weighing the evidence regarding expert opinion. However, here too it is unreasonable to believe that a layperson (or even an expert) can make a reasonable determination regarding the degree of independence of various opinions of the matter. For instance, can we really tell how independent two expert bioethicists' opinions on physician-assisted suicide are? It seems unlikely. Worse still, on matters of great interest it is not merely the independence of two expert opinions that must be determined, but the independence of all the relevant expert opinions. This is an overwhelming task. Since the independence of opinions could be a determining factor in the balance of expert opinion on a matter and since a layperson is not reasonable in making such a judgment, the layperson is not reasonable in believing that the balance of expert opinion lies on any one side of the issue when it is in a state of disarray. So, a topic being in a state of disarray comes with skeptical consequences.[14]

Given the difficulty in determining the exact epistemic position of the relevant parties and the independence of their opinions, one might wonder why these same factors do not make determining the balance of expert opinions in cases of partial consensus also unfeasible. In cases of partial consensus, too, individual differences in epistemic position of the relevant experts will matter and so will the independence of their individual opinions. However, in cases of partial consensus, the balance of expert opinion is sufficiently settled so that even those these other factors may be unknown, it is sufficiently unlikely that they will change the balance of expert opinions on the matter. Returning to the political race analogy, in some races we can declare a winner without knowing all the details about how a number of counties voted. Yes, votes in those counties sill matter, but there is sufficient information elsewhere to declare a winner. So, too, in cases of partial consensus, an inability to determine the particular epistemic positions of the experts or how independent they are will not affect our ability to declare that the expert opinion clearly lies on one side of the issue. For instance, the vast consensus regarding climate changes renders the difficulties in determining the relevant exact epistemic posi-

[13] For some criticism of this claim, see Coady (2012).

[14] For an argument in greater detail to this conclusion, see Carey and Matheson (2013) and Matheson (2015).

tions of the experts and the independence of their opinions obsolete. In this case, it is clear that expert opinion is firmly on the side of believing that climate change is occurring, and this judgment can reasonably be made without having first sorted out these details. These details matter, but in cases of partial consensus, they won't make enough of a difference to tip the evidential scales. This is why cases of disarray are importantly different from cases of partial consensus.

So, in cases where one is reasonable in believing that the expert opinion is in a state of disarray, that person is reasonable in suspending judgment toward the target moral proposition. However, even if suspension of judgment is the rational doxastic response, we often nevertheless need to act. Suspension of judgment does not have a parallel with regard to actions. While for any proposition we have three options (believe, disbelieve, or suspend judgment), the same is not true for actions. Regarding actions we have two options: do it or don't do it. So, even if one is rational in suspending judgment regarding some moral claim, this does not determine how they should act, and regarding moral matters how we should act is of paramount importance. For instance, suppose that Syd is considering whether it is morally permissible to do a certain action, A. Suppose further that Syd knows that it is controversial amongst the relevant experts whether doing A is morally permissible. Given what we have said above, Syd should suspend judgment about whether doing A is morally permissible. While this verdict may settle her doxastic reaction, there remains the issue of what Syd should do. Even if she should suspend judgment she is still forced to either do A or not do A. Given what she knows about the expert opinions on the matter, what should she do?

To answer this question, it is again important to distinguish several kinds of scenarios:

ASYMMETRY: While S is rational in suspending judgment about whether A is morally permissible, S knows that an alternative to A is morally permissible.

SYMMETRY: S is rational in suspending judgment about whether A is morally permissible, and S does not know of any alternative to A that is morally permissible.

Asymmetry and Symmetry give an exhaustive set of options that one may find themselves in when they are rational in believing that the state of expert opinion on some moral matter is in a state of disarray. In cases of Asymmetry, there is an important asymmetry in the subject's options. While one potential action is epistemically cloudy (the subject should suspend judgment as to whether it is morally permissible) another potential action is epistemically clear (the subject knows that it is morally permissible). In such a situation, it is plausible that the subject should exercise moral caution and should not do the epistemically cloudy action. This is to affirm the principle of Moral Caution defended in Matheson in (2015) and (2016).

> MORAL CAUTION: Having considered the moral status of doing action A in context C, if (i) subject S (epistemically) should believe or suspend judgment that doing A in C is a serious moral wrong, while (ii) S knows that refraining from doing A in C is not morally wrong, then S (morally) should not do A in C. (2015 p. 120)

According to Moral Caution it is immoral to take unnecessary moral risks. Put differently, we are morally required to take morality seriously, and taking morality seriously requires avoiding unnecessary moral risks. Moral Caution has it that when one is rational in believing that expert opinion on some moral matter is in a state of disarray, there may nevertheless be moral prescriptions on behavior.

An example may help. Suppose that Stella is considering whether it is morally permissible for her to keep most of her disposable income to herself. Call the proposition 'It is morally permissible for Stella to keep most of her disposable income' proposition 'A'. Stella seeks to find out what the relevant moral experts believe about A, and finds that expert opinion is fairly evenly divided. Stella discovers that a number of experts think that there is nothing wrong with enjoying the fruits of your labor, but she also finds a number of experts who think that given the vast amounts of suffering due to poverty it would be seriously morally wrong for her to do so. Stella comes to reasonably believe that on this matter the expert opinions are in a state of disarray.

Stella, however, notices an important asymmetry in the expert opinions. While the relevant experts are pretty evenly divided about A, very few think that she would do something morally wrong by giving away all of her disposable income to help those in poverty. So, while Stella is reasonable in believing that, regarding A, the expert opinion is in a state of disarray, she is also reasonable in believing that the experts are in a state of partial consensus regarding the permissibility of her giving away all her disposable income. So, given what Stella knows about the relevant expert opinions, she should suspend judgment about A, and she plausibly knows that giving away all her disposable income is permissible. If we then apply Moral Caution to this scenario, we get the verdict that it would be morally wrong for Stella to keep most of her disposable income. So, even in cases where the subject should suspend judgment about the permissibility of some action, there may be factors that are still action guiding.

The factors that were action guiding in cases of disarray had to do with an important asymmetry in the expert opinion. While regarding one action the expert opinion was in a state of disarray, regarding an alternative action, the expert opinion was in a state of partial consensus. It is this asymmetry in the expert opinion that grounded Stella's moral reasons in the case above. However, in many moral matters such an asymmetry does not exist. That is, in many moral matters moral risk is inevitable. Sometimes no alternative is known to be permissible because the expert opinion is in a state of disarray regarding each of the options. For instance, the expert opinion *might* be in disarray regarding these issues, and many more:

- elective limb amputations;
- elective (later-term) abortions for sex-selection, and for certain disabilities;
- whether advance directives should be respected for people who have advanced dementia;
- the permissibility of posthumous reproduction;
- the specific limits of parental authority in pediatric cases;

There are many more contenders for possible disarray. Whether there is a genuine disarray requires major investigation to determine that the parties to the disagreement deeply understand the issues and arguments are indeed likely reasonable in their differing views: not all moral disagreements concerning controversial issues have those features. Given what we have said above, we should suspend judgment about what is permissible in such cases. Since these alternatives are exhaustive, we should suspend judgment about permissibility of any action we can take on this matter. This is a case of symmetry.

In cases of symmetry there is no clear moral path for you to take. Every option you have involves taking a moral risk. Nevertheless, you must act. So, what should you do in a case of symmetry given that each of your alternatives requires taking a moral risk? Unfortunately, we have no advice to give in such a situation.

5.10 Conclusion

We often do not know what to think or do about challenging moral issues. This is especially the case with healthcare decisions that impact ourselves and loved ones. Difficult circumstances can make it hard to see what should be done in a difficult case. This is true for both medicine and morality: a physician should not be the doctor for his or her own child because those emotional connections can cloud clinical judgment and so that physician-parent needs outside help. And a difficult case can make for difficult moral decisions, decisions that are too hard for a family to make on their own, at that time. Ethics consultants, as moral experts, can help with that.

Here were have characterized moral expertise and how to try to find it. We have discussed how merely relying on a moral expert's guidance – deferring to an expert – is less than ideal, insofar as deference does not entail the kind of understanding and display of cognitive and moral virtues that are best. What's best, though, is not always practically necessary in the moment, however, and can be pursued later: at least, deference can result in justified moral beliefs, which are surely better than unjustified moral beliefs.

Finally, we have discussed what to think and do when there are moral disagreements about what to think and do about difficult moral issues. Not all such disagreements are among genuine experts – not all parties to heated moral debates are indeed informed and rational on those debates – but we have discussed what to think and do in a variety of kinds of disagreement. Not surprisingly, however, we have found no magic formula to resolve all such cases of disagreement: morality is often difficult, even for any experts. So, what can we do? At least, we can always do our best to keep thinking critically about the issues, discussing the issues and arguments with other people, especially people who we might disagree with, and doing our best to act on that thinking, when we must act.

References

Anderson, Elizabeth. 2011. Democracy, public policy, and lay assessments of scientific inquiry. *Episteme* 8: 144–164.
Carey, Brandon, and Jonathan Matheson. 2013. How skeptical is the equal weight view? In *Disagreement and skepticism*, ed. Diego Machuca, 131–149. New York: Routledge.
Coady, David. 2012. *What to believe now: Applying epistemology to contemporary issues*. Malden: Wiley-Blackwell.
Collins, H.M., and Robert Evans. 2007. *Rethinking expertise*. Chicago: University of Chicago Press.
Collins, Harry, and Martin Weinel. 2011. Transmuted expertise: How technical non-experts can assess experts and expertise. *Argumentation* 25: 401–413.
Crisp, Roger. 2014. Moral testimony pessimism: a defense. *Aristotelian Society Supplementary Volume* 88 (1): 129–143.
Decker, Jason, and Daniel Groll. 2014. Moral testimony: One of these things is just like the others. *Analytic Philosophy* 54 (4): 54–74.
Driver, Julia. 2006. Autonomy and the asymmetry problem for moral expertise. *Philosophical Studies* 128: 619–644.
Enoch, David. 2014. A defense of moral deference. *Journal of Philosophy* 111 (5): 229–258.
Gelfert, Alex. 2011. Expertise, argumentation, and the end of inquiry. *Argumentation* 25: 297–312.
Goldman, Alvin. 2001. Experts, which ones should you trust? *Philosophy and Phenomenological Research* 63 (1): 85–110.
Hazlett, Allan. 2015. The social value of non-deferential belief. *Australasian Journal of Philosophy* 94 (1): 131–151.
Hills, Alison. 2009. Moral testimony and moral epistemology. *Ethics* 120 (1): 94–127.
———. 2010. *The beloved self: Morality and the challenge from egoism*. Oxford: Oxford University Press.
———. 2013. Moral testimony. *Philosophy Compass* 8 (6): 552–559.
Hopkins, Robert. 2007. What is wrong with moral testimony? *Philosophy and Phenomenological Research* 74 (3): 611–634.
Howell, Robert J. 2014. Google morals, virtue, and the asymmetry of deference. *Noûs* 48 (3): 389–415.
Jones, Karen. 1999. Second-hand moral knowledge. *Journal of Philosophy* 96 (2): 55–78.
Lackey, Jennifer. Forthcoming. Experts and peer disagreement. In *Knowledge, belief, and god: New insights in religious epistemology*. eds. Mathew Benton, John Hawthorne, and Dani Rabinowitz.
Laurence, Thomas. 1993. Moral deference. *Philosophical Forum* 24 (1–3): 232–250.
Matheson, David. 2005. Conflicting experts and dialectical performance: Adjudication heuristics for the layperson. *Argumentation* 19: 145–158.
Matheson, Jonathan. 2015. *The epistemology of disagreement*. New York: Palgrave.
———. 2016. Moral caution and the epistemology of disagreement. *Journal of Social Philosophy.* 47 (2): 120–141.
McGrath, Sarah. 2009. The puzzle of pure moral deference. *Philosophical.Perspectives* 23 (1): 321–344.
———. 2011. Skepticism about moral expertise as a puzzle for moral realism. *Journal of Philosophy* 108 (3): 111–137.
Miller, Boaz. 2013. When is consensus knowledge based? Distinguishing shared knowledge from mere agreement. *Synthese* 190: 1293–1316.
Mogensen, Andreas L. 2015. Moral testimony pessimism and the uncertain value of authenticity. *Philosophy and Phenomenological Research* 92 (3): 1–24.
Nickel, Philip. 2001. Moral testimony and its authority. *Ethical Theory and Moral Practice* 4 (3): 253–266.
Sliwa, Paulina. 2012. In defense of moral testimony. *Philosophical Studies* 158 (2): 175–195.

Vavova, Katia. 2014. Moral disagreement and moral skepticism. *Philosophical Perspectives* 28 (1): 302–333.

Zagzebski, Linda. 2012. *Epistemic authority*. New York: Oxford University Press.

Chapter 6
Credentials for Moral Expertise

Eric Vogelstein

6.1 Introduction

In debates about moral expertise, two questions dominate. First is the issue of whether moral experts do or could exist; second, the question of whether, given that moral experts exist, they can be identified, in some justified or reliable manner. It is the second issue I shall address in this chapter.[1]

The question of whether we can identify moral experts is particularly important for practical applications of moral expertise; indeed, there are people whose professional obligations seem to encompass the role of moral expert. Putative moral experts are most conspicuous in the field of medical ethics, in which there are ethics consultants (whose job it may be to recommend or otherwise pronounce upon the morality of certain actions in particular cases) and ethics committees (which are often charged with institutional, organizational, and public policy formation and ethical oversight of such policies). For example: most major hospitals have ethics consultants and ethics committees to deal with difficult cases and ethically-charged policies; organizations of health care professions such as the American Medical Association and American Nurses Association employ ethicists to devise their position-statements and codes of ethics; federally-sponsored research on human subjects must past ethical muster as determined by Institutional Review Boards, and research on non-human animals must be approved by Institutional Animal Care and Use Committees; and governments sponsor committees aimed at ethical evaluation

[1]There is a third, interesting question that's *related* to issues of moral expertise—that being whether, given that we can identify moral experts (or at least believe we can), there is something problematic about *deferring* to their judgment, i.e. adopting their moral views simply because they, *qua* experts, hold them. As I interpret it, that's fundamentally a question about moral deference, not moral expertise; I will not attempt to answer that question here.

E. Vogelstein (✉)
Duquesne University, Pittsburgh, PA, USA
e-mail: vogelsteine@duq.edu

J. C. Watson, L. K. Guidry-Grimes (eds.), *Moral Expertise*, Philosophy and Medicine 129, https://doi.org/10.1007/978-3-319-92759-6_6

and recommendation of public policy, such as the bioethics committees commissioned by recent U.S. presidents (e.g. the Presidential Commission for the Study of Bioethical Issues under President Obama). To a significant degree, those who work as ethics consultants and on ethics committees and boards are asked to take up the mantle of moral expert; implicit in that is the assumption that we can identify who is and who is not a moral expert. If that is mistaken, then we ought to significantly realign or even abandon those professional roles.[2]

The concern whether moral experts could be reliably identified faces what has been called the *credentials problem*. According to this worry, there is no independent basis from which to verify that the moral judgments of putative experts are in fact *true*, and thus moral experts, even if they exist, lack any satisfying credentials of moral expertise—credentials that would allow people to know who they are (Nehamas 1998, 78–82, LaBarge 2005, Cholbi 2007, McGrath 2008, 2009, 2011). The credentials problem is often framed as a problem only for *non-experts*—after all, one might think, it is plausible that moral experts will themselves know who is and who is not such an expert. But that is not entirely clear—after all, moral experts too seem to lack an independent check of the truth of their moral judgments, thus, if an independent check is required, moral experts would lack sufficient grounds for believing that they themselves are experts. That said, it might be possible to know *a priori* that one's own moral judgments are true, and thus that one is an expert. In any case, for purposes here the credentials problem will be framed as a problem for moral non-experts—the issue is how those who are not moral experts are to identify those who are.

I shall argue that the credentials problem is no problem at all—lacking the ability to independently verify the truth of putative moral experts' judgments does not prevent a moral non-expert from ascertaining who is indeed a moral expert, given a plausible understanding of moral expertise and a plausible view of what moral experts might be like. Thus, absent new concerns, there is no fundamental impediment to moral non-experts identifying moral experts.[3]

[2] That is not to say that if there were no moral experts there would be no professional roles in the vicinity of those that currently imply some degree of moral expertise. For example, we could still preserve roles that involve clarification and explanation of various moral concepts and arguments, but do not involve evaluating those arguments, offering advice, or taking any position on controversial first-order ethical issues. The existence of experts in that sort of clarification and explanation is compatible with the absence of moral experts, in the sense of 'moral expert' relevant here. The important point is that those activities are different in significant respects from what many ethics consultants and those who serve on ethics committees are often expected to do.

[3] It is, of course, a separate question whether those who serve as ethics consultants and on ethics committees are indeed moral experts, and thus whether we have the *right* people serving in those roles.

6.2 What Moral Expertise Is

For purposes here, I shall make the following (plausible) assumptions about moral expertise. First, as I take it, moral expertise is *contrastive* rather than *absolute*. This is a feature of expertise in general. Expertise is always expertise in comparison with some set of people. For example, a professor of atomic physics is an expert in atomic physics compared with those who have significantly less knowledge and skill *vis-à-vis* atomic physics (e.g. ordinary people, physics undergraduates, etc.). However, the professor is *not* an expert in atomic physics compared with a race of aliens all of whom have massively more in-depth knowledge and far greater skill in atomic physics than does any human—compared with the aliens, the professor is essentially a novice. Expertise is also *graded*—the professor is *more* of an expert compared with those who have never studied physics than compared with a Ph.D. student in atomic physics (Goldman 2001).[4]

Furthermore, as I take it, the sort of expertise at issue—the sort that a *moral* expert would exhibit—involves expertise in arriving at *true beliefs* within the domain of expertise.[5] The expert has a certain degree of reliability in arriving at such beliefs. This can be achieved by relying on a stock of domain-specific true beliefs the expert already possesses, or by relying on the ability to form new true beliefs within the relevant domain (e.g. via the employment of certain investigative and/or inferential methods). Thus, a moral expert, as I understand it, is a *cognitive* expert, as opposed to a *performative* expert, which would be someone who is an expert with regard to a particular skill—an expert basketball player, for example. That said, as we shall see, cognitive expertise can itself involve skill—but for the cognitive expert, the skill involved will be *instrumentally* valuable in arriving at true judgments in the relevant domain, whereas for the performative expert, sufficient skill is *constitutive* of her expertise.[6]

---

[4]We can also note that contrast classes are *abstract*, thus there need not be any actual members of the class in order to employ it in determinations of expertise. For example, even if everyone without a doctoral degree in physics were to perish, we can still say of the survivors that they are experts in physics, the implied contrast class being that of ordinary (non-physicist) people (even if no such people exist). That does not rule out *also* saying of such people that they are *not* all experts in physics, in comparison with everyone who currently exists—if that is the contrast class (and there are indeed contexts in which that is the *relevant* contrast class, i.e. contexts in which that is the class we would be interested in) then perhaps only the *best* physicists are experts, or perhaps there are no expert physicists (in that context).

[5]We might also add that an expert is able to provide coherent justifications for her judgments, although that criterion will be controversial. For example, it is an open question whether someone who reliably "sees" the truth within a domain, without being able to articulate (or perhaps without even grasping) relevant justifications or reasons, should be considered an expert.

[6]Some might suggest that cognitive expertise involves the ability to reliably arrive at *justified* beliefs within one's domain of expertise, as opposed to true beliefs *per se*. I am amenable to that understanding of cognitive expertise, and adopting it would not affect the core arguments in this chapter. That said, given that justified beliefs are more likely to be true than unjustified ones (*mutatis mutandis*), it is not clear just how much of a difference exists between truth-based and justification-based views of cognitive expertise. For additional discussion of the relative merits of truth-based and justification-based views of cognitive expertise, see Goldman (forthcoming).

But how much reliability does an expert need? Some have suggested that experts must be *highly reliable* judges of truth in their field of expertise (Cholbi 2007, Rasmussen 2011). I think this cannot be right. Suppose that in a certain domain, it is extremely difficult for ordinary people to arrive at correct judgments; perhaps there are many *prima facie* plausible views on any given issue in the domain, highly complex and conflicting considerations that bear on the truth of such views, competing methodological approaches, easy ways for natural and cultural biases to distort our beliefs in the domain, etc. Let us assume that within that domain, ordinary people get things right 5% of the time. But, suppose that there are some people who judge correctly within that domain 60% of the time. It seems right to say that those people are experts in that domain, compared with ordinary people, despite the fact that they are not *highly* reliable judges. In any case, I shall assume here that high reliability is not a necessary component of expertise. Rather, expertise requires only that the expert be able to judge correctly *significantly more reliably* than those in the contrast class (McGrath 2009). Just what should qualify as a *significant* difference is open to debate, but we can say at least this much: that the *greater* the difference in reliability, the greater the degree of expertise.

So with this general view of expertise in mind, a *moral expert*, as I understand the notion, is someone who, in general, arrives at true (first-order) moral beliefs significantly more reliably than *ordinary or average people* (taken as a group), when such a person attempts to do so. The question I am concerned with here is, can moral non-experts identify who has that feature?

It is worth noting that a moral expert, on this view, is not necessarily more likely to do what's right, or even more likely to be morally motivated, than is anyone else. Of course, on some views, in order have a full understanding of what's right one must also possess such motivations—on a certain kind of virtue-theoretical approach, for instance, the kind of emotional sensitivities and other traits of character that would allow one to fully grasp what's morally important are the very same features that would motivate one to be moral. On other views, however, even perfect moral knowledge might be entirely divorced from the motivation to act accordingly. In any case, the kind of moral expert I shall describe and rely upon in my argument need not be morally motivated—that is, nothing I shall say *requires* that she is. That is compatible with there being other ways (perhaps even better ways) in which someone might have moral expertise that do involve tight connections between moral knowledge or understanding, and moral motivation or action.

Finally, I should note that this chapter relies upon an important metaethical assumption: that there are in fact correct answers to first-order moral questions (e.g. about the morality of particular acts or practices). After all, if nothing is morally right or wrong, then there's nothing for someone to be a moral expert *about,* no domain in which someone *could* be such an expert, given the way I understand the notion of moral expertise—if there are no moral truths, then no one would be able to arrive at true moral beliefs at all, let alone do so more reliably than others, and thus there couldn't be moral experts. Thus, if the question is whether moral experts could be identified *given* that they exist, then the question's condition entails the realist metaethical assumption at issue (given the relevant understanding of moral expertise). For those who accept the sort of anti-realist view on which there are no moral truths, the possibility of moral expertise, and thus the problem of identifying moral experts, does not arise.

6.3 The Proper Credentials

According to the credentials problem, moral non-experts are unable to identify moral experts because non-experts have no way to independently check whether an expert's moral judgments are in fact true, and thus cannot know how reliable such a person's moral judgment is in general. This is not the case in other domains of expertise, because there is ordinarily some readily identifiable independent check or verification of a person's judgment. An expert in weather forecasting, for example, can be identified by checking how often her predictions about the weather are correct—the fact that we can know without relying on expertise what the weather is allows us to know if someone is indeed an expert weather forecaster; and without that kind of verification, we would have no way of knowing who is such an expert.

As proponents of the credentials problem have noted, the problem implies that moral non-experts cannot identify moral experts based on the content of their judgements—according to those proponents, that implies that moral non-experts cannot identify moral experts *at all*. But might there be other criteria by which experts can be identified? Consideration of certain other domains of expertise strongly suggests that the answer must be 'yes.' We know, for example, that mathematicians are experts in mathematics without independently verifying the truth of their mathematical judgments (indeed, independent verification makes sense empirically—it is hard to see what it would mean in *a priori* domains like mathematics).[7]

But why don't we need an independent check in order to identify mathematical experts? How do we identify experts in mathematics? Why doesn't the credentials problem apply to mathematicians? Is there a relevant difference between the moral and the mathematical, or between the respective sorts of expertise? I will suggest that there are no such differences, and I'll argue that the proper explanation of why we don't need an independent check for mathematicians reveals that, for similar reasons, we don't need an independent check for expertise in the moral domain.

The reason why non-experts don't need independent checks of the truth of particular judgments in order to identify experts is that non-experts have ways of identifying when someone has a mastery of *truth-conducive methods* of arriving at judgments within a domain. This is so even if non-experts do not themselves possess

[7] It might be suggested that if mathematical savants were able to intuit the truth of mathematical judgments, then that could serve as an independent check—but, of course, we don't need to enlist the services of savants in order to identify experts in mathematics. Moreover, reliance on savants would be unable to resolve the credentials problem because the credentials problem would apply to savants, as well. Indeed, we are more justified in believing that the mathematician is a math expert than we are believing in the abilities of a putative savant—after all, we would verify the savant's abilities *via* the mathematician's judgment.

a complete understanding of such methods. In the case of the mathematician, for example, we know that such a person is an expert in mathematics because we know that the methods by which she arrives at mathematical judgments are truth-conducive. Specifically, we understand the truth-conducive nature of mathematical methods *in general* (e.g. the method of mathematical proof), and understand that those methods can be extended to more and more complex problems (including ones we do not understand). And since we can trust that her education has inculcated a mastery of such methods (i.e. that university mathematics departments are not fraudulent in that regard), we can know that she is significantly more likely to arrive at true mathematical judgments than are those who lack mastery of those methods, i.e. the ordinary or average person.[8]

Furthermore, if we know that someone's method is *not* truth-conducive, then we know that that person is *not* an expert, irrespective of the person's accuracy. Consider, for example, a weather forecaster who throws dice to determine the weather. Suppose that this person has, up to this point, gotten things right significantly more reliably than the average person (we may even suppose that she has never erred). We would not consider this person an expert, despite the fact that we have independently verified her high and consistent accuracy, because we know that her method is not truth-conducive—she has simply gotten lucky. Thus, if we can know that someone's method isn't truth conducive then we know they're not an expert; and, if we can know that someone's method *is* truth conducive, we can know that they *are* an expert without independently checking the accuracy of their individual judgments (indeed, we can know that such a person is an expert even if we know she has been consistently *wrong*—if we know her method is highly reliable, but not infallible, we should believe that past errors have been the result of bad luck). Thus, as long as we can determine the soundness of the method a person is using to arrive at judgments within the relevant domain, we have no use for independent checks of her particular judgments in determining whether she is an expert.

But can moral non-experts identify a method for reliably arriving at *moral* truths? I believe they can. The method is simply that of *moral reasoning*, broadly conceived. And we know the criteria of *good* moral reasoning. They include criteria of good reasoning in general, e.g. the ability to identify and construct valid arguments, to spot fallacies, to identify ambiguity and make relevant distinctions, to anticipate and offer objections, etc. They also include criteria of good *moral* reasoning *per se*, including an understanding of a wide variety of moral concepts and their

[8] It might be thought that non-experts cannot know whether a putative expert is indeed an expert because they cannot know if such a person is in fact *applying* the relevant truth-conducive method in cases beyond the non-expert's understanding. In response, it is important to note that once the non-expert recognizes that a certain truth-conducive method (e.g. that of mathematical proof) can be applied in ways that might outstrip the non-expert's present understanding, she need only identify *who,* in all likelihood, has mastered that method (including its more complex applications), in order for her to identify (useful) experts in the relevant domain. The way of doing that is to identify reputable *institutions* (e.g. universities), those who are trained therein, and who have been acknowledged in various ways to have mastered the relevant method (e.g. via advanced degrees, academic positions, and publications in high-quality academic journals).

interrelations (including moral principles and theories), "stock" ethical arguments as well as their primary strengths and weaknesses, and methodologies and forms of argumentation most fruitfully applied to the evaluation of first-order moral claims (e.g. argumentation from analogy, reflective equilibrium, etc.). Finally, good moral reasoning requires relevant background knowledge that would inform the premises of one's arguments, e.g. an understanding of various morally relevant psychological, social, and political facts. It is plausible that this sort of knowledge is only satisfactorily acquirable via sufficient "life experience", in which case becoming a good moral reasoner might not be something one can do simply from the armchair, or something that is realistically achievable in relative youth.

The traits and skills I've described will be familiar to academic philosophers, especially those within the analytic tradition and those familiar with the fields of normative and applied ethics. Indeed, developing the skills thus described is the essential point of training in normative and applied ethics. Let us refer to those with such training as 'moral philosophers'. Moral philosophers, then, will have the core skills that form the basis of moral expertise. If some moral philosophers also have relevant background knowledge (likely informed by life-experience), they will, in all likelihood, be moral experts.

Thus, the credentials problem is resolved: everyone can understand (1) that moral reasoning is a truth-conducive method of moral belief-formation, and (2) that moral philosophers are particularly adept at moral reasoning, given sufficient background knowledge. That understanding requires only familiarity with the relevant fields, and a basic understanding of the methodology of moral reasoning and argumentation. Return to the analogy with mathematics. Non-mathematicians know that methods of mathematical reasoning are truth-conducive because we understand their less complex variants (e.g. simple proofs), and understand that that very general method is indefinitely scalable (to more complex problems). Moral non-experts can legitimately make the same sort of inference about moral reasoning. If moral non-experts understand what moral reasoning can do, i.e. that it is a truth-conducive method of moral belief-formation (even if an imperfect one), as well as the expanding complexity of its subject matter and application—and a basic college-level course in applied ethics should be sufficient to supply that understanding—then they can understand the scalability of moral reasoning, the high degree of skill and knowledge required to do it well, and that moral philosophers are much more likely than others to possess those skills and knowledge.

Given that everyone can identify who the moral philosophers are with a reasonable degree of accuracy (e.g. via published papers, talks, academic degrees, and academic positions), that leaves only the issue of identifying those moral philosophers who have the appropriate type and degree of background knowledge. That might be somewhat difficult to determine, but there is no in-principle barrier to moral non-experts figuring that out. For example, whether a certain moral philosopher is sufficiently knowledgeable might be known indirectly, by examining the content of his or her work (and seeing if there is extant or missing pertinent background information), or directly, through interviews or discussion designed to draw out relevant knowledge or experience. Because the sort of background knowledge that would be relevant to moral reasoning is not itself something that is within the purview of the moral philosopher's expertise, moral non-experts are in a position to criticize a putative moral expert on such grounds. For example, if a certain moral philosopher supports the view that the death penalty should be legal, while denying the fact that the criminal justice system in the U.S. routinely involves differential treatment vis-à-vis findings of guilt and sentencing based on race and socioeconomic class, then a moral non-expert can rightly conclude that the philosopher fails to understand key morally-relevant features of the phenomenon in question. That would plausibly disqualify that philosopher from moral expertise, at least about the issue of whether the death penalty should be legal in the U.S.—the philosopher's arguments would thereby be flawed in a way that is transparent to those with no training in moral reasoning.[9] Similarly, moral non-experts can proactively confirm, within reason, sufficient background knowledge by seeing if these sorts of facts are indeed known by moral philosophers. Again, it is far easier to know the sorts of facts that are *relevant* to resolving a controversial moral issue—indeed, easy enough for moral non-experts to know—than it is to employ ethical analysis and moral reasoning in *adjudicating* such issues. The latter requires the sort of skill and knowledge that moral philosophers have acquired through years of study and training.[10]

Thus, insofar as we accept that moral truths can be reasoned to, i.e. that moral reasoning is a truth-conducive method of forming moral beliefs—something I think we should accept, and which I'll address and defend shortly—then there is a way for anyone to know who stands to be significantly better than the average or ordinary person at arriving at true moral beliefs: we can identify who is adept at moral reasoning, and such people will most often be sufficiently knowledgeable (and probably life-experienced) moral philosophers.

To be sure, the notion that moral philosophers are moral experts in virtue of their expertise in moral reasoning is not an original insight—many philosophers have argued along similar lines with the aim of defending the notion that moral experts *exist* (Singer 1972, 1988, Szabados 1978, Brink 1989, Crosthwaite 1995, Nussbaum 2002). My claim here is that same points can be leveraged in response to the creden-

---

[9] That is not to take any position on whether the death penalty should be legal, only on whether the facts thus mentioned *bear* on whether the death penalty should be legal. It is obvious that profound inequities in the doling out of a certain system of punishment count against such a system's implementation, morally speaking, even if they are strongly outweighed by other considerations in favor of that system. Thus, if such inequities exist, anyone who fails to acknowledge that they do, yet argues in favor of the punishment in question, fails to take into account information that bears significantly on that argument, and thus the argument has an obvious flaw.

[10] It is important to stress that it need not be the case that *all* or even *most* moral non-experts have the ability to verify the background knowledge of putative moral experts. It need only be the case that some people are able to verify such knowledge in a way that can be trusted by the general population (and note that those who act as verifiers might require some degree of expertise in *other* areas of inquiry). The precise way in which this is best accomplished is open to debate, of course—but there is no obvious reason why such procedures would inevitably fail to be pragmatically feasible.

tials problem—that if those philosophers' arguments are sound, then the basis for believing that moral experts exist *just is* the basis for identifying them. But my argument does not depend on the claim that moral experts indeed exist. Rather, it depends only on the claim that *if* moral experts of the type I've described exist—something that is at least plausible—then they can be reliably identified by non-experts. In that sense, then, my solution to the credentials problem is a qualified one—it does not entail that *any* moral expert could be identified by moral non-experts, only that a certain sort of moral expert can be so identified.

My view is thus consistent with the possibility of other sorts of moral experts, i.e. other ways in which one can be a moral expert besides being a highly skilled moral reasoner, who are *not* readily identifiable by moral non-experts. For example, some people might be so morally "tuned in" that they have no need to reason morally (at least not explicitly) in order to be significantly more likely to form true first-order moral beliefs than the average person. Such people might have no identifiable signs of moral expertise—and if those were the *only* possible (or plausible) sort of moral expert, then the credentials problem would perhaps be irresolvable. Indeed, apart from the ability of moral non-experts to identify skilled moral reasoners, it is unclear how the credentials problem might be resolved, because it is unclear how there might be another truth-conducive method of moral belief-formation that would be identifiable by moral non-experts.

Therefore, not only is the solution I have described dependent upon the claim that moral reasoning is a reliable way of arriving at true moral beliefs, but, I would suggest, the *only* solution to the credentials problem is so dependent. In that way, the credentials problem is instructive—it presses us to see how moral experts *can* be identified by excluding a way in which they cannot. In developing a response to the credentials problem, we must conclude that the sort of independent check at issue is unnecessary (á la the comparison to experts in mathematics), and that what is required for identifying experts is just identifying mastery of a truth-conducive method of domain-specific belief-formation. We are thus led to look for a truth-conducive method of *moral* belief-formation whose mastery can be identified by moral non-experts, of which moral reasoning is the only plausible sort.

With that in mind, it is important to note that there are some who are, to varying degrees, skeptics about the degree to which moral reasoning is indeed truth-conducive. That said, a moderate skepticism is compatible with the solution I have sketched—such a view would imply only a mitigation of the *degree* of moral expertise possessed by, and justifiably imputed to, moral philosophers or others skilled in moral reasoning. But it is worth considering stronger versions of skepticism about moral reasoning, indeed thoroughgoing versions on which moral reasoning provides *no* (or essentially no) justificatory basis for moral beliefs, thus on which a person's mastery of moral reasoning provides (essentially) no reason to believe that they are more likely than others to form true moral beliefs. If such skepticism is warranted, then moral experts of the sort described above could not exist, and moral non-experts would have no basis on which to identify any moral experts that might exist.

6.4 Skepticism About Moral Reasoning

Moral reasoning possesses the virtues of reasoning *per se* to no less an extent than do other sorts of reasoning, i.e., reasoning employed in the service of other domains of inquiry. For example, reasoning applied to moral matters makes sets of moral beliefs more consistent, teases out the moral implications of propositions we're inclined to affirm, and rules out views that are inconsistent with such propositions. It also provides the conceptual resources needed to sufficiently grasp and distinguish between the moral claims we're interested in, and the arguments that bear on them, rendering moral beliefs less confused, moral claims less ambiguous, and moral arguments less equivocal. These are the sorts of tools that stand to make sets of beliefs more likely to be true than they otherwise would be, regardless of the subject-matter. Therefore, the most plausible version of skepticism about the extent of the power of moral reasoning to make more likely the conclusions that result from it derives not from the power of moral reasoning *in itself,* but rather from inherent limitations in our ability to adequately employ it.[11,12]

On the view under consideration, moral reasoning is *per se* a truth-conducive method of moral belief-formation, i.e. it stands to be so if applied properly (e.g. in an unbiased fashion). The problem, on this view, is therefore with *us,* not with the method—but the result is the same: strong reason to doubt whether the moral conclusions that are the outputs of a skillfully employed process of moral reasoning are

[11] I discuss other skeptical arguments in Vogelstein (2015).

[12] That is not to say that the power of moral reasoning might not be limited in significant ways compared with other areas of inquiry, e.g. mathematics. That might depend, in particular, on the extent of any inherent limitations on the *strength* of moral argumentation in general. For example, moral arguments might depend upon plausible yet questionable intuitive premises, or forms of argument that are often not very strong (e.g. inference to the best explanation). Indeed, such limitations are a plausible explanation of any greater extent of disagreement, should it exist, within applied ethics as compared with other domains of inquiry (although, as I discuss below, widespread disagreement is compatible with expertise). And perhaps, one might think, those limitations are *so* great that they prevent even perfect moral reasoners (limited only by the power of moral reasoning *per se*) from being *significantly* more likely to reach true moral conclusions than others. It is difficult to see, however, how anyone sufficiently acquainted with the complexity of applied moral philosophy, and the work that is done within that field, can reach such a conclusion. It should be obvious, I would submit, to those so acquainted that the level of detail and rigor required to reason well about moral matters is very high, and that there are significant pitfalls and easy errors to be made if one fails to carefully employ the relevant intellectual tools. Indeed, careful, rigorous reasoning is a powerfully truth-conducive tool in any sufficiently complex domain of inquiry irrespective of any limitations on just how certain we can be of conclusions reached within that domain. A key reason is that in complex domains there will very often be multiple positions to take on any given issue and subtle yet important distinctions between those positions. Indeed, simply identifying and distinguishing between possible views might itself be a key component of the relevant reasoning-process—those who are untrained in reasoning about the relevant domain might not even know what positions are possible, let alone the arguments that support or count against them. For these reasons, moral reasoning stands to be a significantly powerful tool for discovering truth within the complex domain of first-order moral inquiry, assuming that it is employed properly and faithfully.

indeed more likely to be true than those that are not. A somewhat common contention that undergirds such a view is that all (or almost all) moral reasoning is subconsciously *post hoc* justification (i.e. rationalization) of antecedently held moral convictions, which are moral beliefs based on intuition (i.e. brute, non-reflective "seemings") that do not derive from, and are immune to, moral reasoning (Haidt 2001). And because moral intuition is not within the purview of the master moral reasoner—skilled moral reasoners are no better at intuiting than others, we may presume—we have no way of identifying people whose moral beliefs are more likely to be true than are the beliefs of others, even if such people exist (a possibility that is not ruled out by the view under consideration). This is equivalent to claiming that the *kind* of moral expert that could be identified by moral non-experts—the expert moral reasoner who *thereby* gains moral expertise—cannot exist (at least if we're restricting the potential pool of moral experts to beings like us).

Notably, this objection presses us to clarify the nature of the central claim at issue in this chapter—in particular, it requires a specification of the *modal* features of that claim. I have been arguing that moral experts *can* or *could* be reliably identified, given that they exist. There is an ambiguity here. Such a claim might be equivalent to the claim simply that it is *possible* that moral experts can be reliably identified by moral non-experts. But such a claim seems obviously true, and indeed vacuously so—we can of course imagine a possible world in which conditions are set up to make that possible, e.g. a world in which God makes his presence known and provides us with a clear way of identifying moral experts. What we want to know is whether moral experts could be reliably identified *by beings like us in a world like ours*. But there is a further ambiguity in an affirmative answer to such a question, because we might wonder how *much* like us beings must be, and how much like ours a world must be, in order for moral non-experts to be in a position to reliably identify moral experts. And to be sure, we might not be satisfied by an answer in which the problem is solved only given a significant deviation from the world as it is, or from humans as they are. If human beings indeed use moral reasoning exclusively (or nearly so) simply to rationalize prior moral convictions, and there is no plausible way to remedy our overwhelming tendency to so rationalize, then moral experts, in a quite relevant sense, *cannot* be identified—indeed *that* context, some might claim, is the relevant one, and it is perhaps interesting but not ultimately pertinent that other sorts of rational beings, or human beings in unrealistic circumstances, might be moral experts in virtue of their (proper) use of moral reasoning, and thus be identifiable by moral non-experts.

The question, then, is whether we should grant the plausibility of the central premise of the objection at issue: that moral reasoning employed by humans as they are, and thus human moral philosophers as they are, serves merely as rationalization, if not universally then nearly so. In the remainder of this section I challenge that view. Not only does the view have implausible implications, but it is implausible on its face in light of the sort of dialectical interaction, doxastic revision, and pedagogical activity that moral philosophers engage in on a routine basis.

There is no doubt that many of our moral beliefs are formed intuitively, automatically, and heuristically, with little to no input from conscious reasoning or moral argumentation; it is also highly likely that many (perhaps most) instances of moral reasoning and argumentation among human beings are essentially rationalizations of those more immediately and unconsciously-formed convictions (Haidt 2001). That implies nothing whatsoever about whether moral reasoning and argumentation can significantly affect our moral beliefs in a rational way. For example, reasoning might provide the basis for the *revision* of moral beliefs (even if reasoning *also* serves to rationalize)—indeed, it is commonsensical that people sometimes change their minds based on moral arguments, and anecdotal evidence abounds that this is what actually happens in communities of applied ethicists (e.g. in graduate school or the professional ranks). And, importantly, note that my claim here need only be extremely weak: so long as moral reasoning *can* rationally determine our moral beliefs, the sort of moral expert at issue (the kind that can be identified by moral non-experts) can exist—and that is so even if moral reasoning isn't *particularly* efficacious in determining our beliefs, and thus even if human beings' potential *degree* of moral expertise is significantly limited. Indeed, I need only claim that moral reasoning rationally determines the moral beliefs *of moral philosophers*—and there is reason to believe that moral philosophers have certain cognitive skills and dispositions that help them avoid the sort of rationalization that might make moral reasoning less effective in others (more on this below).

Furthermore, people are often substantially unsure about what to believe, and reasons can provide the grounds on which a person comes to believe one way or the other. The rationalization hypothesis is not applicable to such cases, because in these cases the person has yet to form the belief to be rationalized. Unless we're inclined to accept that people who begin from a point of substantial uncertainty never (or exceedingly rarely) come to form their moral beliefs based on moral reasons adduced for one position or the other, we must conclude that moral reasoning for human beings has the sort of rational power at issue. Indeed, to deny that conclusion we would have to think that people never (or almost never) believe *for* epistemic moral reasons (i.e. reasons to believe moral propositions) that are known consciously. There *are* such reasons—and it would be strange if it were impossible (or extremely difficult, to the point of high improbability) for humans to hold beliefs for such reasons (even if there are biasing forces within us that make doing so a highly imperfect process). Moral beliefs, in virtue of *being* beliefs, are reasons-responsive at least to some degree (even if they are also especially susceptible to bias)—thus if one is aware of the epistemic moral reasons there are, it should at least be possible to respond to those reasons' normative force, by forming and grounding one's beliefs accordingly. The reasons that favor believing one moral proposition or another should thus be able to rationally influence what beliefs we in fact hold, even if we are pushed to and froe to a significant extent by competing unconscious forces.

Moreover, even if almost all human beings use moral reasoning as rationalization almost all the time, that fact does not impugn the value of moral reasoning under ideal circumstances (as even Jonathan Haidt acknowledges (Haidt 2001, 822–3))—and there is strong evidence that moral philosophers are far closer to the ideal. There is, in other words, a significant difference between "folk" moral reasoning (which is far more prone to use as rationalization) and the moral reasoning of philosophers.

Specifically: Livengood et al. (2010) have found that philosophers are in general far more reflective, deliberate, and aware of potentially biasing forces, in particular because philosophers are highly attuned to the distinction between what *seems* true—that which would be the subject of rationalization—and what one is overall *justified* in believing (Livengood et al. 2010). So, to generalize, the relevant difference between philosophers and non-philosophers derives from differences in intellectual values and cognitive skills associated with those groups, generally speaking, that put philosophers in a better position than others not to be led astray by the easy answer, the automatic response, or the comfortable solution. Because they're aware of the distinction between what seems true and what one is justified in believing, and because they are *motivated* to believe what's justified regardless of what seems true, philosophers have a decreased motivation to rationalize—philosophers, more so than others, don't *want* to rationalize, and that is reflected, for the better, in their reasoning. Rather, philosophers want to believe what they're justified in believing even when it seems false, and they are acutely aware of the ways in which seeming true and overall justification can come apart. That particular skill-value combination constitutes a valuable asset in the employment of philosophical reasoning, including moral reasoning—and the research of Livengood et al. (2010) strongly suggests that philosophers possess the relevant skills and values to a significantly greater extent than do others. And this of course comports with the values and methods standardly taught and employed in philosophical training at all levels.

What I've described might be criticized as painting too rosy a picture of philosophers' motives and abilities, and perhaps rightly so. To be sure, the notion of the objective moral reasoner is an ideal, and I do not mean to suggest that philosophers typically embody that ideal or even that they are close to doing so. But regardless of how close to the ideal philosophers are (and there may indeed be good reason to think that philosophers are not very close at all), there is nevertheless excellent reason (based on the research of Livengood et al. (2010)) to believe that they are significantly *closer* to that ideal than are non-philosophers, in general (i.e. comparing both groups as wholes).

Finally, an acknowledgement of the ability of humans to employ moral reasoning in a way that is more likely to result in true moral beliefs is implicit in the sorts of activities moral philosophers engage in on a daily basis. I have in mind here the kind of moral debate and discussion about substantive moral issues one finds in peer-reviewed articles and presentations on normative and applied ethics, and social and political philosophy. If moral reasoning is substantially and universally *not* a truth-conducive process when employed by human beings, then these fields (as they're practiced by us) would fail to be academic disciplines at all—they would be rationalization games (Gert 1992, Meyers 2007). Indeed, moral philosophers *do* form and revise their beliefs in response to moral argumentation found in scholarly venues—and it would be hard to make sense of that phenomenon, especially in light of the sort of intellectual dispositions common to philosophers outlined above, if such belief-revision were merely the result of non-rational processes. Even more importantly, if moral argumentation served merely as rationalization then teaching students to think critically about moral matters would be a fruitless enterprise. Indeed, grading in the relevant courses would be perverse, as we would be giving greater credit to those students who are better able to rationalize than their peers. Of course, it is *possible* that those practices in fact all rest on fundamental errors, that moral philosophers suffer from a sort of mass-delusion about the nature and value of their work, and that the reasoning skills we teach our students can be used only to rationalize prior convictions. But it seems unlikely that moral-philosophical researchers and instructors would carry on in the way they do if their work were essentially useless, *vis-à-vis* uncovering the truth about controversial moral issues. At the very least, it is disingenuous for any moral philosopher to continue what they do (at least with the thought that they are doing something effective and worthwhile) while suggesting that our use of moral reasoning is not truth-conducive. That said, the considerations thus adduced are perhaps not by themselves convincing evidence that we employ moral reasoning in a truth-conducive fashion (and would thus need to be supplemented by the sorts of arguments defended above), but they point in that direction; and in any case, it is instructive to understand the full cost, and the potentially unsettling implications, of the contrary position.

6.5 Disagreement and Consensus

If what I've said thus far is essentially correct, then moral experts of a certain sort could indeed be identified by moral non-experts, and we should believe that moral experts of that sort can indeed exist; we thus ought to wonder whether the credentials problem still has any teeth. However, I think that this isn't quite the end of the story for the credentials problem (although I don't think that the credentials problem ultimately has any bite). In particular, it might be claimed that even if we can identify moral experts, the credentials problem reemerges at the level of *expert disagreement*. In the moral domain there seems to be strong and widespread disagreement among the putative experts—moral philosophers are continuously debating virtually every substantive ethical issue, with no consensus in sight. Thus, one might wonder, how are non-experts to determine who is *really* the moral expert? Of course, the non-expert might try to figure out who has more fully mastered the method of moral reasoning, or who is employing it better—e.g., whose inferences are flawed, whose life-experience is impoverished, whose understanding of relevant distinctions is muddled, etc. But the point here is that disagreement exists among those who have an equal claim to expertise *from the point of view of the non-expert*—the non-expert is thus in no position to know which putative expert is right and which is wrong. In particular, putative experts might have roughly equal degrees of background knowledge, and as a non-expert one cannot reasonably assess who has a greater mastery of the other aspects of moral expertise based on skill in moral argumentation and analysis. Thus, on this line, there will be no way for the non-expert to know who is the true expert and who is merely the apparent one (LaBarge 2005).

This kind of worry, however, rests on a mistake. The mistake is that there cannot be widespread disagreement among experts (about propositions within their domain of expertise). Expertise *per se* does not rule out disagreement or demand convergence of views; and there is no reason why the *extent* or *degree* of disagreement should matter. Strong and widespread disagreement among a group of people is compatible with those people being significantly more likely to get things right than are ordinary or average people. For example, even if the experts are split 50/50 on a host of issues, many of those issues might lend themselves to some degree of *mistaken consensus* among ordinary people due to common errors or biases of various sorts. Or, there might be positions which experts have ruled out, but which ordinary people commonly hold—e.g., ordinary people might be evenly split between five incompatible views, while the experts are only split between two. In these cases, experts get things right significantly more often than ordinary people despite their widespread disagreement. The upshot is that even if there is no way to know who's *right* about such issues, we can still know who is an *expert.*

It might be objected that this would miss the point of identifying moral experts, which is for non-experts to *rely* on them, in some fashion—to use their judgments as guides to moral truth. Thus, even if we can identify moral experts, widespread disagreement seems to obviate the point of doing so—non-experts will be left wondering which experts are the right ones to go to for moral instruction.[13]

However, it is not clear that there is widespread disagreement among moral philosophers on all controversial issues about which non-experts might seek instruction or knowledge. For example, it seems *prima facie* that a large majority of applied ethicists believe that physician-assisted suicide and euthanasia are morally permissible in paradigmatic circumstances (a reading of the relevant literature, at least, suggests as much). Therefore, unless there is reason to suppose that those who oppose those practices fail, for some reason, to publish their arguments to a larger extent than those who support those practices do (e.g. because the apparent minority but true majority self-selects out of the debate, or because there is bias in the process of peer-review) we have reason to think that there is some significant degree of consensus among moral philosophers. And there is presently no evidence that the proportion of pro-assisted suicide or pro-euthanasia arguments in the literature fails to reflect, at least roughly, the views of those who work on the ethics of those practices. Similar consensus plausibly exists about the morality of abortion in ordinary cases, factory farming, and maybe many other issues as well. If this is borne out by, for example, survey data, non-experts could reasonably seek advice and instruction in line with the relevant consensus.

Moreover, non-experts might make use of aggregation of expert judgment *regardless* of the degree of consensus (or lack thereof) among experts. Insofar as deference to expert judgment is appropriate in the moral domain, non-experts may defer to even a *weak* consensus of experts, e.g. a 60/40 divide, and it would be

[13] Notably, moral non-experts could still, under conditions of widespread expert disagreement, gain knowledge of the *best arguments* for the respective positions. But that's not what we want from moral experts (e.g. we could get that from experts in ethics regardless of whether they're moral experts); or at least that's not the entirety of what we want: we want the right (or most justified) answers to first-order moral questions.

appropriate for the non-expert to adjust her credence-level to the degree of consensus.[14] Indeed, in that case, even a 50/50 split among experts would be instructive, as it would warrant credence of 0.5, among non-experts, with respect to the position at issue. And even if simply deferring to expert consensus is *not* warranted, aggregated expert judgment can still serve, and, it seems, should serve, as particularly weighty evidence to factor in when trying to reach one's own moral conclusions—in that case, we might say that some degree of deference is appropriate, even if one's own deliberation and conclusions should have significant weight as well.

In any case, debate about the appropriateness of deference to moral experts is orthogonal to the validity (or lack thereof) of the credentials problem—such issues arise even if non-experts could perfectly identify moral experts, indeed even if they could in addition identify the precise degree of expertise of each moral expert (*viz.* the overall degree of accuracy of moral experts' first-order moral beliefs). Moreover, the *sort* of reliance on expertise we're after—*epistemic* reliance (i.e. using experts' judgments as a significant evidential basis upon which to ground, at least to some extent, one's beliefs and credences) is, I would submit, uncontroversially appropriate in response to the moral judgment of known moral experts even if moral deference is in some way problematic. That is to say, whatever is problematic about deference to moral experts (if there is anything problematic), once we know that someone is a moral expert, there can be no *epistemic* problem with so deferring.[15]

One potential worry about the aggregate judgment of moral experts remains. The concern derives from the possibility that the breakdown of moral beliefs among moral experts mirrors the breakdown of those beliefs within the general population. If moral experts as a group come down on an issue in the same way that moral non-experts do, that entails that the putative moral experts are not any more reliable *on that issue* than are non-experts, in which case they are not moral experts about the issue in question. If, then, that is the case with *all* moral issues, then we have very good reason to doubt that experts in moral reasoning are indeed moral experts (such people might be better at *justifying* or *arguing for* their views, but they would be no better at getting the correct answer than the average person). But absent strong empirical grounds for thinking that's the case we should think that it is not, given that moral reasoning is a truth-conducive method of moral belief-formation, that at least some humans can employ it effectively, and that those with extensive training and practice in moral reasoning will be significantly better at it than others.

---

[14] The moral, if not epistemic, appropriateness of such deference might depend, in addition, on the non-expert understanding, at least to some extent, the moral reasons and arguments proffered by the experts. Furthermore, the appropriateness of "deferring to the numbers" of aggregated expert opinion depends on the extent to which each expert's opinion is independent of that of the other experts, in the sense that the experts form their expert judgments based on their expertise *per se,* and not on the fact that other experts make such judgments. I shall not further explore that complication here; for more on this issue, see Goldman (2001).

[15] For a more extended discussion of this point, see Sreenivasan (2015).

6.6 Conclusion

It has been argued in this chapter that moral non-experts can both *identify* moral experts and *rely* on such experts, even in the face of widespread expert disagreement. Non-experts can identify moral experts because non-experts can know that sufficiently knowledgeable moral philosophers will tend to get the right answers to first-order moral questions significantly more often than the average person, who lacks their expertise in moral reasoning—and independent checks of the truth of such beliefs are unnecessary for identifying such people as moral experts because moral reasoning is, and can be known by non-experts to be, a truth-conducive method of moral belief-formation. Finally, non-experts can make significant epistemic use of the aggregated judgment of moral experts when forming their own moral views, and indeed do so even when the moral experts are significantly divided, although the extent of that division on some major issues is questionable. For those reasons, moral philosophers can have important roles to play in ethics consultation and policy-formation—in certain cases, they can rightly call themselves moral experts, and others may treat them accordingly.

References

Brink, David O. 1989. *Moral realism and the foundations of ethics*. New York/Cambridge: University Press.

Cholbi, Michael. 2007. Moral expertise and the credentials problem. *Ethical Theory and Moral Practice* 10: 323–334.

Crosthwaite, Jan. 1995. Moral expertise: A problem in the professional ethics of professional ethicists. *Bioethics* 9: 361–379.

Gert, Bernard. 1992. Morality, moral theory, and applied and professional ethics. *Professional Ethics* 1 (1/2): 5–24.

Goldman, Alvin I. 2001. Experts: Which ones should you trust? Philosophy and. *Phenomenological Research* 63: 85–110.

———. Forthcoming. Expertise. *Topoi*.

Haidt, Jonathan. 2001. The emotional dog and its rational tail: A social intuitionist. approach to moral judgment. *Psychological Review* 108: 813–834.

LaBarge, Scott. 2005. Socrates and moral expertise. In *Ethics expertise: History, contemporary perspectives, and applications*, ed. Lisa Rasmussen, 15–38. Dordrecht: Springer.

Livengood, Jonathan, Justin Sytsma, Adam Feltz, Richard Scheines, and Edouard Machery. 2010. Philosophical temperament. *Philosophical Psychology* 23: 313–330.

McGrath, Sarah. 2008. Moral disagreement and moral expertise. In *Oxford studies in metaethics*, ed. Russ Shafer-Landau, vol. 3, 87–107. New York: Oxford University Press.

———. 2009. The puzzle of pure moral deference. *Philosophical Perspectives* 23: 321–344.

———. 2011. Skepticism about moral expertise as a puzzle for moral realism. *Journal of Philosophy* 108: 111–137.

Meyers, Christopher. 2007. *A practical guide to clinical ethics consulting*. Lanham: Rowman & Littlefield.

Nehamas, Alexander. 1998. *The art of living*. Berkeley: University of California Press.

Nussbaum, Martha. 2002. Moral expertise? Constitutional narratives and philosophical argument. *Metaphilosophy* 33: 502–520.
Rasmussen, Lisa M. 2011. An ethics expertise for clinical ethics consultation. *Journal of.Law, Medicine, and Ethics* 39: 649–661.
Singer, Peter. 1972. Moral experts. *Analysis* 32: 115–117.
———. 1988. Ethical experts in a democracy. In *Applied ethics and ethical theory*, ed. David M. Rosenthal and Fadlou Sehadi, 149–161. Salt Lake City: University of Utah Press.
Sreenivasan, Gopal. 2015. A plea for moral deference. *Etica & Politica* 17: 41–59.
Szabados, Bela. 1978. On "moral expertise". *Canadian Journal of Philosophy* 8: 117–129.
Vogelstein, Eric. 2015. The nature and value of bioethics expertise. *Bioethics* 29: 324–333.

Chapter 7
Can Moral Authorities Be Hypocrites?

Marcela Herdova

> *"I saw a mink yesterday ... wearing fur!*
> *If the mink population simply cannot*
> *be bothered to set an example, I see*
> *no reason why we should either."*
>
> *Viz.co.uk*

7.1 Introduction

When I teach classes on poverty, specifically on Peter Singer's *Basic Argument*, which concludes that failing to donate to effective aid agencies is morally wrong (so donating is not merely supererogatory[1]), students always ask if Peter Singer *too* donates to charity. An affirmative answer seems to placate them in some important way: they are more attuned to what Singer has to say and weigh his arguments in a way they might not have been ready to if Singer had turned out to be, among other things, an extravagant spender without a reliable track record of generously donating. It is as if Singer's own acts of giving lend him the credibility to speak on this issue and demand of others that they donate.

This reaction points to something important. We want those who demand that *we* act morally to *themselves* act as they ask us to do. If they do not, they lose credibility: if someone who preaches about the virtues of honesty is shown to be a pathological liar, this discredits her position as someone who has the *mandate* to speak on such matters. It seems that we can refuse to listen, as many students are indeed

[1] Supererogatory actions are those that "go above and beyond" the call of duty. In other words, these are actions which are morally good but not morally *required*. For example, daily volunteering in an animal shelter is commendable and morally good but it is not something that is morally required. Failing to do so does not constitute a moral failing.

M. Herdova (✉)
Florida State University, Tallahassee, FL, USA

J. C. Watson, L. K. Guidry-Grimes (eds.), *Moral Expertise*, Philosophy and Medicine 129, https://doi.org/10.1007/978-3-319-92759-6_7

ready to, should it turn out that Singer's handling of his own finances does not reflect his concerns for the poor captured in his writing.

Here is a more general way of putting this point: it seems that moral authorities are not (and cannot be) *hypocrites*.[2] That is, if one is a moral authority, one rarely (or never) acts hypocritically. One acts hypocritically if one fails to follow one's own moral advice (when, for example, one proposes a general standard of conduct with which one does not comply). In short, moral authorities generally follow their own moral advice, while hypocrites often act against their own moral advice.[3]

Similar claims seem to apply outside of the moral domain: whatever your subject of expertise, you are not (perceived to be) a legitimate authority to speak on this subject if you routinely ignore your own good advice in circumstances where this good advice also applies to your own situation. Imagine an investment banker or a doctor who regularly acts against her own advice. If you value your health, will you, given a choice, go to a doctor who smokes heavily or to the one who doesn't? If you are looking to protect and maximize your capital, will you, given a choice, choose to entrust your funds with someone who keeps losing money in schemes that she dissuades her clients from investing in? Or, will you, instead, choose someone who invests in line with her own advice? The latter, in both cases, seems to be the better—more prudential—choice. If one doesn't follow one's own good advice, this opens room for doubts about one's expertise and credibility. It is not merely that smoking and bad investments are, in and of themselves, often an indication of poor judgment in these particular cases. The discord between one's claims about what ought to be done and one's actions might indicate a lack of care, commitment or some more systematic rational failure of sorts such as in the cases of addiction—none of which we might be inclined to associate with a trustworthy source of reliably good advice.

Whatever the underlying explanation (or a set of explanations) of our reluctance to listen to and follow hypocrites, be this in the moral domain or others, the question remains whether we *should* be so reluctant. Despite our initial hesitation to think so, is a person's being a (moral) authority compatible with that person's not following her own advice? This question is particularly relevant since empirical research

---

[2] I shall delay a more detailed explication of what it takes to be a moral authority until later in the paper. It is worth, however, mentioning the essence of my meaning now. By "moral authority" I mean someone who is a trustworthy source of moral advice. Being a moral authority involves not only having a substantial amount of moral knowledge, but also having a certain *social role*—one dispenses, and can be relied upon to dispense, good advice about morality. I do *not* understand "moral authority" in the strong sense of Linda Zagzebski (see her 2012), according to which we are rationally obliged to defer to moral authorities regardless of what other evidence is available to us. My understanding of "moral authority" maps closely onto the meaning of both "expert-as-advisor" (as contrasted with "expert-as-authority") as this term is used by Jennifer Lackey (see her 2018) and "practical moral expert", as this phrase is defined by Guidry-Grimes and Watson in the introduction of this volume. The other types of moral expertise the latter mention (performative and academic expertise) need not involve being a moral authority.

[3] Hypocrites are also often ready to criticize and condemn those that fall short of their demands while not extending such criticisms to themselves for failing in similar ways. I shall further discuss my understanding of hypocrisy below.

suggests that professional ethicists (more specifically, ethics professors) do not exhibit morally better behaviour than other academic professionals (see, for example, Schwitzgebel and Rust 2010, 2014, 2016). These findings are problematic if professional ethicists are indeed to be considered to be *moral authorities*, i.e. those mandated—by their (moral) expertise—to give advice on moral matters and to whose views on such matters we ought to give significant weight and trust.

One argument to the effect that moral authorities *need not* reliably follow their own good advice to others builds on the differences between different types of knowledge. Roughly, there is propositional knowledge, or knowledge-that, and performative knowledge, or knowledge-how (see Ryle 1949, Noë 2005). Perhaps one can have one of the above types of knowledge about a certain subject, but lack the other type of knowledge about this very subject.[4] In other words, propositional knowledge can come apart from performative knowledge, and vice versa. One can thus claim that an agent can have a great deal of moral expertise, know many moral facts, yet not act on them. While we may *know* what is morally right, or *that* something is morally right, we might simply fail to be steered by these moral facts in our behaviour. If an agent does not *do* what is (morally) good or right, this does not entail that the agent does not know the relevant moral facts.

But then, if one knows what is morally right, even without applying this knowledge in practice, then one can also reliably dispense good moral advice. According to this line of argument, theoretical knowledge *alone* can imbue someone with enough authority to dispense such advice. Maybe hypocrites can be moral authorities after all?

One argument against such a conclusion relies on the thesis of motivational internalism (defended, for example, by Smith (1994))—the view according to which one's judgments about what is best (to do) *also* motivate one to do that very thing. For example, on this view, if an agent judges that it's best, all things considered, to be honest, this will motivate the agent to display honesty in her behaviour. Thus, when a moral authority advises a certain course of action in line with her best judgment, she will also be motivated to act in line with this judgment. Note, however, that being motivated to behave a certain way does not yet amount to actually behaving in this way. One's motivations do not always translate to action. In order to get the conclusion that someone who knows moral facts will also act in accordance with them, one needs to rely on a rather strong version of motivational internalism, according to which knowing moral facts reliably translates into acting in accordance with these facts, regardless of any contrary inclinations the person might have.

While I am sympathetic to some (more moderate) versions of motivational internalism, some have raised important objections to this position (Copp 1997, Svavarsdóttir 1999). In what follows, I shall argue that moral authorities are not hypocrites without relying on any commitments to motivational internalism. In Sect. 7.2, I set out the main argument for my conclusion that moral authorities are not hypocrites and lay out qualifications regarding some of the notions employed in it.

[4] This claim is central, for example, in Lewis's defense of physicalism (1988) which is set out in his reply to Jackson's famous Knowledge Argument (1982).

Sects. 7.3, 7.4, and 7.5 are dedicated to defending the different premises of the main argument. In 7.3, I expand on and defend my understanding of moral authorities. In 7.4 and 7.5, I argue that someone who is a trustworthy source of reliable moral advice will not be not a hypocrite based on considerations relating to, respectively, the systematicity of moral motivation and the non-trustworthiness of hypocrites.

7.2 The Main Argument

My argument for the conclusion that moral authorities are not hypocrites rests on understanding moral authorities as having an important *social* role: (legitimately) dispensing reliable moral advice. It also rests on the claim that someone with that kind of mandate is not a hypocrite. The argument can be presented as follows:

1. If one is a moral authority, then one is a trustworthy source of reliable moral advice.
2. If one is a trustworthy source of reliable moral advice, then one is not a hypocrite.
3. (Therefore) Moral authorities are not hypocrites.

Before I defend the individual premises of this argument in the subsequent sections of this chapter, some qualifications are in order. To start with, I should clarify what I mean by the term "hypocrite". Most simply, a hypocrite is someone who voluntarily acts hypocritically.[5] But, as I note in the introductory section, it is not sufficient for being a hypocrite that someone acts, *on occasion*, hypocritically. Instead, a hypocrite exhibits a somewhat *regular* pattern of hypocritical behavior. One acts hypocritically if one acts against the advice one gives out to people (or *would* give out, if asked) regarding how to behave. In other words, one acts hypocritically if, in the absence of compulsion or coercion, one intentionally, or at least knowingly,[6] acts against what one judges to be the (morally) right course of action, which one nevertheless asks *others* to follow.

Let us turn to moral authorities. First, I do not here commit myself to whether there are, or even *can* be, moral authorities—even though I think we have good reasons for believing they do exist (for a discussion of this issue, see, for example, Ryle (1958), Singer (1972), Burch (1974), Cowley (2005), Varelius (2008), Driver (2013), Meyers (this volume)).[7] Second, being a moral authority does not require that one has command of all, most or even significantly many domains of moral

[5] Someone who acts voluntarily is not compelled or coerced to so act. So, someone who acts against her own advice because, for instance, she suffers from a mental disorder, or is manipulated or forced by another person to so act is not a hypocrite on my account.

[6] Philosophers dispute whether foreseen yet unintended side-effects of one's actions are intentional (see, for example, Bratman 1984, Knobe 2004).

[7] For my purposes, I also do not need to engage with has been called the *credentials problem* (e.g. Cholbi (2007), Vogelstein (this volume)) pertaining to identifying moral experts.

knowledge. For instance, one can be a moral authority on issues such as poverty and the death penalty but not on abortion or animal rights. Being a moral authority can be thus relativized to one or more areas of moral knowledge.[8] Relatedly, on my understanding of moral authority, being a moral authority does not require that one *always* arrives at certain moral beliefs in the right way. In other words, being a moral authority does not require that *all* of one's moral beliefs are appropriately justified (or even true). Third, my claim that moral authorities are not hypocrites is not as strong as the claim that a moral authority would never act against her own advice. This is because being a hypocrite, as I sketch out above, requires exhibiting a *pattern* of hypocritical behaviour.

The last two qualifications on what I understand to be a moral authority pertain to my aims of capturing a psychologically realistic moral authority. Perhaps a *perfect* moral authority would command expertise in all moral domains, would only hold true moral beliefs, would arrive at moral truths in an impeccable manner and would always act on these truths and in line with the good advice she dispenses. But moral authorities need not be so perfect. Just as a virtuous person does not need to be *maximally* virtuous (that is, she need not be the kind of person who lacks vices altogether and displays virtue in all her actions), a moral authority needs only a *certain amount* of expertise, and needs only to act on this expertise, in line with her own advice, *most* of the time.[9]

7.3 What Makes One a Moral Authority?

Let me now turn to the first premise of my argument. Below I expand on what I understand a moral authority to be and say a few things in defense of this notion of moral authority.

As I explain above, I take a moral authority to be someone who is a trustworthy source of reliable moral advice.[10] Being a *trustworthy* source of advice means that

---

[8] See Driver 2006 for a similar point.

[9] It would be somewhat odd to have a notion of "moral authority" (or any other moral category) that does not and perhaps cannot apply to any psychologically realistic agents. However, if one insists that a moral authority must be a *perfect* moral authority, this will not affect the conclusion of my main argument: that moral authorities are not hypocrites. This is because someone who is a perfect moral authority would also be arguably morally perfect, and therefore not a hypocrite. (This understanding of a moral authority would, however, make it more difficult for such authorities to exist. As I explain above, whether such authorities exist is not the main concern of this chapter.)

[10] Such characterization is not revisionary. As Cholbi notes in his (2007), "there is general agreement that a moral expert is someone who very reliably, though not necessarily infallibly, provides correct moral advice in response to moral situations and quandaries" (324). There is a similar agreement on moral experts having a mandate or legitimacy to speak on moral matters—moral authorities *ought to be trusted* (compare, for example, Driver 2006, also referenced in Cholbi 2007). Although these authors talk of moral experts rather than of moral authorities, all moral authorities are moral experts, as I shall go on to argue below. Thus, if moral experts have these properties, then, a fortiori, moral authorities will have them too.

other people have *good reasons* to trust this advice (and its source), and trusting the advice of the moral authority is both *justified* and a *good* thing to do. This means that merely being *trusted* to dispense good moral advice does not suffice for being a moral authority—being trusted does not amount to being trustworthy. This also means that being a trustworthy source of advice must be earned in some appropriate way: no one can become a moral authority instantaneously, without first appropriately establishing herself as the kind of source of advice which *can* be trusted (more on this in Sect. 7.4). Relatedly, once one becomes a moral authority, this does not mean that one keeps this status indefinitely and/or does not need to do anything further to maintain this status. Continuing to be a moral authority requires that one *keeps at* being a trustworthy source of reliable moral advice. This includes, minimally, not doing whatever might undercut one's standing as a moral authority. (Some examples of those kind of things that might subvert one's position as a moral authority include not expanding on or not maintaining one's moral knowledge, and partaking in activities that directly undermine one's trustworthiness, such as intentionally dispensing bad moral advice, etc.) Mine is then a "use it or lose it" conception of moral authority: if you do not exercise your authority (or exercise it correctly), you cease to be a moral authority.[11]

I take *reliable advice* to be synonymous with *good moral advice*, i.e. advice which correctly navigates one to morally good behaviour or to moral truths. Of course, "good" and "reliable" are not always to be used interchangeably: a good car, for example, is not the same thing as a reliable car. While a good car is presumably reliable, a good car is also something more than a merely reliable car. Perhaps one might think that this applies to moral advice too: good moral advice is something over and above reliable moral advice. Whichever way one resolves this issue, nothing of importance rests on this for my overall argument—I am happy for the reader to substitute "reliable moral advice" with "good moral advice." The crucial issue is that for one to be a moral authority, one must be a *legitimate* source of advice. Aside from being trustworthy, this requires that one is a source of advice that is robustly

[11] This does not mean that people cannot be moral authorities after their deaths or if they somehow corrupt or tarnish their previous good legacies as moral authorities. In the first case, one's body of moral work and advice can be still "used" after one's death. For instance, people *still* read, engage with and are guided by Seneca's ethical considerations. Seneca can thus be still considered to be a moral authority. As long as one was a moral authority while being alive, and one's advice is still being used in an appropriate way, this is enough for one to be a moral authority. As for the second set of cases, envisage someone who is a moral authority at one stage of her life but later on intentionally starts dispensing moral untruths, defending morally abhorrent views, etc. In that case, one might argue that there is enough of a break or a split between the person who *previously* dispensed the reliable moral advice and the person who is *now* dispensing the abhorrent views so as to consider the former as a moral authority *still* (again, as long as the original body of moral expertise is used in an appropriate way, as a source of good advice, and legitimately so). In cases where such a break is absent, this *might* cast doubt over one previously being a moral authority at all. If I go from condemning acts of ritual murder to defending such brutalities without any mitigating circumstances, this might raise concerns about the soundness of my previous judgments if the premises and reasoning on which I built the previous judgments also give rise to my new abhorrent set of views.

linked to the moral truth. A cult leader, for instance, would not be a moral authority because she would not be a *legitimate* authority—a cult leader is not a trustworthy source of advice in part because advice dispensed by such a person falls short of the required standard on the quality of moral advice (assuming the cult leader spreads moral falsehoods).

While I understand moral authorities to be those who are *in fact* trustworthy sources of reliable moral advice, one might argue that being a moral authority requires that one is merely *disposed* to be a trustworthy source of reliable moral advice. In other words, moral authorities need only to be *poised* to act as trustworthy sources of reliable moral advice, rather than *actually* act in this way. One might suggest that such dispositional understanding of moral authority is preferable because it can help us accommodate certain cases that prove difficult for non-dispositional views of moral authority. For instance, we might wish to say that someone who is unable to communicate altogether is still a moral authority, as long as she has enough moral expertise, and *would* dispense reliable moral advice if she could (had it not been for whatever is preventing her from disseminating good moral advice).

I see the merits of this claim, and I will not take a decisive stand on this issue. The important point is that my argument to the conclusion that moral authorities are not hypocrites goes through even on the dispositional understanding of moral authority. As will become clear in Sects. 7.4 and 7.5, someone who is disposed to or has the potential to be a trustworthy source of reliable of moral advice will not be a hypocrite because this person still acts in line with the moral advice she is *disposed* to give.

Another possible worry concerning my notion of moral authority is that it is too demanding. One could argue that being a moral authority does not require one to be a trustworthy source of good moral advice. Instead, it suffices that one merely possesses the relevant (amount of) moral knowledge, regardless of whether one is indeed a trustworthy source of reliable moral advice. Perhaps one keeps one's moral knowledge to oneself yet still counts as a moral authority. After all, being an expert does not obligate one to dispense one's expertise. Arguably, then, if one fails to disseminate one's knowledge, this does not make one any less of an authority on the subject.

An internalist about motivation can brush off this challenge with relative ease. According to motivational internalism, someone who knows many moral facts *will* also be a trustworthy source of reliable moral advice. Since it is (arguably) morally good to dispense such moral advice, then if one knows this fact, one will be motivated to act on this fact and (given a strong enough internalism) one will also act on this fact.

There are other ways to allay this worry, however, which do not rely on motivational internalism. One such way draws on a distinction between someone who "merely" knows many moral facts and someone who knows moral facts *and* assumes a particular social role (i.e., the role of someone who provides reliable moral advice). The former might be referred to as a *moral expert* while the latter can be called a *moral authority* (see also footnote 2 above). While moral authorities are also moral experts, moral experts need not be moral authorities (they may instead be merely performative experts or academic experts, in the sense of Guidry-Grimes and Watson (see the introduction to this volume)). Having moral knowledge might suffice for being a moral expert, but not for being a moral authority. A moral authority can be thus understood as being a type of moral expert. It is precisely this (lack of) social role that sets these two notions apart.

Again, it is not directly relevant to my considerations here how we end up referring to those who are trustworthy sources of reliable moral advice. The main point for which I wish to argue is that those who legitimately assume this role (if there are any such people) also act in line with their good advice. To this end, I turn now to my defense of premise 2.

7.4 The Argument from Systematicity

One argument in support of the claim that those who are trustworthy sources of reliable moral advice are not hypocrites rests on what may be best described as systematicity about *acting on reasons*. Consider the following. First, it is plausible that *reasons for advising* others to do something are also the *reasons for doing* that very thing the advice is about. In other words, a reason to A is also a reason to advise others to A, and vice-versa. For example, whatever reasons there are for *donating* to charity, these are the same reasons as those for *advising people* to give money to charity. For example, if one such reason for donating is to lower the number of child-deaths due to preventable causes, then lowering the number of these deaths is *also* a reason for advising others to donate. Of course, this is not to say that the reasons for giving advice about A-ing are fully exhausted by reasons for A-ing. There may be extra reasons in favour of giving advice about A-ing which are not, in and of themselves, reasons to A (to morally educate others, for example). The main point is that whatever reasons there are for doing something moral, these will also be reasons to give advice that steers people towards doing this moral thing. The reasons in favour of giving out a particular piece of moral advice are also the reasons in favour of whatever this advice is directing people to do.

Why does this matter? Given that the reasons for A-ing and for giving advice in favour of A-ing are often the same, then those who are responsive, and indeed *respond*, to the reasons for giving advice in favour of A-ing, will also be responsive, and indeed *respond*, to the reasons for A-ing.[12] Someone who is motivated to

[12] Jamie Watson has raised two concerns with this claim. First, that it presupposes motivational internalism, and second, that it overstates the frequency with which reasons to act and reasons to advise coincide. Concerning the first claim, I do not say that people who make moral judgements must be motivated to act on these judgements, but only that people who are already motivated by certain moral reasons (because they advise people to act on them) will be motivated to act on them—I am appealing to a certain systematicity to motivation, rather than its inevitable presence.

Regarding the second claim, Watson points out that a clinical ethics consultant would advise a medical team that it is morally inappropriate to give blood products to a Jehovah's Witness who refuses them, even though that consultant might rightly think that not giving blood products is, in

dispense advice on charity-giving, advising that it is a morally worthwhile, virtuous, or perhaps even an obligatory thing to do, acts on the very same reasons which favour giving to charity. Thus, by advising people to donate to charity, one is already acting on reasons in favour of donating to charity. Giving good moral advice is thus evidence that one is responsive to, and, in fact, responds to the reasons in favour of whatever the advice is about.

Now, someone who is a trustworthy source of reliable moral advice gives out plenty of moral advice based on a variety of moral reasons, and is motivated to act on those reasons (as evidenced by her giving out the advice). The fact that someone systematically gives out good moral advice suggests that she is systematically responsive to moral reasons. If someone is so *systematically* responsive to moral reasons, it is plausible that her systematic responding to these moral reasons will not manifest *merely* in *giving out* such good moral advice, but also in her (generally) *following* the good advice. And if one generally follows one's own good advice, then one is not a hypocrite. Giving out abundant good moral advice which is evidence of one's sensitivity to relevant moral reasons is in considerable tension with frequently failing to follow this moral advice. In other words, being a moral authority conflicts with being a hypocrite.[13]

One way of spelling out this idea is via the Aristotelian idea that if one possesses one virtue, one will possess *all* the virtues (Aristotle 1915). This thesis has two interesting consequences for our purposes. First, if one is virtuous in one moral domain, one will also be virtuous in other moral domains. For example, if one is virtuous in donating money to charity, one will also exercise virtue in other moral domains by, perhaps, being honest with one's loved ones, keeping one's promises,

general, a bad idea. We might wonder: do not one's reasons to act and reasons to advise come apart in such a (not uncommon kind of) case? I do not think so. One's reasons for giving blood products (that it would save a life, that it would contribute to a person's health, etc.) *are* reasons for advising that one gives blood products, and one's reasons not to give blood products (that it is incompatible with the patient's faith, that the patient refuses such treatment) *are* reasons to advise that one does not give blood products. The case is interesting not because one's reasons to act and to advise diverge, but because the views of the patient can weigh strongly against using effective methods of maintaining the patient's health.

[13] The same reasoning applies to someone who is merely disposed to provide reliable moral advice but is somehow incapacitated or otherwise barred from doing so, and thus does not provide such advice. In other words, if one is (merely) disposed to be a trustworthy source of reliable moral advice, then one is not a hypocrite (still). Why? Simply because someone who is disposed to be a trustworthy source of reliable moral advice must have the same attributes as someone who is, in fact, such a trustworthy source of good moral advice. A potential moral authority will also have the relevant moral expertise and will also be motivated to be moral (as shown by her knowing the relevant moral facts in the first place). If someone is disposed to dispense good moral advice (and would do so if circumstances allowed), this is evidence that she is sensitive to moral reasons in favour of giving out the advice, but also to the reasons in favour of whatever the advice is about (as they are the same reasons). And, if someone is systematically responsive to moral reasons such that she is poised to give out the good advice, it is plausible that her systematic response to these moral reasons will not manifest merely in being disposed to give out such moral advice, but also in her (generally) *following* the good advice she is disposed to give. Finally, if she generally follows her own good advice that she is disposed to give, then she is not a hypocrite.

publicly protesting an unjust war, exercising moderation in eating and drinking, providing good education for one's children, etc. Second, within any particular moral domain, one will reliably act on the moral reasons one has by both giving out good moral advice and routinely following this advice. The Aristotelian moral authority, then, would certainly not be a hypocrite.

Of course, the "unity of the virtues" thesis is controversial. Indeed, I do not wish to commit to this thesis in this paper. As explained in Sect. 7.2, I take it that being a moral authority does not require that one possesses all (or even of most of) the moral knowledge it is possible to have. Indeed, one can be a moral authority within a particular moral domain, but not in others. We may say that a *general* moral authority is a moral authority in all or most such domains, while a *specialized* moral authority is an authority only on some select domains. My argument about holistically or systematically responding to reasons can be taken to apply to morality in general, or to a particular domain of morality. Concerning the latter, if you give out good advice regarding, for example, clinical ethics, you are sensitive and respond to reasons involved in the issues of reproductive rights, assisted dying, etc. According to my argument, if you are such a specialized moral authority on clinical ethics, you are not a hypocrite concerning matters pertaining to clinical ethics. However, assuming the "unity of the virtues" thesis is false, being a moral authority in the domain of clinical ethics may not involve being responsive to other types of moral reasons, such as those reasons related to business ethics. A specialized moral authority, then, may be a hypocrite in other areas of morality. A general moral authority will be a hypocrite in no area of morality.

Before I explore an objection to the defense of premise 2 thus far presented, I wish to make clear that my systematicity argument about acting on moral reasons is not intended to show that acting systematically on moral reasons, even within one particular moral domain, requires that one *always* or even *almost always* act on such reasons, and in *all* the possible ways (by both giving out the good moral advice and following this moral advice). Indeed, it is implausible to claim that being systematically responsive to moral reasons entails that one is always or almost always successfully motivated to action by those reasons. A moral authority need not be *perfectly* responsive to moral reasons, even in one specific moral domain. What is required is that one is *generally* responsive to moral reasons. This point underscores how to understand my main claim—it is not that moral authorities are never hypocritical, but that they are not hypocrites (they do not often exhibit hypocritical behavior).[14]

[14]You might wonder what "often" comes to; that is, at what point does hypocritical behavior make one a hypocrite. I want to leave this question largely open, except to say that the nature of my arguments does not support a strong conclusion that moral authorities literally never act hypocritically. My arguments do suggest, however, that moral authorities do not *habitually* act hypocritically, that hypocrisy is not a character *trait* of moral authorities, that such authorities do not *regularly* act hypocritically, etc. Indeed, I am tempted to think my arguments suggest that moral authorities act hypocritically only very rarely, if at all.

One might object to the above picture on the following grounds. Since *giving* moral advice is, typically, *easier* than *following* the advice, the fact that someone might give out reliable moral advice is not enough to conclude that she will systematically respond to the relevant moral reasons by *also* acting in line with this moral advice. If that is correct, it may well be the case that someone who acts on moral reasons to A by advising in favour of A-ing also regularly fails to act to take her own good advice. After all, we often encounter very strong counter-motivation to *doing* the right thing, while encountering motivation against giving advice on that very thing is rarer (and when we do encounter it, such counter-motivation is typically much less forceful). Consider the following example. While I know the reasons in favour of donating to charity, and I also might find it relatively easy to dispense advice to this effect, I might plausibly find it a lot harder to act on this advice, especially when faced with the temptation of spending money on personal luxuries to which I have grown accustomed. Simply put, the obstacles to my acting on (my own) good moral advice may be far more difficult to overcome than the obstacles to giving good advice.

Two answers lend themselves in reply to this worry. First, there are good reasons to think that someone who has lots of knowledge in particular area also *cares* about this particular area of knowledge. For example, someone who knows a lot about birdwatching is likely to have an interest in birds and birdwatching that motivates her to accumulate this vast knowledge.[15] These considerations can be applied to

[15] Of course, there are some exceptions to this. One might know a lot about a particular area because it has been somehow drilled into one, without one having any significant interest in learning the relevant information. My point here concerns those who acquire the relevant knowledge in a (more) voluntary manner, as I take to be the case for experts. Perhaps, however, this is too quick. Laura Guidry-Grimes has pointed out to me that much of our even specialized knowledge is the result of involuntary knowledge-gathering that results from our upbringing in a certain culture. Might not experts in a certain area be so involuntarily? And, if so, might they not fail to care about the subject they know so much about? Perhaps it is true of many areas of (specialized) knowledge that someone may have come about this knowledge involuntarily, and fails to care much about it. Three points should be considered in reply, however.

First, for this knowledge to rise to that of *expertise* strongly suggests a level of agential investment in the topic—one must seek out such knowledge, actively research it, etc. Second, even if one does not originally care about a certain topic, becoming an expert in it can (and often does) *cause* one to care about it. Consider those who find themselves in rather specialized careers without explicitly planning to do so, but who become deeply involved in their jobs after gaining substantial knowledge about the relevant areas. Third, the case of *moral* knowledge is special. There are societal pressures to become experts in certain areas, even if one does not care about those areas. Thus, consider a student whose parents pressure her into becoming a lawyer or doctor, even though she has little interest in these fields. We do not live in a culture, however, that pressures people into becoming *moral* experts. Those that do attain this status, then, likely do so precisely because they care.

Guidry-Grimes also points out that there are different types of caring—one may be invested in a topic intellectually, emotionally, or agentially. She suggests that agential investment might be required for moral knowledge—one can become aware of nuanced moral distinctions and subtle moral facts only by actively engaging in the moral project. This tallies well with my approach to the topic. As I go on to point out, moral authorities must have certain capacities to attain their moral knowledge. These very same capacities are developed and exercised precisely by *being moral*.

moral knowledge as well. It is plausible that someone who knows many moral facts (enough to count as an authority on the topic) also *cares* about morality and moral knowledge—about both knowing and doing what is right. So someone who devotes many of her resources to accumulating moral knowledge will very likely be interested in *being moral*. Further, someone who is interested in being moral is also likely, in general, to follow the reliable moral advice she dispenses. If you care about knowing moral facts and being moral, then you will be motivated to act on moral reasons, and indeed (often) act on them. (This is particularly relevant given that, as mentioned previously, people might mistrust those that do not follow their own advice. If one is genuinely interested in spreading moral knowledge and having people follow one's advice, this adds another reason in favour of following one's own good advice as this may be conducive to having other people do the same.)[16]

The second line of reply to the aforementioned objection builds on the *kind* of moral advice a moral authority gives. Good or reliable moral advice does not (and should not) come in the form of simple commands; a mere list of do's and don'ts. Such advice would not be the most effective or helpful moral advice. Someone who gives out good moral advice will not only be able to advise people as to what is right and wrong, but will also be able to reason with others regarding *why* they should act as they are being advised to, offer *tactics* to being successful at this behaviour, offer *strategies* to counter competing motivation, etc. For example, good moral advice regarding charity-giving does not merely amount to telling people they need to donate—a moral authority can also explain why such donating is required, the effects donating has, strategies one might employ to be able to give, or to be able to give more. Good advice on donating to charity will alleviate worries one might have regarding donating, offer strategies on how deal with situations where one is tempted to needlessly spend funds that can be of a great help elsewhere, etc. Put more generally, a moral authority will not dispense just generic advice (that perhaps others can easily access themselves), but she will have some sophisticated moral knowledge that can genuinely help and inform people.

What's more, such moral knowledge will not be limited to knowledge-that. A moral authority must also possess moral *knowledge-how* (practical wisdom).[17] Why? In order to come-by and sensibly dispense reliable moral advice, a moral authority must be skilled at all of the following: weighing moral reasons, judging what is and is not morally relevant in a certain situation, treating others respectfully and patiently. Without such skills, someone who attempts to be a moral authority will fail. If she cannot weigh reasons well, she cannot advise others concerning which action is favored by the weight of reasons. If she cannot judge what is morally relevant, she cannot inform others of what factors to consider. If she cannot treat others respectfully and patiently, she is in no position to advise them. These are, of course, just some examples among many. In sum, one is good at giving *moral advice* only if one is good at *being moral*.

---

[16] For some related points, please see Sect. 7.5.

[17] See Arjo, this volume.

We are left with the following picture. Moral authorities are trustworthy sources of reliable moral advice. In being so, such people are sensitive to and respond to moral reasons (in giving out such advice). These very same reasons to advise people to act in certain ways are themselves reasons to act in these ways. Moral authorities possess the skills to weigh such reasons, form best judgements on the basis of such weighing, translate best judgments into action, and treat others respectfully. Moral authorities also *care* about morality as experts who are deeply invested in it (see footnote 15 above for discussion of this claim). In essence, to *have what it takes* to be a moral authority, one must possess a host of other qualities related to morality. Such qualities are precisely those of *moral people*—people who, in general, act morally. As such, moral authorities not only give out good advice, they also follow it. Hypocrites simply lack what it takes to be true authorities.

Even if one disagrees with the above, the very fact that someone has this kind of advice *readily available* makes it more likely that she will in fact act in line with her own advice. This is because someone who gives good moral advice will have all the relevant reasons and tactics at the forefront of her mind, or at least accessible (otherwise it would be difficult to readily dispense good advice to others). Having the relevant knowledge in the first place puts one in a very good position to recognize the circumstances in which one should act in line with one's own advice and to motivate oneself in the right way so that one indeed does.

Still, one might claim that the motivation contrary to acting in line with one's own good advice might just be too strong for a moral authority to combat it, however motivated she is. This might result in the moral authority failing to act on her advice regularly, despite her best intentions. I consider such a scenario unusual at best, but even if we allow that some people with the skills and knowledge outlined above systematically fail to follow their own advice, there are excellent reasons to think that such people are *not* moral authorities. This leads us on the next section where I argue that such regular failing undermines a person's trustworthiness in such a way that this person could not be considered a moral authority.

7.5 The Argument from Trustworthiness

Another argument in support of the claim that those who are trustworthy sources of reliable moral advice are not hypocrites builds on considerations about trustworthiness; more specifically, on how people's actions affect their status as being or not being trustworthy. The general idea is the following. If one does not suitably act on the advice one dispenses, be it due to negligence or intentionally acting against one's own advice, such hypocrisy undermines one's standing as a trustworthy source of reliable moral advice. However, being a moral authority requires that one *is* so trustworthy; as explained in Sect. 7.3, being trustworthy is necessarily tied with the social aspect of being a moral authority. Thus, if one acts hypocritically, this undermines one's trustworthiness, and thereby one's standing as a moral authority. A moral authority must act in such a way that she is trustworthy, which excludes failing to act on one's own good advice.[18]

Why does acting against your own good moral advice undermine your trustworthiness? Simply put, because being trustworthy means that others *ought to* (or are, at least, are rationally *permitted to*) trust you; that there are good reasons for their trusting you and that there are few or no good reasons against trusting you. If you (frequently) act against your own good advice, this is a good reason for others not to trust you, and if they still do trust you, their trust is unjustified. Acting hypocritically by not taking your own advice is evidence for others that they *ought not* to trust your advice.[19]

Here are some reasons why this is so. First, a discord between your advice and the actions you perform may indicate a certain *irrationality*. It is not fitting for an agent who is appropriately responsive to different reasons (including moral reasons and attending instrumental reasons[20]) to (voluntarily) act against what she herself believes is the right thing to do. Acting against one's own advice calls into question one's reasoning in general, and, more specifically, one's reasoning that leads to the very advice given out. This fact is in tension with being a trustworthy source of reliable moral advice, as being this kind of source requires that the (moral) reasoning process which underpins one's advice is also (generally) trustworthy. For example, if someone gives advice to the effect that it is imperative we donate to charity, and she herself (almost) never does, this casts doubt on her being appropriately responsive to (moral) reasons in general, and to those reasons which lead her to give this advice in particular. If one responds to moral reasons in one way (by giving advice on donating) but not in another way (by failing to donate), this raises the question of whether one's advice builds on good reasons in the first place. If one is a trustworthy

[18] I do not here commit to any particular account of trust or trustworthiness. As far as I can ascertain, the points I make below hold true regardless of the specific ways these notions are spelt out. See McLeod (2015) for an overview of trust and trustworthiness. Following Jones (1999), McLeod distinguishes "risk-assessment" views, according to which someone puts her trust in someone by relying on that person (on the assumption that the risk of so relying on her is low because it is in the trustee's own interest to be reliable), from "will-based" accounts, according to which someone is trustworthy only if she shows goodwill to the trustee. Hardin's encapsulated interest account (2002) is an example of the former, while Baier (1986) defends a version of the latter. On both such accounts, and others, a moral authority is trustworthy only if she can be relied upon as a source of reliable moral advice. The considerations below show that a hypocrite cannot thus be relied upon.

[19] This remains so even if one has a dispositional view of moral authority, according to which a moral authority is someone who is *disposed* to be a trustworthy source of reliable moral advice. If someone acts against the advice she is simply disposed to give, she undermines even her potential trustworthiness. This is for the same reasons which I discuss below—such an agent is either irrational or disposed to act on untrustworthy motives, and is disposed to break commitments she has made. As I shall go on to explain, such features render a person untrustworthy.

[20] *Instrumental* reasons are reasons which help show the most effective means of executing one's plans. For example, if you want to fly to Europe from America, that there is a cheap flight from Orlando to London shows that one effective and efficient means of doing so is taking that cheap flight.

source of reliable moral advice, then one's reasoning process which leads one to the (contents of the) advice given is not questionable in this kind of way.

Second, if there is a discrepancy between what one does and what one asks of others to do, this casts doubt on one's *motives* for giving the advice. Regardless of whether the advice is good/reliable, if the advisor does not take her own advice, this calls into question *why* she gave it if she (regularly) does not follow this advice herself. This leaves it open that she might not have given out the advice for the right (moral) reasons but instead acted on some ulterior motives (and for reasons which are *not* reasons for whatever the advice is about). For instance, if one is advised to give money to charity, but the person dispensing this advice does not donate herself, this raises a question of whether the advisor gives out her advice based on reasons in favour of donating, or for some other reason(s) altogether. This is a problem especially if we accept the plausible condition on trustworthiness which says that the trustee must act out of *good will* for the trustor (Baier 1986)—if someone does not follow what she claims to be good advice, this is evidence that she is indeed not acting out of good will.

Third, if one does follow one's own good advice, this casts doubt on the *accurateness* of this advice (being *accurate* is part of being reliable or good advice). Whether one's reasoning process or one's motives are called into question because of one's failing to take one's own advice, the audience to this advice are not in a position to know whether the advice given is correct. If there are good reasons to believe that one's advice on our moral obligations regarding charity donations rests on shaky reasoning or that it is given out for the wrong reasons, this makes it questionable that we *ought to* donate to charity after all. Legitimate doubts about the accuracy of advice are too in tension with the advisor being a trustworthy source of good moral advice. Being a trustworthy source of advice requires that the advice itself can and ought to be trusted.

There is a further challenge to trustworthiness if moral advice either *involves* or *is* a kind of a *commitment* to certain actions' being morally good. Why? If one acts contrary to one's own *sincere* good advice, this amounts to a violation of one's commitments: one is violating one's own commitment to a certain course of action being morally good. We can put the point in the following way: good moral advice rests on one's best/better judgement regarding what should be done. Such judgments are best described as *evaluative commitments* (see Mele 2012)—commitments to certain courses of action's being best all things considered. If one acts against one's own advice, one breaks one's evaluative commitments. If one sincerely advises others that it is morally obligatory to donate, one judges it best to donate. If one then fails to donate, one breaks one's evaluative commitment to donating. This renders one untrustworthy. If one breaks commitments by not following one's own good advice, it is fitting to question whether one's other actions—those which also involve making a commitment such as giving advice—are to be trusted. This is akin to someone breaking promises: if someone (regularly) does not fulfill her promises, this gives us excellent reason not to trust her promises in general.

Lastly, if one goes against one's own good advice and, as a result, acts *immorally*, this too calls into question the goodness of one's advice. Acting immorally is evidence that one is ready to embrace immoral behaviour. If one is ready to embrace immoral behaviour in one domain, this gives others reason to be skeptical about whether one acts morally in other domains. Since giving good advice requires that one is motivated by the right moral considerations and is acting morally in giving this advice, it thus becomes questionable whether the advice one gives out is indeed good advice. In essence, on those occasions when acting hypocritically is also acting immorally (which will, of course, be often), one presents oneself as willing to be immoral, and thus possibly willing (immorally) to give bad moral advice, which undermines one's position as being a trustworthy source of reliable moral advice.[21]

All of the above points apply even if it is *not known* that someone who dispenses good moral advice is not following her own advice. If someone acts *secretly* in opposition to her own advice, this *still* undermines her trustworthiness. That is, even if this person is *in fact* trusted to dispense good advice, it does not mean that she *ought* to be so trusted. Just because other people fail to garner or recognize reasons against treating someone as a trustworthy source of reliable moral advice, this does not mean that no such reasons exist. This point draws on a distinction between *having* a reason to do something, and there *being* a reason to do something (see, for example, Kearns and Star 2009). Someone who is ignorant of what it takes to keep in good health might not *have* a reason not to eat too much junk food (she does not know that junk food is generally bad for her), but that does not mean that there *is* no reason for her not to eat junk food. Similarly, while I might not *have* a reason to distrust your advice (I do not know that you often act against your own advice), there still *is* a reason why your advice ought not to be trusted.

The above argument rests on normative considerations regarding when one ought (not) to be trusted. But we can make an argument in support of Premise 2 based on descriptive considerations, too (this, unlike the argument above, requires that others *do* know of one's hypocritical behaviour). If one acts against one's own advice, this might easily make it so that other people in fact stop trusting one (or one will never earn the trust of others in the first place). However, if people do not trust an agent, then this agent is not a moral authority. This is because, by losing the trust of others, one loses the *social standing* that is required for being a moral authority. Such social standing is built in part on *actually being trusted* to dispense good advice: if one is trusted, other people will believe one's advice, will follow it, will rely on it, etc.[22] One's advice is thus being received in a way that's appropriate. If others mistrust a person's advice, this means that this person's advice will not be utilized in the

[21] Conversely, just because someone does follow her own advice (and so does not act hypocritically), this is not evidence that she is reliable in the relevant way: an astrologist may do all the things she recommends, but this does not imply expertise or authority.

[22] This implies that someone who ought to be treated as a moral authority is *not* a moral authority if others have not given her an appropriate social standing. This has consequences for thinking about *epistemic injustice*. If an ethicist is not trusted on moral matters because of her gender or race, for example, this would mean she is not a moral authority. This is, of course, consistent with the view that she *should* be a moral authority.

relevant ways required for being a moral authority, and this person thus will thus fail to be or become a moral authority.[23] All the fur-wearing minks dispensing anti-fur advice as well as others who frequently act hypocritically (including many ethics professors) are not moral authorities.

Acknowledgments I would like to thank the editors of this volume, Laura Guidry-Grimes and Jamie Watson for their extremely helpful comments on this chapter. I would also very much like to thank Stephen Kearns for his numerous suggestions on different drafts of this chapter.

References

Aristotle. 1915. *Nicomachean ethics*. (Ross W.D. Trans.). In *The works of Aristotle*, vol. 9. Oxford: Clarendon Press.

Baier, A.C. 1986. Trust and antitrust. *Ethics* 96: 231–260.

Bratman, M. 1984. Two faces of intention. *Philosophical Review* 93: 375–405.

Burch, R.W. 1974. Are there moral experts? *The Monist* 58 (4): 646–658.

Cholbi, M. 2007. Moral expertise and the credentials problem. *Ethical Theory and Moral Practice* 10: 323–334.

Copp, D. 1997. Belief, reason, and motivation: Michael Smith's *The Moral Problem. Ethics* 108: 33–54.

Cowley, C. 2005. A new rejection of moral expertise. *Medicine, Health Care and Philosophy* 8: 273–279.

Driver, J. 2006. Autonomy and the asymmetry problem for moral expertise. *Philosophical Studies* 128: 619–644.

———. 2013. Moral expertise, judgment, practice and analysis. *Social Philosophy and Policy* 30 (1–2): 280–296.

Hardin, R. 2002. *Trust and trustworthiness*. New York, NY: Russell Sage Foundation.

Jackson, F. 1982. Epiphenomenal qualia. *The Philosophical Quarterly* 32: 127–136.

Jones, K. 1999. Second-hand moral knowledge. *Journal of Philosophy* 96 (2): 55–78.

Kearns, S., and D. Star. 2009. Reasons as evidence. *Oxford Studies in Metaethics* 4: 215–242.

Knobe, J. 2004. Intention, intentional action and moral considerations. *Analysis* 64: 181–187.

Lackey, J. 2018. Experts and peer disagreement. In *Knowledge, belief, and god: New insights in religious epistemology*, ed. Matthew Benton, John Hawthorne, and Dani Rabinowitz. Oxford: Oxford University Press.

Lewis, D. 1988. What experience teaches. *Proceedings of the Russellian Society* 13: 29–57.

McLeod, C. 2015. (Fall 2015 Edition) Trust. *The Stanford encyclopedia of philosophy*, ed. Zalta E.N. URL = https://plato.stanford.edu/archives/fall2015/entries/trust/.

Mele, A. 2012. *Backsliding. Understanding weakness of will*. Oxford: Oxford University Press.

Noë, A. 2005. Against intellectualism. *Analysis* 65: 27–90.

Ryle, G. 1949. *The concept of mind*. London: Hutchinson.

———. 1958. On forgetting the difference between right and wrong. In *Essays in moral philosophy*, ed. A.I. Melden. Seattle: University of Washington Press.

Schwitzgebel, E., and J. Rust. 2010. Do ethicists and political philosophers vote more often than other professors? *Review of Philosophy and Psychology* 1: 189–199.

---

[23] The same idea applies even on the dispositional conception of moral authority. Even if one is merely disposed to give out moral advice (but does not actually do so), one's advice would not be trusted by those aware of one's immorality.

———. 2014. The moral behavior of ethics professors: relationships among self-reported behavior, expressed normative attitude, and directly observed behavior. *Philosophical Psychology* 27: 293–327.

———. 2016. The moral behavior of ethicists. In *Companion to experimental philosophy*, ed. J. Sytsma and W. Buckwalter, 225–233. Chichester: Wiley-Blackwell.

Singer, P. 1972. Moral experts. *Analysis* 32 (6): 115–117.

Smith, M. 1994. *The moral problem*. Oxford: Basil Blackwell.

Svavarsdóttir, S. 1999. Moral cognitivism and motivation. *Philosophical Review* 108: 161–219.

Varelius, J. 2008. Is ethical expertise possible? *Medicine, Health Care, and Philosophy* 11: 127–132.

Zagzebski, L. 2012. *Epistemic Authority: A Theory of Trust, Authority, and Autonomy in Belief.* New York: Oxford University Press.

Chapter 8
If There Were Moral Experts, What Would They Tell Others? Answers for Dilemmas from Early Chinese Philosophy

Ai Yuan

8.1 Introduction

In western philosophical discussions,[1] the debates on moral expertise often concern whether moral opinions come to us with a certain level of guarantee that they are true.[2] Scholars have defended moral experts from the perspective of, for example, the problem of autonomy,[3] the issue of trust in relation to moral teaching,[4] the validity of moral knowledge and its usability,[5] as well as philosophers' abilities to make moral judgments based on their moral training and dedication of time to moral deliberations.[6]

Among those who believe the possibility of moral experts, some scholars claim that expert advice should always be accepted and followed, since experts can replace

[1] I am grateful for the comments and suggestions from editors Jamie Watson and Laura Guidry-Grimes. I also thank the suggestions given by Rens Krijgsman.

[2] A review of debates can be seen in Parker, Lisa S. 2005. Ethical expertise, maternal thinking, and the work of clinical ethicists. In *Ethics expertise: History, contemporary perspectives, and applications*, ed. Lisa Rasmussen, 165–207. Dordrecht: Springer. Moreover, we see Alison Hills calling for the idea of "moral understanding" as a reason to defer to moral experts. Hills, Alison. 2009. Moral testimony and moral epistemology. *Ethics* 120: 94–127. Michael Cholbi questions non-experts' ability to identify and appreciate the testimony of moral experts when facing complex moral issues. See Cholbi, Michael. 2007. Moral expertise and the credentials problem. *Ethical theory and moral practice* 10: 323–334.

[3] Driver, Julia. 2006. Autonomy and the asymmetry problem for moral expertise. *Philosophical studies: An international journal for philosophy in the analytic tradition* 128: 619–644.

[4] Jones, Karen.1999. Second-hand moral knowledge. *The Journal of Philosophy* 96: 55–78.

[5] Hopkins, Robert. 2007. What is wrong with moral testimony? *Philosophy and Phenomenological Research* 74: 611–634.

[6] Singer, Peter. 1972. Moral experts. *Analysis* 32: 115–117.

A. Yuan (✉)
University of Oxford, Queen's College, Oxford, UK
e-mail: ai.yuan@queens.ox.ac.uk

J. C. Watson, L. K. Guidry-Grimes (eds.), *Moral Expertise*, Philosophy and Medicine 129, https://doi.org/10.1007/978-3-319-92759-6_8

one's beliefs on certain issues instead of merely adding reasons for it.[7] In other words, they argue that moral experts possess the justifiable right of compelling us to do X or not to perform Y, from the standpoint of rationality.[8] They treat their expert advice as decisive.

However, drawing on the resources from early Chinese texts, this paper argues, when considering the nature of moral expertise, moral experts do not treat their advice as decisive. On the contrary, early Chinese moral experts treat themselves as recognizable and trustworthy advisers who can help the recipient using their own reflective moral reasoning, understand their situations, all the while respecting the recipient's perspective. Under such conditions, moral experts can provide methods, information and rational deliberations that enable individuals to make morally sound decisions, without demanding how they ought to act.

More importantly, this paper emphasizes that early Chinese moral experts do not see themselves as merely responsible for providing competing values and rational deliberations regarding what one ought to act; they also see themselves as responsible for providing a proper attitude for the agent to deal with a situation, and to live with their choices, namely the attitude of living in equanimity (*an* 安). In other words, they focus on the proper attitude for dealing with a situation, instead of merely focusing on actions.

Such a role for early Chinese moral experts can be vividly illustrated through their expert advice on dealing with cases in which the agent[9] regards herself/himself as facing a moral dilemma. It refers to the case in which the moral agent regards herself/himself as having "moral reasons" to do each of two (or more) actions. The agent can do each of the actions but believes he or she cannot do both (or all) of the actions. The agent thus seems condemned to moral failure; no matter what she does, he/she will do something wrong (or fail to do something that she ought to do).

The idea of "moral reasons" cannot be reduced to moral virtues. One's moral virtue can be in conflict with one's religious belief.[10] One's general obligation can be in conflict with a role-related obligation.[11] Two different roles may pose conflicting choices.[12] All of these conflicts can be seen as conflicting moral reasons.

---

[7] See Zagzebski, linda Trinkaus. 2012. *Epistemic Authority: A Theory of Trust, Authority, and Autonomy in Belief.* Oxford: Oxford university press.

[8] Frey argues that what we want from moral experts are normative decisions instead of sound arguments. See Frey, R. G. 1978. Moral experts. *Pacific Philosophical Quarterly* 59: 47.

[9] This paper only concerns single person dilemmas instead of multi-person ones.

[10] Quinn, Phillip, 2006. *Essays in Philosophy of Religion.* Oxford: Oxford university Press, 6.

[11] For example, once a lawyer finds out that his/her defendant is guilty, one's role as a lawyer may conflict with the general obligation of telling the truth due to the attorney-client privilege. The conflict of values due to a person's multiple roles is more easily encountered in early China. Some scholars argue that early Confucian ethics are in fact "role ethics". See Rosemont, Henry Jr. and Ames, Roger T., eds. 2016. *Confucian Role Ethics, Confucian Role Ethics.* Taipei: National Taiwan University press.

[12] A talented artist who travels around the world to learn new skills and gains? inspiration may not be seen as a responsible spouse, since such an artist is mostly away from for his/her family. In this case, one's responsibility to develop one's talent is in conflict with one's duty as a spouse.

Also, whether the cases are genuine dilemmas is not the concern of this paper. As long as one regards oneself as being caught in a moral dilemma, one will ask for moral consultancy. And it is the role of the moral expert to clarify the case and assist in dealing with it. Taking the self-imposed dilemma as an example, one may argue that self-imposed dilemmas are not genuine dilemmas, since their existence does not point to any deep flaws in moral theory.[13] However, this does not mean that those agents do not deserve advice from moral experts when they make a mistake. As moral experts, simply explaining how this conflict of values should not have happened as a rational moral agent is certainly not enough, and does not help, since a dilemma has been perceived to exist by the moral agent, despite others trying to argue otherwise. The key issue is, how can a moral expert help when the agent regards him or herself to be stuck in a dilemma? Early Chinese moral experts take the alleged dilemmas as opportunities to help others.

The importance of giving moral advice on dilemmas as moral expert is pointed out by Peter Singer. He argues that, "It is when, say, honesty clashes with charity (If a wealthy man overpays me, should I tell him, or give the money to famine relief?) that there is need for thought and argument."[14] This thought and argument does not merely refer to how one ought to act, or which value can override another. According to Chinese moral experts, the expert's advice should also include the proper attitude to deal with the alleged dilemma, and to live with one's decision in peace, namely with an attitude of living in equanimity (*an* 安).

When facing a dilemma, expert advice on "equanimity" can help the recipient in the following ways. (1) It frees him or her from the mental disturbance resulting from what he or she was not able to achieve or has lost. Accordingly, it allows the recipient to direct attention to the new situation that he or she has decided to act on. (2) It requires the recipient to willingly accept the decision that he or she has made with affirmation. Only by keeping the mind in equanimity, can one free oneself from emotional and ethical struggles, and therefore direct attention to what one ought to and can do, and live in peace with a decision without regret, guilt, or fear.

The importance of giving expert advice also focusing on the right attitude for dealing with a dilemma, can be seen from the following reasons. First, such an attitude of equanimity indicates a respect towards the subjectivity of different individuals, and a trust in their ability to make specific decisions based on each individual case. Through deliberation, moral experts allow the agent to understand the possible consequences that each action may result in. With the best possible information available, moral experts help others to make a decision, without forcing the other to act, but rather based on sound reasoning, best information, trust, and respect.

Second, moral experts can make themselves better understood when they can free the recipient from emotional distress. Imagine a moral expert explaining complicated and complex philosophical debates to a moral agent who is not only under stress, but is also burdened with fear and worries. The expert will not be able to

[13] Donagan, Alan. 1977. *The Theory of Morality*, Chicago: University of Chicago Press. See especially Chapter 5.

[14] Singer, Peter. 1972. Moral experts. *Analysis* 32: 115.

deliver the moral message and make the theories understood unless the agent's mind is at ease. Without a clear understanding of the analysis provided by the moral expert, the agent could easily believe he or she were being compelled by the expert, rather than reasoned with and respected as a moral agent.

Third, once moral experts take into account how recipients can live with their decision for the rest of their lives, it means that moral experts also care about the recipients' moral commitment and well-being as a whole. Therefore, moral experts are no longer moral code dictators, but rather recipient-centred moral advisers.

In order to demonstrate the necessity of providing moral expertise, not only on actions but also on attitudes, this paper will be divided into three parts. First, using anecdotes in the *Analects* and *Mencius* (these texts are perceived as a collection of philosophical ideas of Confucius [551?-479? BCE] and Mencius [c. 372—289 B.C.E.] respectively, which are attributed to the classics of Confucianism), this paper discusses how moral experts not only provide the principles for temporarily disregarding a social norm to deal with an alleged dilemma, but also pay attention to how to live in peace with the dilemma, without dwelling on ethical or emotional struggles. Second, using the resources in the *Zhuangzi* (this text is perceived as a collection of philosophical ideas related to Zhuangzi [late fourth century BCE], who is seen as one of the most important figures in classical Chinese Daoism) this paper shows how, with an attitude of living in equanimity, one can practically free oneself from emotional disturbance and therefore fulfil one's moral duty and solve the dilemma. Lastly, this paper will examine how an attitude of living in equanimity can be used as morally expert advice to help with (not necessarily solve) unresolvable dilemmas, such as "Sophie's Choice."

Before examining the early Chinese cases, this paper will first demonstrate the theoretical assumptions and then clarify the idea of morality in early China.

This paper makes the following assumptions. First, it assumes that the possibility of moral expertise is indeed acknowledged within each knowledge community in early China, in the form of sages (*shengren*圣人). To be specific, sages are those who not only command a body of moral values regarding how to live a good life,[15]

---

[15] Morality here refers to a series of codes of conduct that are recognized and accepted by society for living a good life. The early Chinese moralistic view has been discussed by scholars. For example, David Wong calls for a kind of moral relativism in the *Zhuangzi*, which is unseen in the Wester philosophical tradition. See Wong, David B. 2006. *Natural Moralities: A Defense of Pluralistic Relativism*. Oxford: Oxford University Press.

Also, Huang Yong points out the "ethics of difference" in the *Zhuangzi*. See Huang, Yong. 2010a. The ethics of difference in the Zhuangzi. *Journal of American Academy of Religion* 78.1: 65–99. 2010b. Respecting different ways of life: A Daoist ethics of virtue in the Zhuangzi. *Journal of Asian Studies* 69: 1049–1070. Other moralistic ideas for living a good life can be found in Kjellberg, Paul Kjellberg, ed. 1996. *Essays on skepticism, relativism, and ethics in the Zhuangzi*. Suny Series in Chinese Philosophy & Culture. As for Confucian moral ideas, see for example, Olberding, Amy. 2012. *Moral exemplars in the Analects: The good person is that*. New York: Routledge. Olberding argues for emulating moral exemplars, which can make people be moral, and how it is possible for people to identify moral exemplar with their pre-theoretical ability. Also, Confucianism is frequently related to virtue ethics. See Angle, Stephen C. and Slote, Michael. ed. 2013. *Virtue ethics and Confucianism*. London: Routledge.

but are also capable of embodying their obtained knowledge and way of living as an ideal model. Therefore, this paper also assumes that sages in early China can be recognized and trusted as moral experts within a certain community. When people encounter moral difficulties, sages are the ones they look to for advice.

When discussing ethics and morality in general, as mentioned in the introduction of this book,[16] we will follow the majority of academic philosophers and use these terms interchangeably to refer to the study of a series of codes of conduct that, given specified conditions, can be put forward by all rational human beings. This normally includes the study of concepts surrounding right and wrong, good and bad, etcetera.[17]

The study of ethics in early China, however, lies in the questions related to how one ought to live, within which not all of the considered problems can fit into the study of ethics within western philosophical frameworks. For early Chinese thinkers, the study of how to live a good life includes such issues as, "How to weigh duties toward family versus duties toward strangers, whether human nature is predisposed to be morally good or bad, how one ought to relate to the non-human world, the extent to which one ought to become involved in reforming the larger social and political structure of one's society, and how one ought to conduct oneself when in a position of influence or power. The personal, social and political are often intertwined in Chinese approaches to a subject."[18] This means that questions of how to be a filial child and loyal minister, how to follow the ritual codes properly, how an elite female should be virtuous, and how to protect one's physical and mental health are all ethical questions in relation to how to live a good life in early China.

Admittedly, certain conflicting ethical requirements in early China may not make sense to many modern readers. But they are true ethical problems that early Chinese thinkers must deal with. And their expert advice on attitude can be helpful for modern ethical dilemmas. I will examine related cases in the last section. Now let us enter into the conversation with ancient Chinese sages.

8.2 Expert Advice from Confucian Sages

Recently, the importance of moral dilemmas in early China have begun to draw scholarly attention. Paul Goldin reveals the vulnerable moral situations that women encountered in early China through conflicting moral situations recorded in the *Categorized Biographies of Women* (*lienü Zhuan* 列女傳) which is a selection of idealized biographies in the earliest traditional histories of Han dynasty China (2nd ct BC– 2nd c AD).[19] With a focus on the same Chinese anthology, César Guarde-Paz

---

[16] See Introduction footnote 1.

[17] For a detailed analysis on the concept of morality, please see Gert, Bernard. 2016.

[18] For a detailed analysis on Chinese ethics, please see Wong, David. 2013.

[19] Goldin, Paul R. 2016. Women and Moral dilemmas in early Chinese narrative. In *The Bloomsbury Research Handbook of Chinese Philosophy and Gender*, Pang-White Ann A ed., Bloomsbury Academic, 25–33.

compares the different dilemmatic cases from early China with Greek tragedy and maps them under the philosophical theories on dilemma.[20] The previous two articles focused on the issue of moral dilemmas from the perspective of the moral agent. They argue about how one should act when facing a moral dilemma through a discourse on what the agent decides to do.

This paper, however, changes its focus to the role and responsibilities of moral experts and reveals how moral experts should advise moral agents on the alleged dilemma. Now let us look at a different type of answer through the sayings attributed to two early Chinese sages, namely Confucius and Mencius, who were purported to be the founders of a Confucian school.

When the agent encounters conflicting norms of either following ritual or saving his or her life, and has no idea how to act, Mencius advices on the abstract principle of weighing (*quan*權) which "denotes the act of disregarding an otherwise binding norm in exigent circumstances."[21]

> Chunyu Kun said, 'Is it prescribed by the rites that, in giving and receiving, man and woman should not touch each other?'
> 'It is,' said Mencius.
> 'When one's sister-in-law is drowning, does one stretch out a hand to help her?'
> 'Not to help a sister-in-law who is drowning, is to be as cruel as jackals and wolves. It is prescribed by the rites that, in giving and receiving, man and woman should not touch each other, but in stretching out a helping hand to the drowning sister-in-law one uses one's weighing (*quan*權).'
> 'Now the Empire is drowning. Why do you not help it?'
> 'When the Empire is drowning, one helps it with the Way; when a sister-in-law is drowning, one helps her with one's hand. Would you have me help the Empire with my hand?'[22]

> 淳于髡曰:"男女授受不亲,礼与?"
> 孟子曰:"礼也。"
> 曰:"嫂溺则援之以手乎?"
> 曰:"嫂溺不援,是豺狼也。男女授受不亲,礼也;嫂溺援之以手者,权也。"
> 曰:"今天下溺矣,夫子之不援,何也?"
> 曰:"天下溺,援之以道;嫂溺,援之以手。子欲手援天下乎?"[23]

[20] Guarde-Paz, César. 2016. Moral Dilemmas in Chinese Philosophy: A Case Study of the *Lienü Zhuan*. *Dao: A Journal of Comparative Philosophy* 15, 1: 81–101

[21] Goldin, Paul R. 2016, Women and Moral dilemmas in early Chinese narrative, in *The Bloomsbury Research Handbook of Chinese Philosophy and Gender*, Pang-White, Ann ed., Bloomsbury Academic, 29. For a more detailed discussion on this concept of weighing in early China, see also Vankeerberghen, Griet. 2005–2006. Choosing balance: weighing (*quan*) as a metaphor for action in early Chinese texts. *Early China* 30: 47–89. Goldin, Paul R. 2005. The Theme of the Primacy of the Situation in Classical Chinese Philosophy and Rhetoric. *Asia Major* 18.2: 1–25.

[22] For the English translation of the *Mencius*, please see Lau, D.C. 2003 [first edition 1972]. *Mencius*. London: Penguin, p.84.

[23] Yang, Bojun楊伯峻. 2010 [first edition 1984]. *Mengzi Yizhu*孟子譯注. Beijing: Zhonghua shuju, pp.177–178.

Modern readers may question why such an issue even needed to be discussed, when one needs to decide whether to save others or to avoid body contact. This comment is reasonable for many culture codes today. However, in early China, the proper adherence to ritual, such as members of different sexes not touching hands, not only involves the chastity of a woman (a woman with social status would rather die to protect her chastity than be touched), but is also related to the proper ethical cultivation of oneself.[24] Therefore, expert advice to deal with the abovementioned dilemma was necessary.

In this case, we see Mencius advising on a specific application of the principle "weighing," namely temporarily disregarding an otherwise binding norm. Mencius also provided reasoning for his choice. He first acknowledges the validity of the ritual regarding not touching hands while giving and receiving. However, he argues that if one sees another person drowning without giving a hand, such a behaviour is as cruel as an animal's. Therefore, in such an extreme case, one's moral sensibility to save others should override adherence to ritual norms. And only by doing so, can one know the true nature of ritual, and act as a truly morally correct person.

In this case, Mencius directly demonstrates his method of choosing between two conflicting values. But this does not imply that Mencius' expert advice should always be accepted.

When the recipient has a valid reason to choose differently than a moral expert would, the responsibility as a moral expert, according to Confucius, is not to enforce one's expert opinion and demand that others act in a certain way. Instead, as a moral expert, one should not only participate in a proper understanding of the recipient's idea, assisting him or her with rational reasoning, but should also provide the recipient with the idea of living in peace with his or her final decision. This is seen in the following expert advice Confucius gives to Zai Wo as recorded in the *Analects:*

> Zai Wo asked about the three-year mourning period [after the death of a parent], saying, 'Even a full year is too long. If a gentleman gives up the practice of the rites for three years, the rites are sure to be in ruins; if he gives up the practice of music for three years, music is sure to collapse. A full year's mourning is quite enough. After all, in the course of a year, the old grain having been used up, the new grain ripens, and fire is renewed by fresh drilling.' Confucius said, 'Would you, then, be able to live in peace with eating your rice and wearing your finery?' Zai Wo said, 'Yes. I would.' Confucius said, 'If you are able to live in peace with this, do so by all means. The gentleman in mourning finds no relish in good food, no pleasure in music, and no comforts in his own home. That is why he does not eat his rice or wear his finery. Since it appears that you can live in peace with this, then do so by all means.' After Zai Wo had left, the Master said, 'How unfeeling Zai Wo is. A child ceases to be nursed by his parents only when he is three years old. Three years' mourning is observed throughout the Empire. Was Yu not given three years' love by his parents?'[25]

宰我問三年之喪。「期已久矣。君子三年不為禮,禮必壞;三年不為樂,樂必崩。舊穀既沒,新穀既升,鑽燧改火,期可已矣。」子曰:「食夫稻,衣夫錦,於女安乎?」曰:「

[24]For the importance of ritual in relation to ethical cultivation, please see for example Shun, Kwong-loi. 1993. *Ren* and *Li* in the Analects. *Philosophy East & West* 43: 457–79.

[25]The translation mainly follows that of Lau, D.C. 1979. *The Analects---The Sayings of Confucius.* London: Penguin.

安。」「女安則為之!夫君子之居喪,食旨不甘,聞樂不樂,居處不安,故不為也。今女安,則為之!」宰我出。子曰:「予之不仁也!子生三年,然後免於父母之懷。夫三年之喪,天下之通喪也。予也,有三年之愛於其父母乎?」

Again, one may ask, how can we understand a child's mourning duty towards his late father as morally relevant, other than seeing it as either a social convention or a family problem? In fact, in early China, good and right behaviour is largely defined as whether one can be a good son.[26] The value of being a good child, namely the value of filial piety, has "shaped nearly every aspect of Chinese social life".[27] Therefore whether one can act properly towards one's parents, even when they are dead, is closely related to how to live a better life, including one's emotional life, religious worship, ritual performance, as well as being a good human being (*junzi*君子).

In this case, as moral expert, Confucius not only focuses on providing moral reasoning by adopting the principle of "weighing," but also incorporates the attitude of living in equanimity (*an*安) with one's choice as part of his expert advice.

Admittedly, on appearance, when Confucius said, "Would you, then, be able to be at ease with eating your rice and wearing your finery", the idea of living in peace is similar to whatever one feels like doing, namely choosing based on one's preference without reflection.

However, the advice on choosing based on what one can live in peace with is not a choice without rational deliberation. Instead, it involves an anticipation and acceptance of the problems, preparation for an adaptive mind to deal with a changing situation, and a moral commitment to one's choice.

To be specific, on the one hand, Confucius understands the concern of Zai Wo's choice and finds it justifiable. He agrees with Zai Wo's concern that there exists a danger in withdrawing from social affairs and mourning for three continuous years. In other words, Confucius agrees with Zai Wo that the three years' mourning is in conflict with the practice of public ritual.

On the other hand, Confucius helps Zai Wo with the best available information and rational analysis regarding how to weigh this moral choice. According to Confucius, he himself would choose three years of mourning and disregard the public practice of ritual. This is because first, even if one shortened the period of mourning from three years to only one year, normally, for the remaining two years, one would still feel painfully sad and therefore would not be able to practice ritual properly. Moreover, Confucius tells Zai Wo that he would face severe social criticism, since the three-year mourning period is adopted by all gentlemen in society, and therefore acting against it means acting against recognized social protocol. Confucius' expert analysis focuses on the possible emotional difficulties and social criticism that Zai Wo might face if he gives up the mourning ritual.

[26] Knapp, Keith N. 2005. *Selfless offspring: filial children and social order in medieval China*. Honolulu: University of Hawai'i press, pp. 3–4.

[27] Brown, Miranda, 2007. *The politics of mourning in early China*, New York: SUNY Press.

Knapp, Keith N. 2005. *Selfless offspring: filial children and social order in medieval China*. Honolulu: University of Hawai'i press.

Through this context, we can clearly see that as moral expert, Confucius chooses differently from Zai Wo. However, this does not mean that Confucius has the justifiable right to compel Zai Wo to act in the way he thinks he ought to. With an understanding of Zai Wo's concerns, an analysis of the possible moral and emotional problems that allows Zai Wo to anticipate what might lie ahead, finally, Confucius advises Zai Wo to choose based on what he can live in peace with.

The importance of expert advice to choose based on what the agent can live in peace with can be revealed from the following:

First, Confucius makes sure that Zai Wo can follow his choice without regret or emotional conflict. He not only provides possible moral options, but also anticipates the possible emotional and ethical consequences that Zai Wo will face. Therefore, with a double confirmation, and the full explanation of the coming difficulties, Confucius not only assists Zai Wo in reasoning how to make a choice without enforcing it, but also helps him participate in his choice without regret or inner conflict. As moral expert, he not only cares about the agent's moral correctness, but also about the way the agent lives with his right choice.

Second, when expert advice not only focuses on how to act, but also on the attitude that one should live with, moral expertise does not take moral choice merely as a one-off, but as a kind of choice that one can commit to and insist on. In other words, it takes moral commitment into account for giving moral advice. When Zai Wo gives up his mourning duty and returns to society to devote himself to practicing ritual and music, the practical and emotional situation also changes. Equanimity of mind enables one not only to direct one's attention away from social criticism and emotional disturbance, but also to adapt one's mind to how to better practice ritual so as to create value out of giving up mourning. Without anticipation and adaptation of one's mind, one may end up giving up one's original choice while lingering in worry and guilt resulting from a different choice. By taking moral commitment into account, moral experts allow recipients to deal with dilemmas with a proper and lasting attitude.

Last, it shows the trust and respect given by moral experts. With a full analysis of the situation and a complete understanding of the agent's concern, moral experts fulfil their duty by giving the best information and analysis to the recipient. Although they may disagree, when moral experts find the opposing opinion equally valid and are clear that the agent can live with the decision, moral experts ought to show trust and respect for the final choice the agent makes.

8.3 Moral Expertise Advice in the *Zhuangzi*

The previous examples in the *Mencius* and *Analects* show us why the advice on attitude should be regarded as part of moral expertise. The following example suggested by Zhuangzi, on the other hand, shows that only by giving advice on retaining mental equanimity, can moral experts truly make themselves heard and understood. And under certain circumstances, moral experts can only persuade a recipient to act when that recipient has gained mental equanimity.

In the chapter 'Renjian Shi' 人間世 (In the World of Men) in the *Zhuangzi*, which is seen as one of the most important Daoist philosophical work, the text incorporates an anecdote about Duke Xie (葉) receiving a mission to travel to the state of Qi (齊). Once he was appointed to this mission, the Duke regards himself as being stuck in a dilemma. On the one hand, he realizes the importance of this duty and wishes to fulfil it. On the other hand, with self-concern, he knows that proceeding on this mission will endanger his health. Duke Xie is torn by the conflicting values of public duty and personal well-being.

> Zi Gao, Duke of Xie, being about to proceed on a mission to Qi, asked Zhongni, saying, 'The king is sending me, Zhu Liang, on a mission that is very important. Qi will probably treat me as his commissioner with great respect, but will not be in a hurry (to attend to the business). Even an ordinary man cannot readily be moved (to action), and how much less the prince of a state! I am full of apprehension. You, Sir, once said to me that of all things, great or small, there were few which, if not conducted in the proper way, could be brought to a happy conclusion; that, if the thing were not successful, there was sure to be the evil of being dealt with after the manner of men; that, if it were successful, there was sure to be the evil of constant anxiety; and that, whether it succeeded or not, it was only the virtuous man who could secure it not being followed by evil. In my diet I take what is coarse, and do not seek delicacies - a man whose cookery does not require him to be using cooling drinks. This morning I received my charge, and in the evening, I am drinking iced water; am I not feeling internal heat (and discomfort)? Such is my state before I have actually engaged in the affair; I am already suffering from conflicting anxieties. And if the thing does not succeed, (the king) is sure to deal with me after the manner of men. The evil is twofold; as a minister, I am not able to bear the burden (of the mission). Can you, Sir, tell me something (to help me in the case)?'
>
> 葉公子高将使于齐,问于仲尼曰:"王使诸梁也甚重,齐之待使者,盖将甚敬而不急。匹夫犹未可动,而况诸侯乎!吾甚栗之。子常语诸梁也,曰:'凡事若小若大,寡不道以欢成。事若不成,则必有人道之患;事若成,则必有阴阳之患。若成若不成而后无患者,唯有德者能之。'吾食也,执粗而不臧,爨无欲清之人。今吾朝受命而夕饮冰,我其内热与!吾未至乎事之情,而既有阴阳之患矣;事若不成,必有人道之患。是两也,为人臣者不足以任之,子其有以语我来!"

When Duke Xie asked Confucius his opinion to help with the dilemma, Confucius gives his expert advice in relation to how to adopt a new perspective and a new attitude to deal with the given situation, so that a seemingly unsolvable dilemma can be seen as a situation within one's control.

As Confucius said,

> 'In the world, there are two great admonishments: one is fate and the other is duty. A son's love to his parents is fate. You cannot untie this from the mind. A subject's serving his ruler is duty. There is no place where a subject can go and be without his ruler; there is no place where he can escape to between heaven and earth. These are what I mean by the great admonishments. Such being the case, to serve your parents, to be in peace with anywhere they are; this is the perfection of filial piety. To serve your ruler, be at peace with serving him everything that could be; this is the peak of loyalty. Those who serve their own heart-mind are not swayed in the face of sadness and joy. To understand what you can do nothing about and to be at ease with it as if it were fate – this is the perfection of virtue. As a subject and a son, although there are things you cannot avoid, if you act in accordance with the state

> of affairs and forget about yourself, then how will you have time to be concerned about loving life and hating death? Act in this way and you will be all right.'[28]

> 仲尼曰:「天下有大戒二:其一,命也;其一,義也。子之愛親,命也,不可解於心;臣之事君,義也,無適而非君也,無所逃於天地之間。是之謂大戒。是以夫事其親者,不擇地而安之, 孝之至也;夫事其君者,不擇事而安之,忠之盛也;自事其心者,哀樂不易施乎前,知其不可奈何而安之若命,德之至也。為人臣子者,固有所不得已,行事之情而忘其身,何暇至於悅生而惡死!夫子其行可矣!」[29]

The expert advice from Confucius shows the importance of mental equanimity (*an* 安) so as to deal with the dilemma. The expert advice focuses on two aspects. First, a full analysis of the problem that Duke Xie is facing. Second, advising on dealing with the situation with mental equanimity so that the expert advice can be heard and understood, and that understanding can be put into practice without emotional struggle.

To be specific, on the one hand, Confucius analyses the source of Duke Xie's problem, namely his loving life and hating death. This is the reason for his unhealthy situation, and it is the reason that hinders his confidence and dedication in fulfilling his duty. In other words, Confucius understands that, unlike someone who treats death as an honour or as a part of duty, to Duke Xie, protecting his own life is a crucial value when facing duty.

Second, Confucius reveals a fact that Duke Xie does not realize, namely that such a kind of duty is inescapable and unchangeable (*bukenaihe* 不可奈何). As Confucius said, 'There is no place that a subject can go without his ruler', and therefore if one does not face this problem and find a way to deal with it properly, one will still have to suffer from this sooner or later. Not to mention that this mission is a direct order from a ruler. This implies that one may endanger one's life by rejecting this order. In other words, practically, Duke Xie is in fact facing a given, unavoidable and inescapable situation, although theoretically one can risk one's life and escape from such a situation for the moment.

From the previous analysis, Confucius points out that what Duke Xie regards as a dilemma is in fact an inescapable situation if he wants to keep himself safe and alive. With a true understanding of Duke Xie's valuing his life, the nature of his mission, and a rational analysis of the situation, moreover, Confucius points out the importance of mental equanimity for Duke Xie to truly accept his advice, and act on his moral duty without endangering his life.

To be specific, when Duke Xie can keep his mind in equanimity, it means that he can direct his attention away from thinking about why it was he who was appointed, why he was appointed such a difficult job, whether he should quit this mission and how he can keep himself alive. Once he can free himself from constantly worrying

---

[28] The translation is adapted from that of Watson, Burton trans. 1968. *The Complete Works of Chuang Tzu*. New York: Columbia University Press.

[29] Guo, Qingfan郭慶藩. 2007 [first edition, 1961]. *Zhuangzi jishi* 莊子集釋. Beijing: Zhonghua shuju. pp. 152–155.

about potential death, his personal safety, or the unknown factors of the mission, and direct his attention to the duty given from above, he can deal with it directly.

In other words, the rational expert advice from Confucius is truly heard when and only when Duke Xie is free from fear and worry. The mission cannot be fulfilled within one day. The moral advice given by moral experts is not one-off. This means that, without helping Duke Xie with his emotional and mental equanimity, the expert advice cannot truly be understood and put into continuous practice. Suppose an expert only informs him in terms of how he should fulfil his duty; without paying attention to his mental state to support the fulfilling of duty, then Duke Xie may end up dealing with it inattentively, with constant worry for and risk of his life.

8.4 Conclusion by Way of Example: "Sophie's Choice"

The stories in the *Analects* and *Zhuangzi* inform us that, when the agent regards himself as facing a dilemma with two conflicting values, Chinese moral experts not only provide rational deliberations on competing values, but also provide the agent with an attitude to better deal with the situation, and to live with his or her choice in commitment and in peace.

By providing the agent with the attitude of making a choice, on the one hand, the moral expert gives the agent the respect and trust to make his or her final decision without forcing it. Such a kind of moral expert advice does not demand that the recipient unconditionally accept it. Moral experts do not treat their advice as decisive. When the agent holds a different choice with a valid reason, as does the Confucian disciple who reasonably argues against three years of mourning, the moral expert's duty is to provide a thorough analysis of competing values, to evaluate the recipient's choice, to provide the attitude that allows the agent to follow through on his or her ultimate choice without regret, and finally to respect the agent's choice with trust.

On the other hand, by enabling equanimity of mind free from emotional disturbance, a moral expert can ensure that a moral agent truly understands the act that he decides with a clear mind. In the decision-making process, in order for the agent to truly listen to and understand the analysis of ethical values, the agent needs to direct his/her attention away from his/her own emotional and ethical struggles. Only in so doing, can he/she adopt a new perspective and deal with the issue with an adaptive mindset.

Most importantly, giving advice on living in peace with one's choice shows that the moral expert cares more about the agent's well-being than about moral codes of conduct. When facing two conflicting values, the decision one makes is not merely about his or her actions at a certain moment, but is more concerned with how one can live with that choice. For example, when Zai Wo chooses to give up the three-year mourning ritual, the key problem is how he can live a normal life without being crushed by the pain of loss, as well as by criticism from the other nobles. Zai Wo has valid reason to choose either way, but this does not mean he can easily live with his decision.

One may question how such morally expert advice on attitudes from early China can help us with modern dilemmatic situations, such as the case of "Sophie's Choice". How can this kind of expert advice contribute to modern philosophical discussion? Now let us examine the application of moral advice from early Chinese sages.

Admittedly, moral experts from early China cannot "solve" all dilemmas. As in cases like William Styron's *Sophie's Choice*,[30] the principle of "weighing" cannot provide answers for Sophie in terms of which of her children she ought to save, and why. But the attitude of living in peace with the choice can help Sophie to live a better life with her remaining child. Sophie's Choice is as follows:

> 'Sophie and her two children are imprisoned in a Nazi concentration camp. A guard confronts Sophie and tells her that one of her children will be allowed to live and one will be killed. But it is Sophie who must decide which child will be killed. Sophie can prevent the death of either of her children, but only by condemning the other to death. The guard makes the situation even more excruciating by informing Sophie that if she chooses neither, then both will be killed. With this added factor, Sophie has a morally compelling reason to choose one of her children. But for each child, Sophie has an apparently equally strong reason to save him or her.'[31]

Suppose one is facing such a situation and asking for moral advice on what to do - what should a moral expert tell her? As a theorist, one can explain and compare ethical rules such as deontology or act-utilitarianism to help in this case. One may also tell Sophie that a theorist should not provide answers regarding why and what she ought to do. Or, a moral expert may tell Sophie, as Michael Zimmerman suggests, that she should act to save one or the other of her children, since that is the best that she can do.[32]

However, none of the previous moral advice focuses on the deeper issue of how she can live with her choice. If we think of this issue in a practical way instead of theorizing it, we know that the pain and difficulty that Sophie faces is which child she should decide to save, and which to kill, instead of whether she should save her children at all. No doubt, ideally Sophie wants to save both. But practically, she cannot. Therefore, the problem she faces is how she can live her life facing the fact that she did not choose the one who died, and how she can still live a good life with the saved child while carrying guilt (both to herself and possibly to the saved child), remorse, and anger.

Concerning the difficulty of how Sophie can still live a good life with the saved child, apart from giving theoretical analysis on competing moral rules, early Chinese moral experts will advise on the attitude of living in peace while helping Sophie

---

[30] Greespan, Patrica S. 1983. Moral dilemmas and guilt. *Philosophical Studies* 43 (1): 117–125.

[31] The summary of this case is quoted from McConnell, Terrance. 2014.

[32] Zimmerman, Michael J. 2008. *The concept of moral obligation*, Cambridge: Cambridge University Press, Chapter 7.

with her choice. To a Chinese ethicist, when Sophie chooses, the attitude she chooses to adopt is more important than which child she chooses.

Only by willingly accepting the fact that she cannot save both but only one of them, can she not punish herself for not being able to save both and therefore, free herself from dwelling in guilt, pain and anger. Accordingly, only by affirming her own choice of the one that she saves, can she direct her attention to the saved child and her own life without being restrained by doubts because of her decision. This does not mean that Sophie should no longer feel pain or sorrow. The direction of attention enables one to focus on what one can still do, without being overly emotional and stuck within the previous mind-set.

Suppose that after saving one of her children, without being able to live in peace with her choice and at ease with the undesirable situation, Sophie might have been thinking of what else she could have done. She could have spent her entire life blaming herself and living in pain (which also in turn harm the saved child), instead of focusing on how she could live better together with the saved child. Therefore, in the case of Sophie's Choice, a moral expert's advice not only analysing the competing values of how she should act, but also advising on the attitude of how she can live in peace with her choice, shows another level of moral caring for the recipient.

Chapter 9
Moral Experts, Ethico-Epistemic Processes, and Discredited Knowers: An Epistemology for Bioethics

Nancy Nyquist Potter

This chapter presses against the concept of moral expertise—but not for the standard reasons. I argue that the very question of whether or not there are 'moral experts' is misguided—not because of arguments from disagreement or other objections to the possibility of moral experts—but because the concept of moral expertise itself disguises a number of mistaken assumptions—namely, that knowers are individuals; that knowledge is a success term based on truth and/or certainty; that one must choose between objectivist or relativist ethics/ epistemology; and that ethics and epistemology are independent branches of philosophical and clinical theory and practice. This essay, therefore, aims to shift the terrain and the premises of the debate. Drawing on feminist and decolonialist critiques of epistemology and ethics, I argue that both the aims and the model of ethical reasoning need to be reconceptualized.

I divide this essay into two parts, each of which identifies masked assumptions. In the first part, I discuss the equation of moral expertise as a kind of knowledge—a 'knowing-how', which is an achievement. This idea assumes a particular kind of knower, the autonomous, disinterested, objective knower of mainstream epistemology and science. This picture of the reliable knower is now widely contested. In its place, I offer the work of Lorraine Code, Heidi Grasswick, and Naomi Scheman, all of whom challenge that view with a nuanced account of relational or participatory, ongoing knowing as an activity. Thus, a model of moral knowing as inclusive and cognizant of something participants do together—not something that is done, accomplished—calls upon character traits of open-mindedness and humility that the status of 'moral expert' seems to close off.

The second part focuses in on some of the effects that mainstream epistemology has on people's everyday lives—especially as they apply to the field of bioethics: that of who counts as a knower. Recent work by Miranda Fricker (2009), José

N. N. Potter (✉)
University of Louisville, Louisville, KY, USA
e-mail: nancy.potter@louisville.edu

J. C. Watson, L. K. Guidry-Grimes (eds.), *Moral Expertise*, Philosophy and Medicine 129, https://doi.org/10.1007/978-3-319-92759-6_9

Medina (2013), and Kristie Dotson (2014) show the myriad ways in which people's credibility and voices are denied, silenced, and erased by structurally-encoded conditions for listening and giving uptake to 'different' and 'other' people than the advantaged and privileged in society. I end with the idea that an adequate ethico-epistemic model in bioethics needs actively to create and sustain the space for those whose voices are discredited to bear witness to their experiences, needs, and suffering. In using the term 'ethico-epistemic', I highlight the way that ethics and epistemology are co-constitutive and are integrally bound up with one another, a position that challenges the historical and still-prevailing view that those fields are separate and distinct.

In coming to think differently about the idea of moral experts and expertise, I have been deeply influenced by my ten-plus years of experience participating in a local emergency psychiatry department and in our mood disorders clinic. As an ethics-trained philosopher whose research primarily is in philosophy and psychiatry, including psychiatric ethics, I am not called upon only when conflict or uncertainty arises; I am not brought in as a consultant. Instead, as an Associate with the Department of Psychiatry and Behavioral Science. I have been working with Dr. El-Mallakh for over 10 years now, either in the Emergency Department or in the Mood Disorders Clinic. Currently, I am in the Clinic and meeting with patients alongside Dr. Rif El-Mallakh one day a week, whether or not ethical conflict or uncertainty arise. While we do discuss ethical issues as they arise, we also talk about patients' progress, their medication needs, diagnostic questions, and other matters relevant to caring for each patient; ours is a teaching hospital, so such discussion with residents is vital to clinical education. I provide a case in Sect. 9.4.

9.1 Ontological and Epistemological Assumptions. Knowers, Experts, and Achievements

In this section, I explain why the debate about whether or not there are moral experts is misguided. There are three intertwined assumptions I call attention to: what sort of being a moral expert is; the ontological-political context in which moral experts exist; and what the aim of moral expert/ise is.

A moral expert is a kind of knower, or has a particular kind of knowledge, or reasons in a particular kind of way (cf. Grasswick 2014; Scheman 2011; Code 1991). To identify the assumptions that undergird the concept of 'moral expert,' one must look to the broader project of epistemology. Mainstream knowledge is envisioned as an individual enterprise in which the knower has certain features: he is autonomous, rational, disinterested, and detached. His objectivity makes possible his rationality, in that he is not tainted by emotion, passion, concern for outcomes, and other subjective stances. This is to say that conventional norms of knowledge require that knowledge be discovered through processes that eliminate bias and subjectivity. To be clear: by 'mainstream knowledge' I do not just mean philosophical epistemology but knowledge-production more broadly—specifically, in the sciences and other bodies of knowledge where the value of what is known is tied to how that knowledge is produced. The expectation that scientific knowledge, including psychiatric knowledge, is objective and value-free has not been replaced despite epistemological challenged in the academy.[1]

A central concern in clinical ethics consultations is the question of whether normative claims actually can have prescriptive force in the absence of justification. Dien Ho argues that it is not necessary to have a solid ethical foundation in order to do ethics; in fact, this is not how we actually live our moral lives. 'When we try to determine what we ought to do, we do not take some broad ethical theory, plug in the particulars of the situation, and see what recommendation falls out. Moral problems, unlike calculus, are usually not solved by filling in the values for the variable' (Ho 2016, 374). Yet Ho proposes that we deal with disagreement and value conflicts in clinical decision-making by appealing to reasons, with the important caveat that the side that wants to restrict another's autonomy holds the burden of proof—what he calls the Default Principle (DP). Specifically, the DP holds that, when an asymmetric moral conflict arises between two parties, one is permitted to do what one wants unless the party who wants to restrict one's actions offers justification as to why one is not permitted to do that action (Ho 2016). Ho argues that DP offers a way to resolve moral conflicts by requiring that one is free to do what one wants unless the opposing party can offer good reasons as to why one should not be permitted to do that action. Since DP is value-neutral, it offers a deliberative framework that does not require any particular normative framework. Nevertheless, Ho argues, it does give us a way to hold onto the notion of ethical expertise: ethical expertise involves skill and adeptness at formal and informal reasoning, including facility with analogical reasoning, and clinical knowledge. While he admits that his model emphasizes a rational ideal, he still holds to the centrality of reasons-giving to resolve ethical conflicts. Thus, he concludes, for the most part, and following his model of appealing to the Default Principle, ethical expertise is indeed possible.

Ho's view on the centrality of reasons-giving is not uncommon: in fact, it has been the prevailing ethical and political view for centuries now. Christopher Meyers, for example, holds that expertise is connected with justification, where justification is a matter of using 'the standard philosophical repertoire of deductive and inductive reasoning' (Meyers 2003, 259; see also Cross 2016). David Archard points out that we do not know what counts as success in ethical reasoning, in particular as it applies to clinical decision-making. Yet, he argues, even if we do not ascribe to an ethical framework of truth and justification, we can still identify experts because some judgments are better than others (Archard 2011). I agree that some moral responses are better than others; in eschewing the notion of moral expert/tise, I am not claiming that just any person will do in clinical ethics consultations or the sort of participation I do in psychiatric work. Yet I want to point out the assumption that the sort of responses we are talking about are judgments: moral judgments are taken

[1] For recent work in social epistemology, see Zabzebski 1996; Goldman and Whitcomb 2000; Sosa 2009; Grasswick 2012.

to be part of moral reasoning, with all of the baggage that comes with Enlightenment rationality.

However, recent work by Lisa Tessman (2015), Karen Jones (2004), Jesse Prinz (2004), Alison Jaggar (1989), and many others have challenged the view that our best ethical reasoning arises from reason alone. Eschewing rationality as the foundation of knowledge will require that we think differently about who knowers are and what the aims of knowing are—entailing that we reconceive what characteristics are involved in knowing well.

One epistemological implication of a reductionist 'view from nowhere' (Nagel 1986) is that only those individuals who are capable of having the specialized characteristics required of knowers—rationality and objectivity—can be experts in various knowledge practices (science, yes, but also law and ethics). It follows, from that view, that not everyone is, or even can be, knowers. Some groups of people, the prevailing epistemology goes, are not rational enough and too subjective (too emotional by nature, too politically interested in certain outcomes) to be knowers. It is true that, with respect to particular domains of knowledge, not everyone can be a knower. In making the stronger claim that whole groups of people can be discredited as knowers just in virtue of their social groupings, I am drawing on extensive literature from feminist and critical race theorists' writing, some of which I discuss below. The entire question of whether there can be moral experts, or what such expertise consists in, is ensnared in mainstream epistemology.

However, feminists and decolonizing epistemologists have long argued that the ontology[2] of what a knower is, and the demand of objectivity entailed in it, does not map onto reality. What it has done is to silence members of many groups on the grounds that they are not capable of being knowers—they are, by nature or by position in society, not credible. As noted above, the impossible demand for objectivity requires that we bracket off emotions. Epistemic achievement/success is measured by how well we do this, but the loss is significant: if we try to become good knowers, we undermine our ability to attend well to emotional responses, which arguably are an important source of knowledge (Jaggar 1989; Scheman 2011; Haidt 2001; Grasswick 2012). Finally, existent epistemic norms structure the very subject of knowledge—meaning, what counts as knowledge, what the scope of knowledge is, and what the important questions are. So the problem is not just a matter of 'how well do epistemic norms map onto the world' but which world, and whose world, we are talking about here—and who is the 'we'? (Scheman 2011). If we want to produce bodies of knowledge that map onto the truths of everyday lives (including ethical problems in living as well as our scientific and ecological world), we need a different epistemology.

If we take seriously the idea that different subject-positions in society produce different knowledges and values,[3] then we see that we would do well to draw upon

---

[2] Ontology is the study of being, or what exists in the world. In philosophy, this ranges from questions such as 'what is time?' to 'what is real?' to 'what is race'?

[3] A subject-position is a concept that shifts the understanding of self-identity from an individual being separate from context and autonomously discovered or created, to that of self-identity as

practices of knowing that bring together the ethico-epistemic perspectives of diverse groups. This is because assumptions may be invisible or may seem unquestionable to those inside a given scientific or psychiatric practice. This calls both for the recognition of, and the normative and active practice of, epistemic interdependence.

Note that this is not an ethico-epistemic theory of relativism: it is a recognition of epistemic pluralism (cf. Code 1991; Torcello 2011; Costa 2015). The difference between pluralism and relativism often is misunderstood, but many scholars who reject theories of epistemology that arose out of Enlightenment rationality are not relativists. A quick way of distinguishing between them is that, 'pluralism lies in the inability to defend any position categorically; whereas relativism, I [Torcello] will maintain, lies in the inability to condemn any position categorically' (Torcello 2011, 2). Similarly, Lisa Tessman argues for moral value pluralism. By this, she means that there exists a great range of values that humans have constructed because they have been found to be 'necessary in order for us to live good enough lives together' (Tessman 2015, 25). One implication of this is that some moral requirements can be negotiated away and others not, and there is not one single unified theory that will settle this for all relevant parties. (I say more about this in Sect. 9.2.1).

Nor is mine a claim to standpoint epistemology: the idea that some class of people has a privileged epistemic standpoint. Instead, it is a claim that objectivity and pluralist values and knowledges are compatible. What we need in theories of clinical ethics 'experts' is a kind of objectivity that allows for ongoing critique, challenge, and negotiation.

A good place to start, in rethinking the concept of experts, is found in the work of Moulton et al. (2007). They argue that expertise comes not from knowing well and with confidence, but from being the sort of knower who handles unclear situations slowly and carefully. The experienced clinician comes to know so well the sorts of things he needs to reason about that he internalizes the reasoning process, thus becoming more efficient in treating patients (Moulton et al. 2007, S109). Yet the epistemological process of learning to internalize assessments and decisions about what a patient needs does not aid clinicians when things occur out of the ordinary. Moulton et al.'s review of the literature on clinical expertise points, instead, to the idea that 'expertise is achieved when one constantly and intentionally engaged with one's environment during the routines of daily practice' (Moulton et al. 2007, S112; emphasis added). In novel, unclear, or messy contexts, they argue, we require additional cognitive resources because our own may be constrained by usual practices of relying on tacit knowledge. However helpful it may be in ordinary situations, it is not sufficient to deal with indeterminate cases. One source of additional resources they call for is situational awareness—the 'perception of the elements in

---

necessarily implicated in cultural and historical contexts. More technically, it is the subjectivity of self that is produced by and through various discursive practices, where 'discourse' is not only linguistic but includes the ways our bodies are ordered in space and time and are gendered, raced, and classed according to norms, values, laws, and so on. As Jukka Törrönen says, 'Subject positions are relational categories that obtain their situational meaning in relation to other possible subject positions and discourses' (Törrönen 2001).

the environment within a volume of time and space, and comprehension of their meaning, and the projection of their status in the near future' (Moulton et al. 2007, S112). Simply put, it is the constantly evolving understanding of the environment. This requires a process of slowing down. 'At any moment in time, attention needs to be allocated to monitoring the environment for unexpected and unanticipated cures, as well as for assessing results of actions already taken' (Moulton et al. 2007, S 113). Expertise isn't so much what one knows, therefore, but having the 'ability to respond effectively in the moment to the limits of his or her automatic resources and to transition appropriately to a greater reliance on effortful processes when needed' (Moulton et al. 2007, S114).

Yet when we reach the limits of our experience and knowledge, and are uncertain about what to do, this is precisely the moment when we need others with whom to sort things out. Contrary to mainstream epistemology, many scholars have argued that, in reality, our knowing is always socially situated—meaning that we necessarily know the world from particular perspectives (Lugones and Spelman 1983; Code 1991; Collins 2003; Dotson 2014; Grasswick 2014). As Heidi Grasswick points out, 'feminist [and decolonizing] epistemologists have overwhelmingly rejected the vision of the self-sufficient and atomistic knower who must acquire knowledge on his own and is undifferentiated from other knowers in his capacity to know. Instead, they have drawn attention to the epistemic relevance of differently located knowers' (Grasswick 2012, 308).

In fact, as Lorraine Code characterizes it, knowledge is the ongoing relational work of epistemic communities. Indeed, Code argues that knowledge is commonable (Code 1987, 196). It 'is an intersubjective product constructed within communal practices of acknowledgment, correction, and critique' (Code 1991, 224). Knowledge production is a reciprocal process:

> Cognitive commonability is dependent for its existence upon community, which, in turn, is sustained by commonability. In fact, for something to count as an item of knowledge, it must be possible for at least some members of an epistemic community to locate it within the context of what one might call a 'communication system.' (Code 1987, 171).

The point is that our different social situatedness places us as necessarily epistemically dependent on one another. Grasswick thus emphasizes the 'uneliminable role of the social in shaping knowledge,' while at the same time upholding a commitment to knowledge that is oriented toward truths about the world (Grasswick 2014, 8).

Furthermore, this epistemic dependence is inseparable from our moral lives and the ways we relate to and treat one another. In the words of Grasswick:

> Accordingly, we are deeply epistemically interdependent. We must rely on those who are differently situated for insights that we cannot obtain on our own and which need to be integrated with our own perspectives. As a result, relationships and interactions between knowers form an integral part of epistemic inquiries. Such social interactions necessarily have an ethical dimension to them, and the ethical quality of our relationships with others will in many cases affect our possibilities for epistemic success in the long run. (Grasswick 2014,16).

Not only are epistemic and ethical lives inextricably interwoven, but also the process of seeking moral knowledge often requires that we give moral and epistemic deference to others—when our own capacities for perception and understanding are limited by our positionalities. 'We need to be able to perceive how things are with other people, becoming aware of the relevant moral detains of the situation so that we will be able to respond appropriately' (Grasswick 315).

While I cannot address these issues in detail here, in the next section I will elaborate on why this is important and describe some of the characteristics a good ethico-epistemologist would have.

9.2 Characteristics of an Ethico-Epistemology for Bioethics

9.2.1 Testimonial Injustice, Epistemic Violence, and Democratizing Ethico-Epistemic Processes

In this section, I will argue that an adequate model of moral epistemology in bioethics needs to actively create and sustain the space for those whose voices are discredited to bear witness to their experiences, needs, and suffering. The idea that being a clinical ethicist is like being an architect is articulated by Margaret Urban Walker, who argues that the task of the clinical ethicist is best thought of as 'mark[ing] and open[ing] up moral-reflective spaces' (Walker 1993, 38). This metaphor serves to remind us that such actual spaces need to be created—that they are not 'natural' in institutional settings—and that these spaces must be designed as inclusive and diverse ones. Walker eschews the idea that mastery of ethical matters consists in the abstract knowledge of epistemological and moral-theoretic ethics:

> Could full moral competence really consist entirely in intellectual mastery of codelike theories and lawlike principles? What of skills of attention and appreciation, of the practiced perceptions and responses that issue from morally valuable character traits, of the wisdom of rich and broad life experience, of the role of feelings in guiding or tempering one's views? (Walker 1993, 34).

Walker's questions raise the issue of epistemic authority, a central concern of this paper. The question must be pressed, whose authority and which epistemology are we using in these debates? In clinical ethics practices? As I suggested in Sect. 9.1, some people, just in virtue of group memberships, are routinely treated as less credible, less trustworthy, and as not-knowers. Medina puts the point clearly: 'Credibility never applies to subjects individually…' (Medina 2013, 61). His reasoning is that credibility, like many other epistemic qualities, is interactive, and the interactive nature of credibility judgments is comparative or contrastive; 'being judged credible to some degree is being regarded as more credible than others, less credible than others, and equally credible than others' (61). Attributions of credibility affect everybody in an interaction, listeners and speaker alike. And when members of a group are treated as untrustworthy testifiers just in virtue of their group membership, when one speaker in that group is discredited, the implication of untrustworthy speakers can spread to the whole group via stereotyping. Medina also argues that some speakers, assigned excessive credibility via group membership, may harm everyone involved; they are given too much epistemic authority.

Miranda Fricker (2009) argues that treating some groups of people as less credible is a kind of testimonial injustice. To be epistemically unjust means that the listener could have done otherwise and that his or her failure to attend appropriately to the speaker results in distorted beliefs. It occurs when the listener holds (most often, socially based) biases and prejudices that influence his or her assessment of the speaker's telling. It is ethically unjust because it is unfair: the listener does not accord the speaker the credibility that is warranted, because he or she holds biases and prejudices that influence his or her assessment of the speaker's telling. While it might seem that the problem epistemic injustice lies only in regarding people as less credible for epistemically irrelevant reasons, this would miss the larger point I am making about how mechanisms of knowing the production of knowers works in practice: the contested issue precisely is what it means to call something 'epistemically irrelevant' and who gets to decide these matters: historically and now, the lines of knowledge, knowers, and unknowers has been drawn in relation to structures of power. Belief, knowledge, justification, credibility—these are normative notions that are never free from real-life facts of power and oppression.

Kristie Dotson makes the point more sharply: epistemic dismissal of a person's credibility is a form of violence when it is rooted in prejudices and stereotypes of members of marginalized and oppressed groups. As she puts it, epistemic violence is 'a failure of an audience to communicatively reciprocate, either intentionally or unintentionally, in linguistic exchanges owing to pernicious ignorance' (Dotson 2011, 239, 242). Pernicious ignorance is a reliable ignorance that harms another person. By 'reliable,' she means that it 'is consistent or follows from a predicable epistemic gap in cognitive resources' (Dotson 2011, 238). Reliable ignorance is not necessarily harmful; it is the pernicious kind that we need to worry about. Whether or not reliable ignorance is pernicious and what kinds and degrees of harm have been done depends on context. To determine this 'requires not only identifying ignorance that would routinely cause an audience to fail to take up speaker dependencies in order to achieve a successful linguistic exchange, but it also requires an analysis of power relations and other contextual factors that make the ignorance in that circumstance or set of circumstances harmful' (Dotson 2011, 239),

To understand Dotson's claim about a 'predictable epistemic gap in cognitive resources' requires an understanding of feminist and critical race theorist work on situated knowledge. The idea of situated knowledge is that intersecting oppressions, and the social locations we occupy within cultural and political domains, shape the experiences we have, the kinds of concepts we form, and the beliefs we hold regarding what is true about the world. It presents certain epistemic, social, cultural, sexual, and ethical norms as options for us, and others as off-limits, depending on social location(s). Such experiences are formed in patterns of differentially situated people, wherein the locations one occupies have real-world material, epistemic, ethical, sociopolitical, and economic consequences. Social location shapes our very perceptions. Thus, social location shapes (but does not entirely determine) situated knowledge and situated ignorance.

As Dotson explains, 'situated ignorance follows from one's social position and/or epistemic location with respect to some domain of knowledge' (Dotson 2011, 248). It produces or sustains a kind of unknowing (not-knowing, not needing to know, needing not to know) that fosters significant epistemic differences among diverse groups. Epistemic difference is 'the gap between different worldviews caused by differing social situations (economic, sexual, cultural, etcetera) that produce differing understandings of the world, differing knowledges of reality' (Bergin 2002, 198, as quoted in Dotson 2011, 248). When linguistic exchanges take place between people where epistemic gaps are present, reliable ignorance of the pernicious kind can occur—meaning that that ignorance is harmful. Pernicious ignorance may be unintentional but still be due to what Dotson calls a kind of incompetence that concerns a maladjusted sensitivity to the truth with respect to some domain of knowledge. Testimonial injustice and epistemic violence caused by pernicious ignorance are both epistemically and ethically wrong to engage in because they cause harm.

I apply these ideas about situated ignorance and epistemic violence to questions of moral expertise in the clinic. As do Archard and Ho, I advocate an ethico-epistemic democratic process—although the ways that we think this can occur are distinct (Archard; Dien Ho). The asymmetry between an expert and a layperson requires that we expand the epistemic asymmetry to recognize not only that patient-clinician power differentials exist, but what those asymmetries do to our conception of knowers and how we engage in epistemic communal practices of knowledge-production. As Meyers explains about clinician-patient encounters, 'those who hold power—especially power that has historically been established within a hierarchical institution like medicine—characteristically dominate conversations. They control the tone, they control the content, and they usually control the outcomes' (Meyers 2003, 235). Nevertheless, ethico-epistemic processes in clinical practice require that clinicians learn to listen to patients and to take seriously their voices (cf. Potter 2016).[4] Writing on decolonizing epistemologies, Walter Mignolo says that 'decolonial thinking is what colonial subjects do when they do not want to assimilate and are not happy with remaining colonial subjects… [It] means engaging in knowledge making and transformation at the edge, in and of, the disciplines' (Mignolo 2012, 42). A democratizing epistemology in the clinic does not aim for equality; it calls for the decentering of power that clinicians and any people who claim to be experts need actively to promote. Indeed, it calls for a decentering of the self. Roughly, this means that we learn to shift from our world-view, values, assumptions, and epistemic commitments to be able to grasp what the world is like as the other person experiences it. This involves animatedly and lovingly shifting out of the comfort of 'home' to be open and receptive to another's world where we may not be comfortable

[4]A discussion of the importance of service user voices and inadvertent mechanisms of silencing that may be at play in clinician-service user encounters is beyond the scope of this paper, but it is a topic I am writing on.

and things are unfamiliar to us (cf. Potter 2003 for a full discussion of this issue as it applies to psychiatry. See also Mendieta 2012 on a decolonizing relationship between ethics and knowing well.)

9.3 Acceptance of Uncertainty and Moral Remainder

A second characteristic of an adequate ethico-epistemic practice in clinical work is a resistance to the ontology and the value of certainty. Lisa Rasmussen argues against ethics consultants/specialists aiming for epistemic certainty or moral truth. 'A frequent assumption is that only absolute certainty would justify an 'ought' statement' (393). The notion of moral certainty, Rasmussen argues, assumes a particular and contestable understanding of what counts as a justified knowledge claim. As I have argued, this picture of the knower comes out of a rationalist epistemology that has been challenged for many decades. Instead, Rasmussen says, 'the justification for a particular clinical ethics decision has to be understood within a context of uncertainty,' so the real question is not whether there are moral experts, or even moral expertise, but what we ought to do in the face of uncertainty (Rasmussen 2016, 392). This is not to say we just 'wing it,' because even in the absence of certainty, there are better and worse ways of making decisions. Yet most of us are highly uncomfortable with uncertainty and ambiguity, and we tend to claim certitude both in order to decrease uncertainty and a feeling of loss of control, and in order to silence or repudiate others who challenge our need to be correct.

In embracing a stance of uncertainty, we also develop an attitude of humility. It is true that I stated in Part I that I'm suggesting a model of ethico-epistemic integrative work in the clinic that is pluralist. But I also made clear that pluralism does not require that we think that all positions are equally right. However, given power differences among members of different groups, including relationships between patients and clinicians, we who hold more authority and social or professional power would do well to recognize our epistemic shortcomings (see Meyers 2003; Medina 2013, 51). In admitting what we don't know (in the ethico-epistemic sense), we exercise humility. But, in addition, we may need to not only recognize and accept the limitations of our own knowings, we may need sometimes to rely upon others to fill in gaps, or even to call us on our errors, revealing truths we may not have seen. Grasswick puts the point this way: 'If testimony offers a legitimate form of knowing, the privileged could gain access to knowledge of social relations through testimony of the underprivileged…'(Grasswick 2012, 309). Once again, we must exercise caution in interpreting this claim. Grasswick argues that sometimes our epistemic and social positionalities prevent us from perceiving an undistorted picture of what is going on or what is called for in a particular situation. In such cases, our relative ignorance may require us to give epistemic and moral deference to others who are better placed to perceive more clearly. However, this is not to say that those others have special epistemic privilege. It is, instead, to suggest that, in conceding that our own perspectives our limited, we give prima facie credibility to those who are marginalized or less socially privileged than we are and sometimes to defer to their judgment. (See Grasswick 2012 for a full analysis and example.) The 'we' here includes those trained in philosophical ethics—precisely those whose experience with clinical ethics consultations raise the question of whether or not moral experts truly exist.

A related point is the experience of moral remainder even in the face of choosing. Lisa Tessman argues that genuine, irresolvable moral dilemmas exist and that it is a fact of actual ordinary life that we sometimes experience deep loss and moral failure when we must choose between two non-negotiable moral requirements. Moral requirements arise not out of rational deliberation but our emotional life and relationships with others: as Tessman puts it, 'love and care are sources of value and of requirement' such that love makes some actions quite literally unthinkable (Tessman 2013, 50, 51). The 'I must' that we feel toward one we love stands as a moral requirement to protect that person and that is non-negotiable. This can make the moral atmosphere of clinical work anguishing: as Tessman suggests, caring for the vulnerable sometimes means that we suffer even when we fulfill the 'I must.' We sometimes feel morally responsible to different people at the same time, and the fact that we cannot fulfill conflicting responsibilities does not absolve us of the responsibility we did not fulfill. To think it does is to subscribe to an account of ethical theory that holds that no genuine moral dilemmas exists—that an apparent conflict between two moral requirements, on that view, is always resolvable by rationally prioritizing one over the other. Yet, as Tessman argues, theory aside—in real life—we sometimes do feel the pull of two conflicting moral requirements, neither of which can be overridden by the other. Thus we experience a moral remainder. The illusion of ethico-epistemic certainty might defend against the experience of loss at what we cannot do under the circumstances, but moral remainders are like an internal bleed; they cannot be reasoned away.

9.4 Ethico-Epistemic Vices and Virtues

Lorraine Code says that belief and understanding require effort on the part of would-be knowers or believers (Code 1987, 177). That is, it takes ongoing work on our part in order to be epistemically responsible: 'To strive for insight into the extent of one's own cognitive capacities, to distance oneself as much as possible so one can be critical of one's own knowing, is a crucially important aspect of epistemic competence' (Code 1987, 176). I would add (and this is consistent with Code's later work) that epistemic and ethical responsibility are intertwined—they are not independent branches of knowing, as the history of epistemology has held.

Let me dwell on these ideas.

José Medina argues that epistemic flaws are grounded in and exhibit our character (Medina 2013, 29). Vices (and virtues) are not temporary or one-off flaws or strengths but are partly constitutive of who we are and how we perceive, respond to, and help shape the world and the various people within it. Thus, they are not only individual flaws or strengths, but also systemic and structural ones: as I have discussed, epistemology is a social endeavor that involves others in deciding what counts as knowledge and knowers, and so on. Medina explains that social injustices leave people—both the privileged and the disadvantaged—liable to have epistemic deficits that affect our ability to hear and to be heard correctly (Medina 2013, 28).

Cognitive and social development work together to cultivate our epistemic schemas for navigating the world—schemas that simultaneously create our characters. These schemas or blueprints shape our bodies of knowledge: who we count as knowers, what we count as evidence, who we count as credible, and who determines the structure of various practices. These schemas may not be conscious: although we typically are not aware of our everyday attitudes, beliefs, and assumptions, and usually are not critically evaluating our own epistemic frameworks, we are responsible for them because we can be critically aware, we can evaluate and change our own epistemic character, and we can learn to understand who we are and who others are in a more epistemically and socially accurate way. We can engage in strong objectivity, both in science and in ethics (cf. Potter 2016; Harding 1993). And because epistemic vices are integrally tied to social injustice, we not only can, but should make the necessary cognitive corrections in order to cultivate more virtuous characters. Thus, Medina argues that one form epistemic character flaws take is a resistance to self-correction and openness to correction from others (Medina 2013, 31). This is a vice when it becomes a habit, part of our disposition, because "letting one's perspective go unchecked results in an unavoidable, mundane accumulation of oversights, errors, biased stereotypes, and distortions. In this way, racist and sexist biases become undetectable and incorrigible blind spots" (Medina 2013, 32).

Regardless of where we are situated in relation to structures of domination and subordination, we need to develop a character with epistemic virtues in order to serve social justice and fight against injustice.[5] But the road to epistemic virtue is, in many ways, more challenging and more difficult for the privileged. In particular, it presents a challenge to psychiatrists and other mental health professionals.

Epistemic arrogance, for example, hinders one's ability to learn from others and from the facts, and decreases the capacity to self-correct and to be corrected by others. 'Thinking whatever one wants to think, without resistance, does not lead to the development of good epistemic habits' (Medina 2013, 32). Another flaw of the privileged is epistemic laziness. Epistemic laziness, the main epistemic vice of the privileged, manifests in a lack of curiosity that is carefully orchestrated to reproduce the world as the privileged experience it. Epistemic agency, Medina explains, requires a minimum of diligence, which is undermined by arrogance and laziness. The third epistemic flaw or vice of the privileged is closemindedness, a kind of epistemic hiding: attitudes of not knowing (culpable epistemic ignorance), not needing to know, and even, needing not to know. The path to becoming open to epistemic [and ethical] counterpoints to these ethico-epistemic vices, then, are the

[5] I do believe that psychiatrists and other clinicians should be engaged with such matters, as they are integrally bound up with clinical practices and patients' lived experiences.

cultivation of the character traits of humility, intellectual curiosity/diligence, and open-mindedness.

9.5 A Case Study from Psychiatry

A few months ago, Sylvan, a white man in his mid-40s who has been diagnosed with dysthymia,[6] expressed despair that, even though the psychiatrist had tried numerous medications over the years, he would never get better. After reviewing the patient's medication history with him, and discussing his current medications, the psychiatrist responded with a comment that he was not sure there was much more he could do for the patient. I did not respond at the time with a question about that comment because I did not want to challenge the psychiatrist in front of the patient. However, a vigorous discussion ensued afterward, both between the two of us and the 3rd year residents.

I wondered out loud whether it was harmful to the patient not to offer hope to him. I posited that hope is an ethical concept and that, from my perspective in psychiatric ethics, hope is necessary to patients' well-being. One of the sharpest and most confident residents objected to my characterization of hope as ethical; Dr. El-Mallakh[7] remarked that water is also necessary to health but is not ethical. While nothing was decided about whether or not it was good for the patient not to hold out hope that he would experience relief from his dysthymic life, the conversation continued. (For example, I brought in for Dr. El-Mallakh a postcard reproduction from the Courtauld Museum in London depicting hope as one of the three virtues.)

When a different patient's mood was described on a later clinic day as hopeless, we returned to the question of whether or not hope should be given to patients when they are feeling ready to 'give up.' While we no longer debated whether hope is an ethical concept, this time we went more deeply into when hope is appropriate and when it is not. Dr. El-Mallakh expressed what I initially took to be a strictly fact-based approach to talking with patients, in which he sees his role as, in part, to help patients face reality. I expressed the worry that that may not be a sufficiently empathetic and ethical approach to helping patients who are despairing to have a reason to go on living. But Dr. El-Mallakh's point was that having realistic expectations is a central component of facing reality, and the degree of hopefulness to offer must be mapped onto a world in which the patient has realistic expectations about his or her illness. Additionally, he remarked that a complicating factor in my view (that ethical treatment requires that we give hope to the patients) is that patients present with

[6] Dysthymia is a mood disorder where the person is chronically mildly depressed for a period of at least 2 years. It includes feelings of hopelessness and self-doubt and can affect energy levels and general functioning. In the DSM-V, it has been combined with Chronic Depressive Disorder and is now called Persistent Depressive Disorder

[7] Dr. El-Mallakh has read and commented on this write-up of the case study and has agreed to being identified as the psychiatrist.

different attitudes towards, and expectations of, their psychiatrist. If a patient expects his psychiatrist to solve his life-long misery with medications, it may not be appropriate merely to offer hope that medication adjustments will accomplish that. It turned out that Dr. El-Mallakh held a more nuanced view of the relationship between giving hope and giving a realistic picture of appropriate patient expectations than I had thought prior to our second conversation.

Nothing has yet been settled for us about this topic. But when Dr. El-Mallakh met with this second patient he was, as always, very empathetic and helpful.

This case illustrates the integrative approach that Dr. El-Mallakh and I take. Our collaborative approach, for over 10 years now, is an on-the-ground way of working through the ethico-epistemic and clinical issues as they arise in the moment and evolve over time. In Sylvan's case, an ethical and clinical discussion of the role of hope was prompted, and this exploration was sustained over time and included discussion of treatment and behavior involving a second patient. While there is literature that addresses philosophical and ethical dimensions of hope, it can only serve to direct our thinking in a general way; it cannot help us sort out what is best for an individual patient (see Kadlac 2015; Garrard and Wrigley 2009; Martin 2008; Waterworth 2003; Bovens 1999). Indeed, it is not clear that there always is a 'best' approach for a psychiatric patient, as I suggested in Sect. 9.2. Instead, Dr. El-Mallakh, the residents, other students, and I were thinking together about what hope is and how and when it is necessary for good care of a given patient. We weren't looking for the right answer but, instead, were taking our understanding to a deeper level by examining it from a number of levels.

The ethico-epistemic model I advocate in this chapter is one that Dr. El-Mallakh and I strive to achieve and that is most beneficial to patients, to residents who learn from us, and to us. It is true that he particular patient whose clinical visit gave rise to ethico-epistemic issues may not be the beneficiary of such talk, and our aim is not to find 'the moral truth.' But it isn't just idle time spent: our discussions aim at understanding problems in the context in which a patient's needs come to light. Additionally, in the case of Sylvan, I was struggling to understand the patient's needs from his own world-view—to attend to his voice as a mentally ill patient; the voices of people living with mental illnesses sometimes are not given credibility, so in clinic we work together to understand how to listen more attentively. (See Potter 2018 for more on patient/service-user voices and stigma against service-users.) Similar to Walker's emphasis on the creation of regular and actual spaces for open, moral-reflective cooperative engagement, these kinds of discussions occur over time and in clinic when we have time between patients, and are almost always prompted by particular questions or problems that one of us identifies while meeting with a patient in clinic that day. Yet sometimes, decisions have to be made, and we do not always have the luxury of time to examine a topic from different angles. When this occurs, Dr. El-Mallakh or other clinicians may be less than fully confident, or they may act on an 'I must' and be left with a moral remainder for a moral requirement that was left unaddressed. (See Potter and El-Mallakh forthcoming for a discussion of our collaborative approach.)

9.6 Conclusion

This argument aims to undermine the authority that an assertion of 'moral experts' in bioethics is seeking to assert. That is, I deny that people with academic ethics training have the right to assert special authority in bioethics settings. However, I do believe that those with experience in disenfranchisement, feminism, decolonialism, and activist politics can provide an important perspective on problems that arise in psychiatry and elsewhere. Thus, the voices of people with nuanced perspectives on ethics are vital in clinics, hospital ethics, committees, and IRBs. With respect to psychiatry, that especially includes the voices of patients with mental disorders. In Dr. El-Mallakh's and my current model, patients are not yet included in discussions; this is a weakness of our otherwise exceptional model. Nevertheless, I believe that it is diverse voices and inclusivity of patients' perspectives that will give clinicians the resources for a pluralism of goods to be recognized, and a commitment to the good of the patient and the cultivation of the virtues of humility, intellectual curiosity, and open-mindedness prevent us from sliding into relativism.[8]

References

Alvin, Goldman., and Dennis Whitcomb, eds. 2000. *Social epistemology: Essential readings*. Oxford: Oxford University Press.

Archard, David. 2011. Why moral philosophers are not and should not be moral experts. *Bioethics* 25 (3): 119–127.

Bergin, Lisa. 2002. Testimony, epistemic difference, and privilege. *Social Epistemology* 16 (3): 197–213.

Bovens, Luc. 1999. The value of hope. *Philosophy and Phenomenological Research* 59 (3): 667–681.

Code, Lorraine. 1987. *Epistemic responsibility*. Hanover: University Press of New England.

Code, L. 1991. *What can she know? Feminist theory and the construction of knowledge*. Ithaca: Cornell University Press.

Collins, Patricia Hill. 2003. Some group matters: Intersectionality, situated standpoints, and black feminist thought. In *A companion to African-American philosophy*, ed. Tommy Lee Lott and John P. Pittman. Malden: Blackwell.

Costa, Paulo. 2015. Realism, relativism, and pluralism: An impossible marriage? *Philosophy and Social Criticism* 4 (4–5): 414–432.

Cross, Ben. 2016. Moral philosophy, moral expertise, and the argument from disagreement. *Bioethics* 30 (3): 188–194.

Dotson, Kristie. 2011. Tracking epistemic violence, tracking practices of silence. *Hypatia* 26 (2): 236–257.

———. 2014. Conceptualizing epistemic oppression. *Social Epistemology* 28 (2): 115–138.

Fricker, Miranda. 2009. *Epistemic injustice: Power and the ethics of knowing*. Oxford: Oxford University Press.

[8] I have not said enough in this chapter to illustrate and make good on this claim. Full treatment of this matter in the context of psychiatry will have to wait until my next project.

Garrard, Eve, and Anthony Wrigley. 2009. Hope and terminal illness: False hope versus absolute hope. *Clinical Ethics* 4 (1): 38–43.

Grasswick, Heidi. 2012. Knowing moral agents: Epistemic dependence and the moral realm. In *Out from the shadows: Analytical feminist contributions to traditional philosophy*, ed. S. Crasnow and A. Superson, 307–338. Oxford: Oxford University Press.

———. 2014. Understanding epistemic normativity in feminist epistemology. In *The ethics of belief*, ed. J. Amtheson and R. Vitz. Oxford Scholarship Online. 1093/acprof: oso/9780199686520.003.0013.

Haidt, Jonathan. 2001. The emotional dog and its rational tail: A social intuitionist approach to moral judgment. *Psychological Review* 108 (4): 814–834.

Harding, Sandra. 1993. Rethinking standpoint epistemology: What is 'strong objectivity'? In *Feminist epistemologies*, ed. Linda Alcoff and Elizabeth Potter, 49–82. New York: Routledge.

Ho, Dien. 2016. Keeping it ethically real. *Journal of Medicine and Philosophy* 41: 369–383.

Jaggar, Alison. 1989. Love and knowledge: Emotion in feminist epistemology. *Inquiry: An Interdisciplinary Journal of Philosophy* 32 (2): 151–176.

Jones, Karen. 2004. Emotional rationality as practical rationality. In *Setting the moral compass: Essays by women philosophers*, ed. C. Calhoun, 333–352. Oxford: Oxford University Press.

Kadlac, Adam. 2015. The virtue of hope. *Ethical Theory and Moral Practice* 18 (2): 337–354.

Lugones, María, and Elizabeth Spelman. 1983. Have we got a theory for you! Feminist theory, cultural imperialism, and the demand for "the woman's voice". *Women's Studies International Forum* 6 (6): 573–581.

Martin, Adrienne. 2008. Hope and exploitation. *Hastings Center Report* 38 (5): 49–55.

Medina, José. 2013. *The epistemology of resistance: Gender ad racial oppression, epistemic injustice, and resistant imaginations*. Oxford: Oxford University Press.

Mendieta, Eduardo. 2012. The ethics of (not) knowing: Take care of ethics and knowledge will come of its own accord. In *Decolonizing epistemologies: Latina/o theology and philosophy*, ed. Isasi-Diaz and Eduardo Mendieta, 247–264. New York, Fordham University Press.

Meyers, Christopher. 2003. A defense of the philosopher-ethicist as moral expert. *Journal of Clinical Ethics* 14 (4): 259–269.

Mignolo, Walter. 2012. Decolonizing western epistemology/building decolonial epistemologies. In *Decolonizing epistemologies: Latina/o theology and philosophy*, ed. Isasi-Diaz and Eduardo Mendieta, 19–43. New York: Fordham University Press.

Moulton, Carol-anne, Glen Regehr, Maria Mylopoulos, and Helen MacRae. 2007. Academic Medicine. *Supplement* 82 (10): S109–S116.

Nagel, Thomas. 1986. *The view from nowhere*. Oxford: Oxford University Press.

Potter, Nancy Nyquist. 2003. Moral tourists and world-travelers. Some epistemological issues in understanding patients' worlds. *Philosophy, Psychiatry, and Psychology* 10 (3): 209–223.

———. 2016. *The virtue of defiance and psychiatric engagement*. Oxford: Oxford University Press.

Potter, Nancy. 2018. Reform and revolution in the context of critical psychiatry and service-user/survivor movements. *Szasz: Appraisal of his legacy*. Ed. Chaitanya Haldipur. Oxford: Oxford University Press, forthcoming 2018.

Potter, Nancy Nyquist, and El-Mallakh. forthcoming. The interface of ethics and psychiatry: A philosophical case consultation on psychiatric ethics on the ground. In *Philosophy, Psychiatry, and Psychology*.

Prinz, Jesse. 2004. *Gut reactions: A perceptual theory of the emotions*. Oxford: Oxford University Press.

Rasmussen, Lisa. 2016. Clinical ethics consultants are not 'ethics' experts—but they do have expertise. *Journal of Medicine and Philosophy* 41: 384–400.

Scheman, Naomi. 2011. *Shifting ground: Knowledge and reality, transgression and trustworthiness*. Oxford: Oxford University Press.

Sosa. 2009. *Reflective knowledge: Apt belief and reflective knowledge*. Vol. II. Oxford: Oxford University Press.

Tessman, Lisa. 2015. *Moral failure: On the impossible demands of morality*. Oxford: Oxford University Press.
Torcello, Lawrence. 2011. Sophism and moral agnosticism, or, how to tell a relativist from a pluralist. *The Pluralist* 6 (1): 87–108.
Törrönen, Jukka. 2001. The concept of social position in empirical social research. *Journal for the Theory of Social Behavior* 13 (3): 313–329.
Walker, Margaret Urban. 1993. Keeping moral space open: New images of ethics consulting. *Hastings Center Report* 23 (2): 33–40.
Waterworth, Jayne M. 2003. A philosophical analysis of hope. Palgrave-Macmillan.
Zabzebski, Linda. 1996. *Virtues of the mind: An inquiry into the nature of virtue and the ethical foundations of knowledge*. Cambridge: Cambridge University Press.

Chapter 10
The Nature of Ethics Expertise in Clinical Ethics and Implications for Training of Clinical Ethics Consultants

Johan Christiaan Bester

What exactly is expertise in clinical ethics? We know that experienced clinicians are supposed to have it to the degree that they may practice ethically and professionally. We know further that clinical ethicists are supposed to have it to the degree that they can help navigate uncertainties and ethical questions in clinical practice. But what does such expertise look like, and what does it consist of? When can one be thought of as an expert in clinical ethics, and what sort of training is needed to become one?

In this essay, I consider the nature of expertise in clinical ethics by way of philosophical reflection. I will present an extended argument that stipulates the components of clinical ethics expertise, the competencies of the expert derived from these components, forms of pseudo-expertise that represent threats to true expertise, and implications of these ideas for the training and practice of clinical ethics consultants. In essence, I describe and defend a theory of clinical ethics expertise that may be used to inform thinking about clinical ethics practice and the training and accreditation of clinical ethics consultants.

This essay will proceed in four parts. The first will consider the existence of ethical conflicts and ethical uncertainty in medicine. The second will examine the nature of ethical expertise required to manage the ethical uncertainties and conflicts found in clinical practice. The third will present possible threats to ethical expertise, ways of practice that obscure instead of clarify ethical reasoning and judgments. The fourth will distill these reflections into recommendations for training of clinical ethics experts and consultants as well as the accreditation of clinical ethics consultants.

J. C. Bester (✉)
University of Nevada, Las Vegas, Las Vegas, NV, USA
e-mail: johan.bester@unlv.edu

J. C. Watson, L. K. Guidry-Grimes (eds.), *Moral Expertise*, Philosophy and Medicine 129, https://doi.org/10.1007/978-3-319-92759-6_10

10.1 Ethical Conflicts and Ethical Uncertainty in Medicine

10.1.1 The Moral Foundations of Medical Practice

Medicine is in its essence a moral practice, with an integral ethical code and a central set of values. I take it that the grounding justification of medicine is found in its moral commitment towards the wellbeing of individual patients and communities of patients. Medicine exists to help and serve the sick by bringing comfort and healing, the well by maintaining good health and preventing illness. This makes medicine essentially a moral endeavor, with its goal to benefit patients and communities. This point has been argued by different writers in different ways.

Leon Kass (1983) argues that medicine is inherently a moral profession in view of the fact that the practitioner professes her adherence and commitment to a moral way of life that has its focus and endpoint in a moral good. Edmund Pellegrino (1990) argues that medicine is a moral community with its practitioners committed to a moral way of life centered around serving the health and well-being of their patients.[1] Pellegrino's moral community is composed of members who are bound to one another by a shared a set of ethical commitments and an overarching purpose beyond mere self-interest. In the case of medicine, this purpose is to serve the health of patients and society, placing the well-being of patients and the best interests of the sick as the foundation of medicine. Pellegrino views these moral commitments as inherent and central to the practice of medicine, and it is not possible to conceive of medicine without it. Some authors have cited the Hippocratic tradition and the Hippocratic Oath to demonstrate that medicine was founded on moral considerations and always had central ethical commitments (Jotterand 2005).[2] The Oath of Hippocrates reads like a document in medical ethics, an account of the moral commitments of medicine to benefit the sick while refraining from abuses of power. Although the values of medicine continue to be reviewed, reinterpreted and revised, the point remains that medicine has since its inception been rooted in its obligations towards patients and communities of patients and its mission to help those who need it.

Medicine is practiced in the context of a clinical relationship – a professional relationship that exists between practitioners and patients. An interesting feature of this relationship is that there are inherent asymmetries and power differentials pres-

[1] Pellegrino argues in many of his works that medicine is a moral endeavor and its practitioners a moral community, committing the practitioner to a set of virtues. As an example, see: Pellegrino, Edmund D. The Medical Profession as a Moral Community. *Bulletin of the New York Academy of Medicine* 1990;66(3): 221–232.

[2] For an analysis of this idea, the relationship between contemporary ethics and the Hippocratic tradition, and an argument for the need for a philosophical exploration of the values internal to medicine, see: Jotterand, Fabrice. The Hippocractic Oath and Contemporary Medicine: Dialectic between Past Ideals and Present Reality? *Journal of Medicine and Philosophy* 2005;30(1): 107–128.

ent. The patient has many interests at stake, which are completely dependent on the conduct of the physician. Patients make themselves vulnerable, decide to trust the physician with intimate personal details and personal vulnerabilities, and make themselves dependent on the conduct of the physician. The physician's medical knowledge and skills, as well as the position of trust afforded to the physician, gives the physician a measure of power over the interests of the patient. This means that physicians can act in ways that compromise the goals of medicine; through their behavior they can harm patients, fail to act in patients' best interests, or compromise patients' trust. Ethical behavior that seeks to protect the interests and wellbeing of patients and the integrity of the clinical relationship is therefore a necessary feature of medical practice. The goals of medicine presuppose ethical behavior in its practitioners. Given the kind of thing medicine is, the nature of the clinical relationship with its inherent power, knowledge, and interest differentials, and the basis of medicine being the moral call to help and to serve, it follows that ethical considerations are central and foundational to the practice of medicine.

Thus, I think of medicine as inherently a moral practice, with an integral ethical code and set of values interwoven into the practice of medicine. However, things are not always straightforward. Given the complex nature of medical situations and the ethical considerations involved, difficulties can arise in the form of ethical conflicts, including value conflicts or dilemmas, and ethical uncertainty.

10.1.2 Ethical Conflicts, Ethical Uncertainty, and Ethical Dilemmas

Ethical values, ethical commitments, and ethical obligations can be (and often are) in conflict with one another.[3] Those who practice medicine may frequently encounter such moral conflicts in practice. Consider as examples the following. The physician is committed to being honest with her patient, but also does not want the patient to give up hope and sink back into major depression in the face of a devastating diagnosis. The physician feels compelled to tell the truth and give the patient an honest diagnosis, but simultaneously feels she needs to somewhat bend the truth in order to avoid the patient's giving up hope. As another example, the physician feels she has to provide the benefit of life-saving treatment, but her competent patient refuses a life-saving treatment. Her perceived obligation of having to save the patient's life and preventing harm to the patient conflict with her perceived obligation to respect the self-rule of the individual with regards to healthcare decisions.

[3] Although I take this as obvious, it is worth noting that some may see this as a controversial claim. For example, an act utilitarian may say that although there seems to be a moral conflict, once we've weighed all the relevant consequences and moral content, there is actually only one right thing to do, and that the appearance of moral conflict is illusory.

In such ethical conflicts, competing ethical obligations, values, or directives may guide us to different (and perhaps mutually exclusive) courses of action. Sometimes such conflicts are resolvable. For example, some ethical approaches conceive of our obligations as being *prima facie*, often in conflict with one another, and requiring resolution. Resolving the conflict requires weighing the different competing ethical obligations against one another to arrive at one's ethical duty *all things considered*, or one's *actual duty* (Beauchamp and Childress 2013; Ross 1930).

Resolving such conflicts may be possible in some cases through weighing and deliberation. But sometimes conflicts cannot be satisfactorily resolved. Take as an example the various well-known trolley thought experiments, with a train trolley barreling down the track upon which five people are walking. You can save the life of these five by pushing a button that diverts the trolley to the other track, but there are three people on the other track. You either must push the button or not. If you push the button, three people die. If you do not push the button, five people die. Here we have a truly intractable dilemma, with competing obligations creating an intractable situation. It is no use here to feel one can arrive at a satisfactory moral resolution after weighing these duties against each other; both choices are bad, and whatever one does one would have acted contrary to one's duty. Yet, one must act and must therefore try and make the best of the situation.

Let us expand on my second medical example above. Consider a physician confronted with a 35-year-old patient, with an active and flourishing life, who develops an acute abdominal condition that can be corrected through surgery. The surgery is fairly straight-forward, is expected to be low risk, and has a high chance of providing a cure. Without surgery, the patient is likely to die. The patient has no desire to die, and values her life. But she absolutely refuses the surgery because she has an objection to the practice of surgery, and she also feels that she cannot bear to live through the post-operative recovery period in the Intensive Care Unit. On the one hand, the physician may recognize duties to provide benefits, to heal, to remove suffering and disease. Consequently, the physician may feel she has a duty to provide life-saving treatment, remove harms, and restore the patient to full health and function. On the other hand, the physician also has a duty to respect her patient's wishes and values. But to compound this conflict, the patient expresses two different sets of values. She values her life and does not wish death; she wishes to be well. At the same time, she finds surgery and the process of recovery repugnant. Here, then, is a complex dilemma where duties to respect patient wishes conflict with one another, as well as with duties to provide beneficial treatments and remove harms. However the physician acts, something a doctor is obligated to do is not done. This does not lend itself to easy resolution by saying what our actual duty is, and then feeling justified that one is doing the right thing. It is clear that something of moral significance is being lost whichever way one acts. This, of course, does not mean that all ways of action are equal – we may well have to work hard to identify the option that is least repugnant or leads to the least amount of moral loss.

Rosalind Hursthouse (1999) distinguishes between tragic moral dilemmas and non-tragic moral dilemmas. In the former, we have what I have described as a true dilemma. Whatever we do, we are left with the failure to discharge some obligation that we are duty-bound to discharge. We are left with a tragic choice; whichever way we decide we leave something morally compelling undone. This leaves us with moral regret and moral loss, something called a moral remainder. This remainder, says Hursthouse, places on us further obligations to respond, such as an apology or some effort to mitigate the moral loss as best we can. Even when we recognize we only have bad options, we should strive for the least bad option and be mindful of how we should respond to the moral remainder. To continue our second example, let us imagine the physician decides to withhold life-saving treatment and allow the patient to die. The moral loss here is great and tragic; the preventable loss of life is searing to a physician. Even if one presents a justification that it is the best among the available decisions in the circumstances, there is moral loss here. The moral remainder here would require steps such as acknowledging the moral loss, ensuring the patient is comfortable in her last moments, and providing emotional support to the patient and her family. In contrast to such tragic dilemmas, non-tragic moral dilemmas may admit to a resolution, and do not necessarily leave us with the same sense of moral loss and moral regret. With resolvable dilemmas, working through the dilemma to arrive at a satisfactory and ethically supportable resolution is called for.

Another way of thinking about ethical conflicts is in terms of value conflicts. In work done by the American Society for Bioethics and Humanities, for example, the goal of ethics consultation is stated as to respond to and/or resolve value uncertainties and conflicts that occur in healthcare (ASBH 2011).[4] They conceive of values as being diverse normative components of relevance to health care, which can have many different sources such as health law, professional obligations, morals and ethics, and different conceptions of the good life. Value conflicts may often arise and can drive uncertainty as to how to proceed.

Thus, medical practitioners may face a variety of ethical conflicts, which may be resolvable or may not be resolvable. These may include tragic and non-tragic ethical dilemmas, and value conflicts. In all cases, those facing ethical conflicts or dilemmas can experience uncertainty as to how to act or how to resolve the ethical conflict (Hurst et al. 2007; Kälvermark et al. 2004). Uncertainty can be introduced at either end of the process of working through a conflict or dilemma. First, it may not be clear what the underlying values or obligations are that drive the conflict or uncertainty. This refers to the first step of the process, where one perceives the various normative components involved in the case and how they conflict with one another or drive uncertainty. In this instance, those who have to make a decision may feel uncomfortable or uncertain about what to do in a given situation without knowing

[4] See the discussion on value conflicts and uncertainty in: ASBH Core Competency Task Force. 2011. *Core Competencies for Healthcare Ethics Consultation.* Chicago: American Society for Bioethics and Humanities (p. 2,3).

exactly which competing values or obligations drive the uncertainty. Second, it may not be clear how to resolve the identified conflict. Here, persons who have to act know what the underlying conflict is, know which values and duties drive the uncertainty, but they are not sure how to choose between different competing obligations and values to resolve the conflict, or how to decide which is the best action choice in a dilemma. This refers to the second part of the process, using reasons to assign relative weight to conflicting obligations and values in order to arrive at an ethically supportable way forward in the face of ethical conflict. Finally, in true tragic dilemmas, it may be unclear to physicians how to state the conflicting values and obligations, or how to proceed in a situation where moral loss is unavoidable. In such a situation, it is important to be clear about what is at stake, what drives the conflict, and what may be the least bad decision among the available decisions.

As I will show in the next section, these considerations form the foundation of the ethical expertise of clinical ethics experts. They can perceive and state the conflicting duties well, and they can weigh conflicting duties well, providing ethical judgments supported by justifying reasons. I will illustrate this by reflecting on the type of expertise to which clinical ethics consultants lay claim to respond to ethical uncertainty or conflicts.

10.2 The Nature of Ethics Expertise in Clinical Ethics

10.2.1 Two Components of Ethics Expertise in Clinical Ethics

Let us reflect on the expertise of clinical ethics consultants to identify what is meant by ethics expertise in clinical ethics. While I use this as spring board to reflect on the different components of ethics expertise, I by no means think that only clinical ethics consultants are ethics experts. I trust (and hope) that many experienced clinicians, educators, and leaders in clinical medicine have the kind of ethics expertise I unpack here. But reflecting on the role of the clinical ethics consultant in identifying and responding to ethical uncertainty in clinical situations is meant to help identify those components that underlie ethics expertise.

Ethics consultation in healthcare aims to help patients, family members, and practitioners resolve or work through ethical questions and uncertainties that may arise during medical care (Tarzian et al. 2013, p. 3–13). Let's unpack this further. Healthcare ethics consultation has the aim of responding to and resolving ethical questions, ethical uncertainty, and ethical conflicts (ASBH 2011, p. 2). Clinical ethics consultants, therefore, should have expertise that can help with situations in which ethical uncertainty exist: to make clear what is unclear, to facilitate resolution where possible, to find the best way forward when resolution is not possible. That is of course not the only expertise that clinical ethics consultants have. Some have claimed that clinical ethics consultants are also mediators or conflict managers as in the sense of resolving interpersonal conflict and facilitating communication (Dubler and Liebman 2011; Fiester 2012). Clinical ethics consultants should therefore also be good communicators, and have expertise in facilitating interpersonal conflicts. While that is no doubt important, central to clinical ethics consultation remains the "identification, analysis, and resolution" of ethical questions, uncertainty, and conflict (ASBH 2011, p. 3). And in my view, this is what sets the ethics expert apart from just the expert communicator or interpersonal conflict resolver. This is what makes one an ethics consultant rather than just a difficult-situation consultant. This unique ethics expertise is the ability to identify and clarify (and where possible resolve) ethical uncertainties and ethical conflicts.

Thus, experts in clinical ethics should have mastery over a set of skills and a body of knowledge that allows the clarification and resolution of ethical conflicts and ethical uncertainty in clinical medicine. As I've argued before, those who are uncertain may be either unable to clearly identify and state the duties, actual duties, or values that impinge on the situation or they may be unable to proceed with weighing and deliberation to decide on satisfactory next steps. The expertise of the ethics expert therefore is to identify and clearly state the various obligations or values involved in the situation and to provide assistance and guidance in how to balance and weigh these obligations. The ethics expert can also provide recommendations with justifying reasons as to ethically supportable courses of action in the face of such conflicts or dilemmas.

Ethical conflicts are of central importance here. If uncertainty is merely because of lack of knowledge of straight-forward ethical obligations or established norms, these are presumably easily corrected. But when it comes to complex value-laden situations, ethical conflicts, messy situations, or tragic dilemmas, the expertise of the ethics consultant is really of value, and it is in these situations that ethics expertise comes to the fore. The ethics expert can identify the values and obligations driving the conflict and assist with a reasoned deliberation in weighing the conflicting values and obligations.

In short, the ethics expertise of clinical ethics experts is that they can clarify and analyze ethical conflicts and provide an ethically supportable way forward for those experiencing such conflict. Such expertise must have two components:

(a) Identifying and clarifying the underlying ethical conflict. This involves moving from discomfort and uncertainty to a clear statement of which underlying values and obligations drive the conflict, what the nature of the conflict is, and what is ethically at stake. After this process, it is clear which values and obligations drive conflict and uncertainty, and what has to be weighed in order to arrive at ethically supportable next steps.
(b) Provide guidance as to ethically supportable actions in the face of ethical conflicts or uncertainty, supported by appropriate justifying reasons. This means that ethics expertise involves weighing different available actions in the light of the conflicting obligations/values to provide recommendations with justifications. Reasons should be provided for how one action better preserves value or better satisfies conflicting obligations than another, or why one obligation or set of obligations or values should be preferred over another. This also involves giving attention to moral remainders, those obligations incurred because of duties that cannot be discharged in a dilemma.

My component (b) may be viewed by some readers as controversial, particularly those who feel uncomfortable with the implication that clinical ethics consultants could provide recommendations to clinicians. In my view, and in my theory of ethics expertise, it is pivotal that an ethics expert can provide an ethically supportable way forward in the face of ethical conflicts, and even in tragic ethical dilemmas. While there may be some who feel that clinical ethics consultants should not provide recommendations, and only clarify values and normative components involved in conflict, this seems to me to only do half of the job of ethics expertise. If we then ask how the clinician will move forward in deciding what to do, one would have to say that the clinician would engage in the kind of weighing and deliberating that I envision here. This would imply that the clinician has the kind of ethics expertise that enables her to successfully do this. But note what this means: this acknowledges that a second step is necessary in order to remove ethical uncertainty and provide an ethically supportable way forward. It attributes to the clinician mastery over the components of ethics expertise. I have no objection to that, if the clinician truly has the ethics expertise to successfully engage in weighing and deliberation to manage the uncertainty. In such a situation, one would also expect from the clinician to provide supporting reasons for her judgments. However, if the clinician lacks the expertise necessary, guidance from someone with ethics expertise seems called for. Thus, whether one agrees that a clinical ethics consultant should provide a set of recommendations or not, at the very least one has to see that a second component of ethics expertise after the initial identification of normative content is the weighing of such content and arriving at ethically supportable actions with accompanying justifying reasons.

Thus, I view ethics expertise as consisting of two components. Component (a) I will call *analysis* and component (b) *weighing*. Each of these components presupposes a set of competencies mastered by the ethics expert.

10.2.2 Component (a): Analysis and its Relevant Competencies

To engage in analysis (a), one must have the ability to recognize or intuit the various moral components that impinge on a given situation in clinical practice and provide descriptions and explanations of said moral components that make the conflicting normative aspects of the case clear to others. But this, on reflection, requires at least two competencies.

One is a practical skill, a practical moral judgment, which can be described as a kind of praxis or practical knowledge. W. D. Ross (1930 p. 29–30) argues that the *prima facie* duties present in a situation are self-evident in the way that mathematical axioms are self-evident to practitioners in mathematics. However, one needs "sufficient mental maturity" to understand and see their truth (Ross 1930, p. 29). It is similar to the study of mathematics: the capacity of the student to see the validity of mathematical axioms must be practiced and sharpened; it is a practical judgment that must be honed. Just like the practitioner of mathematics must be taught how to recognize and apply self-evident axioms, so the ethics expert must be trained and must develop a type of moral judgment that allows her to see *prima facie* duties present in clinical situations.

Kathryn Hunter (1996) argues that moral knowing within clinical medicine is a kind of practical knowledge that is akin to clinical judgment. Clinical judgment is developed through exposure to clinical cases, practical and theoretical. Clinicians build up a store of practical case experience that inform pattern recognition and practical judgment. Clinical judgment is a kind of practical judgment, based on integrating the patient's narrative history and symptoms with the expert's knowledge of prior cases and theoretical medical knowledge into a case statement. This case statement makes clear what is at stake, what the essential points are, and what the issues are the practitioner has to address in order to proceed with clinical management. Moral judgment, Hunter argues, is inextricable from this and develops in the same way. Moral judgment in the clinical sphere is a practical judgment that presupposes experience with a set of cases. Such case experience affords the practitioner the opportunity to sharpen her judgment through reflection, inculcation of habits, pattern recognition, and familiarity with the underlying nature of clinical medicine and clinical cases with its internal moral structure. Through experience, Hunter argues, the ability of the expert to recognize the moral components of a case is honed, and the ability to pick out what is morally relevant from a narrative account is developed, in a way that is like the development of clinical judgment. Whereas clinical judgment focuses on the diagnostic and treatment components of a case, moral judgment is concerned with the moral components and various values internal to the case. Both are methods of practical knowing, both are developed through case experience and reflection, and both scrutinize the narrative structure and facts found within a case. They are somewhat related, but have different ends in mind. One seeks to bring diagnosis and offer treatment options. The other seeks to identify the different values and moral components involved, rearranging them and framing them to the point where it is clear what is morally at stake within the case, and what questions the practitioner must answer in the weighing stage. Thus, we can see that clinical judgment and practical moral judgment develop in similar ways and work together to help the practitioner determine a course of action: clinical judgment illuminates what can be done, moral judgment determines what should be done.

In the case of the clinical ethics expert, this practical judgment or praxis includes knowledge of things such as medical practice, different clinical roles, and the clinical environment. The expert has to be familiar with the clinical and moral structure of medicine, as well as having experience with case narratives, so that she can see the relevant facts and moral content as it is contained in the case narrative presented by patients and clinicians. It requires the ability to recognize patterns and construct "ethics cases" from different strands of information within specific clinical situations. As with the ability to understand and manipulate mathematical axioms, or the ability to understand and manipulate symptoms and clinical signs to reach a diagnosis and construct a clinical case, this competency is developed and constructed through experience, education, reflection, and exposure.

The second competency is familiarity with a body of theoretical ethics knowledge. In order to name and recognize ethical concepts and values, the expert needs to have mastery of ethical language, ethical theories, *prima facie* duties of different kinds, and the process of framing and presenting ethical dilemmas. This is a theoretical, building-block type of knowledge that forms the knowledge basis of ethics expertise. Once again, this is not dissimilar to other facets of the practice of medicine. To perform surgery, the surgeon requires theoretical knowledge of relevant anatomy, physiology, and therapeutic techniques. This is then applied to the practical situation and employed through the practical judgment already discussed, together with case and practical experience in the provision of patient care. Similarly, the ethics expert needs to know the theoretical aspects and methods of bioethics as is found in the theoretical learning of philosophical clinical ethics. This includes knowledge of various ethical theories and methods of justification, differing approaches to bioethics and the merits of each, a systematic approach towards identifying and describing moral content within clinical situations and how to frame and communicate these in understandable and consistent ways.

10.2.3 Component (b): Weighing and its Relevant Competencies

Weighing is the process of evaluating the various normative components in the clinical situation, such as ethical obligations and values, considering the various action options available in the light of these normative components, and then deciding on or providing guidance on an ethically acceptable way forward. Such a way forward would usually seek to optimally preserve values against value loss and seek to discharge obligations in the best way possible. In cases where there is no true dilemma, or where weighing can lead to a resolution, this would mean providing guidance on the resolution to the conflict with a set of supporting reasons. In a true dilemma, where resolution is not possible without value loss or leaving some duty undone, it would mean providing reasons for acting on one obligation over another, while being mindful of and responsive to moral remainders created in such dilemmas. Whatever form the ethical conflict or uncertainty takes, this component typically means guidance is provided by the ethics expert in the form of recommendations with accompanying justifications which provide clarity as to an ethically supportable action or actions given the ethical conflict or uncertainty.

This component similarly requires two competencies. First, a measure of practical knowledge is required. The ethical consultant needs familiarity with the clinical environment, the roles of different stake-holders within the clinical situation, and what is actually practically workable in clinical situations. It includes the ability to problem solve and to take into account various practical considerations that are necessarily part of the clinical environment.

Second, a sound basis of theoretical ethics knowledge is required. This refers to knowledge about theory as well as methods of justification. When engaging in weighing, the clinical ethics consultant provides reasons for prioritizing one moral obligation over another, or preferring one action over another. Such reasons are presented as objective reasons. They are not based in the authority or the personality of the consultant, nor in the individual preferences or relationships of the consultant. Rather, valid reasons in the process of weighing are objective normative reasons that are meant to be convincing as grounds for justification irrespective of the personal preferences or authority of the ethics consultant. The ethics consultant therefore justifies recommendations through impersonal and objective criteria, using ethical norms. These objective criteria are the norms of medicine, the integral moral components of medicine. Thus, being an expert in clinical ethics necessarily involves the study of, interaction with, and deliberation around the objective moral norms of medicine. The clinical ethics expert must therefore have knowledge of ethical language, theories, methods of weighing, and objective norms. For these we can use the catch-all terms of philosophical ethics or theoretical ethics, a foundational body of knowledge required to engage in weighing.

If you think these two competencies seem the same and therefore repetitive, you would be right. Reflecting independently on the two components of ethics expertise, we can see that each component leads us to the same set of competencies. The point is, therefore, that both components of ethics expertise require similar sets of competencies.

10.2.4 Summary: The Competencies of the Ethics Expert

The expertise that allows the clinical ethics consultant to respond to and clarify ethical questions and uncertainty, then, has two components: (a) Analysis, the recognition of the moral components within a clinical case, and framing normative components that drive the uncertainty and conflict; and (b) Weighing, considering the various normative components and available actions in the clinical situation to arrive at an ethically supportable way forward supported by justifying reasons.

Each of these components require two competencies, which are similar to one another. (1) A practical moral judgment, which is developed through exposure to and reflection on the clinical environment and clinical cases, and (2) Theoretical ethical knowledge, which is a theoretical knowledge of ethical concepts, method, and objective norms within medicine, acquired through the study of theoretical ethics.

This can be represented by the following diagram.

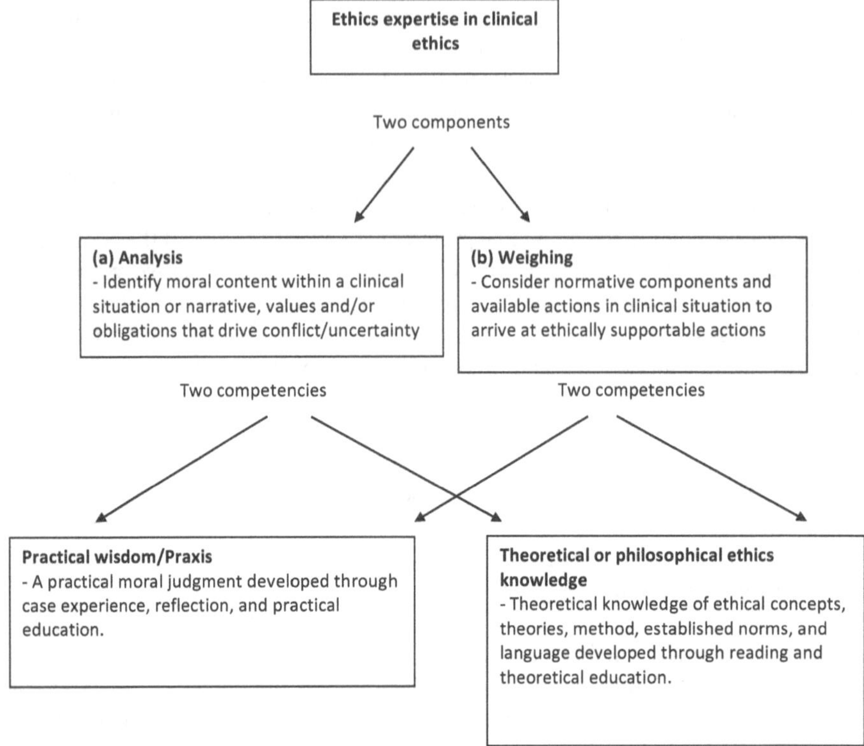

10.3 Threats to Claims of Ethical Expertise

Here I highlight two methods of practice within clinical ethics that may have the outside appearance of ethics expertise but are not fully consistent with the model of ethics expertise I have defended. These methods of practice undermine the consultant's claim to having ethics expertise and introduce a deficiency and unreliability in the ethical conclusions and methods of justification. Such conclusions are therefore flawed and do not represent fully coherent ethical justifications. The danger is that pseudo-expertise may mask the deficiencies present and portray the image of expertise while not in reality providing clarity and ethically justifiable courses of action in situations of uncertainty. Instead of illuminating, these methods of practice obscure ethical components and ethical reasoning within a case. They are based on the two competencies required to lay claim to ethics expertise, with each form of pseudo-expertise representing an absence of one of the components I have identified.

10.3.1 Lack of Practical Clinical Moral Judgment

Someone may purport to be an expert in clinical ethics due to their vast knowledge of theoretical ethical concepts, legal and moral norms, and philosophical method, and yet lack the practical moral judgment that is an important part of clinical ethics expertise. This may be due to an inadequate case load or clinical experience, leaving the practitioner unfamiliar with the practical aspects of clinical medicine and the clinical environment. The practitioner therefore will not be able to appreciate the context of the moral content fully and will not be able to distill out important nuances and relevant unspoken ethical aspects often present in clinical situations. She will also not always know what is workable and what is unworkable in the clinical environment, and may offer "solutions" that would leave clinicians exasperated at its lack of practical application. She will not have the case experience to know what the central issues are in a given narrative, missing the point of what it really is that clinicians are actually struggling with.

Such an ethics consultant is in danger of completely missing the point when confronted with an ethically complex clinical situation. She will not be able to identify all the moral content present, will miss important aspects, and will ultimately not advance recommendations that are relevant to clinical practitioners. She will often not actually clarify or resolve the conflict and uncertainty for those who have to engage in clinical decision-making. She is like the medical student who has memorized all the latest research on antibiotic resistance, has knowledge of microbiology and pharmacology, and read about all the diagnostic tests available, but cannot in the clinical situation elicit important points from a patient's history, identify important signs, create a case description, diagnose the infection correctly, and ultimately prescribe the correct antibiotic. The practical skill in employing theoretical knowledge in the context of a practical problem has not been developed.

Because the non-expert practitioner has mastery of the theoretical knowledge, philosophical method, ethical language, and some established norms, she may at times provide the correct answers. The clinical novice who has all the theoretical knowledge may also sometimes identify the correct antibiotic to prescribe. But there will be times when important clinical nuance is overlooked, and the point of the case is missed. In the case of clinical ethics, lack of true expertise may be obscured by the use of ethical language and methods drawn from theoretical knowledge. However, to the true expert (and clinicians), the defect in the pseudo-expert's advice should be apparent, as offered solutions would appear divorced from what is practically feasible. If there was an impression among some clinicians that ethics consultants are really not of much help or use, one has to wonder whether this non-expert mode of practice may be a contributing factor.

10.3.2 Lack of Ethics Knowledge and Method

The opposite is also possible. It is possible to have practical knowledge or practical judgment that may assist one in navigating some cases, although one may lack the necessary ethics theoretical knowledge to provide thorough analyses and engage fully in philosophical weighing. It certainly seems plausible that some experienced clinicians may have the ability to provide guidance in some dilemmas by drawing on their rich experience and practical judgment, arriving at ethically justifiable conclusions through a type of casuistry. There are difficulties here, though. It may not be possible for such practitioners to illuminate their reasoning and thinking, and they may not be able to provide justifications for their conclusions that are open to review and scrutiny. In essence, we have to take the recommendations at the practitioner's word, justified by their confidence they know this is the right thing to do given their experience. There is no systematic reasoning to display. Furthermore, the practitioner may not have the necessary knowledge and language to frame and present conflicts, justifications, and ethically supportable actions in ways that makes it clear to all involved parties what is ethically at stake. Although there is obvious value in the practical judgment developed through case experience, this has to be supplemented by knowledge of theoretical ethics and philosophical ethics method in order to lay claim to clinical ethics expertise.

Scott Yoder (1998) argues that an important aspect of ethics expertise is the ability to provide a reasoned justification for one's ethical judgments and actions. For Yoder, this is even more important than the question of whether there are objective moral facts or not. Yoder draws heavily on the work of Jan Crosthwaite (1995), who argues that ethics experts have ways of reasoning and justification that allows them to provide moral judgments supported by reasons. Crosthwaite identifies different kinds of philosophical abilities relevant to this expertise: the ability to analyze and clarify conceptual issues and problems, knowledge of ethics theories and methods, philosophical assumptions and the like, and a commitment to values, which means an ability and commitment to understanding and questioning, and reasoned inquiry. To Yoder and Crosthwaite, the provision of objective reasons in support of ethical judgments and recommendations are key components to ethics expertise. Expert judgments are, for them, moral judgments supported by reasons.

These ideas are further explored and clarified by Michael Cholbi (2007). Cholbi also argues that the provision of reasons to account for and justify moral conclusions is an essential component of ethics expertise. Because ethics is not aimed at some instrumental end, expertise cannot be evaluated by outcomes. Compare with the financial expert – even if we are not financial experts, we can evaluate whether someone else is by judging the outcome of their investments. But ethics, argues Cholbi, cannot be evaluated in the same way, because ethics expertise consists partly in choosing what proper ends and outcomes would be. The correct way of thinking of ethics expertise is therefore that moral judgments and recommendations conform to the proper shape, not whether the recommendations conform to a specific set of ends. Similarly to Yoder and Crosthwaite, Cholbi considers it an important component of ethics expertise that ethical judgments are supported by objective reasons. This is the proper shape of the ethical judgments of the ethics expert. But Cholbi goes further, and distinguishes between a merely performative expert and a descriptive expert. A performative expert can perhaps work through some ethical conflicts and problems through some personal moral insight, and provide moral judgments, but cannot provide justifications that make their reasoning apparent. Non-ethics experts should rightly be suspicious of such purported experts; after all, what have they to go on but trust in the integrity and authority of the purported expert? In contrast to this, Cholbi describes a descriptive expert, one who has true ethics expertise. This expert does not just issue correct judgments, but also provides clear and accessible reasons in support of these judgments. This allows non-experts to scrutinize such reasons, to learn from the process, and facilitates moral decision-making among non-experts in complex situations. Cholbi argues that ethics decisions have sizeable consequences, and non-experts cannot be expected to take seemingly ungrounded recommendations from purported experts purely based on the experts' authority. At least some justification is required in order to ground expert advice; and clear support for moral judgments is a feature of ethics expertise.

In my view, a specific issue to be mindful of in clinical ethics is the unsystematic use of ethical theories or approaches, a smorgasbord type approach to ethics. The practitioner starts off with a desired ethical conclusion in mind and then scours the ethical marketplace to see which ethical theory or method of justification can be used in order to reach her desired conclusion. For example, a consultant may appeal to care ethics, utilitarian, rights-based, and deontological approaches within the same argument as if they are equivalent to one another in an ad hoc fashion, without its being clear how they are related to one another. No systematic theory or method of justification is offered by the consultant who makes this error; whatever method of justification seems right for a particular section of the analysis and weighing processes are used as the consultant sees fit. Such a consultant has a superficial understanding of ethical theory and the philosophical underpinnings of different approaches to ethical decision-making without appreciating deep conflicts between these differing approaches. What we then have, in effect, is the consultant's desired conclusions offered with some loosely associated arguments based in different and often conflicting ethical paradigms. This introduces a subjective element to ethical reasoning. The desired values of the consultant are inserted into the clinical situation, and because there is no systematic or consistent reasoning offered, the reasoning cannot be reviewed or challenged. In effect, the "ethics expert" engages in a disguised form of value imposition, and she may even be blind to this herself, convinced that she is practicing legitimate ethical reasoning. Furthermore, it introduces a relativism into ethical discourse, where one can find any ethical theory or argument to justify any fancied position. The ethical consultant that practices in this way severely undermines the claim to having expertise in an objective body of knowledge or the expertise to generate objective reasons supporting recommendations that are open to review and scrutiny. Instead of clarifying ethical reasoning, this mode of ethical practice obscures it.

10.4 Implications for Training and Accreditation of Ethics Consultants

10.4.1 Training Ethics Consultants

Since there are two components to ethics expertise, each requiring a robust development of two related competencies, it follows that the education of clinical ethics consultants should develop both competencies.

To develop practical moral judgement, it is necessary to have an experience-based component to training, immersing the student in a clinical environment together with the optimal blend of supervision, mentorship, and graduated responsibility that allows for reflection on cases within the clinical environment. This may combine observation of practice, engagement in practice, and mentored reflection on consultation experiences. This is nothing other than a type of apprenticeship, a form of on-the-job training. Indeed, this type of training is becoming ever more available within the United States. It would be important then, given that such a fellowship or apprenticeship aims at developing identifiable competencies and skills, that some form of standardization among fellowship programs takes place, ensuring the right blend of case experience and mentored reflection in order to develop the practical skill.

There is also the need for a type of traditional, theoretical or philosophical ethics education. The format is less important – seminars, lectures, assigned readings, discussions – these may all suffice. What is important is that the student absorbs a body of ethical knowledge, which includes core theoretical ethical knowledge, knowledge of important concepts and norms within medicine, and knowledge of philosophical ethical methods. These are critically important skills in systematic argument, framing a dilemma and conflicting ethical norms, and the process of weighing.

Training for ethics experts, then, looks quite similar to the model employed in general medical training: education in basic core knowledge combined with a practical apprenticeship such as residency or fellowship.

10.4.2 Standardization and Quality Attestation in Clinical Ethics Consultation

I would suggest the implementation of some mechanism to ensure standardization and sufficient quality in ethics consultation. This would include some way in which it would be possible to determine who is an ethics expert. Accreditation or credentialing of ethics consultants has been considered as a way to do this. For example, Martin Smith and colleagues suggested a four-step certification process for clinical ethics consultants (Smith et al. 2010), and a variety of other authors supported the idea of credentialing or accreditation in some fashion (Acres et al. 2012; Dubler et al. 2009). But there are barriers to such processes, as pointed out by Eric Kodish and colleagues (Kodish et al. 2013). For one thing, it is not always clear what the standards should be that clinical ethics consultants should be held to, or how to adequately measure performance. Another concern is that the multidisciplinary nature of ethics consultation would be adversely affected by an accreditation process. It is also not clear how training requirements would be standardized, given the multidisciplinary nature of the field, and how or by whom ethicists who do not conform to standards would be held accountable. In their work on behalf of the American Society for Bioethics and Humanities, Kodish et al. (2013) advance the idea of quality attestation for clinical ethics consultants through a peer-review process instead of formal accreditation.

The matter is evidently quite controversial. Be that as it may, I would suggest some sort of process is required to ensure quality in ethics consultation and practice in accordance with ethics expertise competencies. Such a process would ensure that consultants have the required competencies that make up ethics expertise, and that they do not practice according to the models that undermine ethics expertise. Of course, only those who have the necessary competencies of ethics expertise are really in a position to judge whether some individual also shares the competencies of ethics expertise. This is similar to the practice of clinical medicine; only those who are experts in family medicine, for example, can really judge whether an individual shares the expertise and practices of this specialized discipline accordingly. This means that the clinical expert has to be submitted to some form of accountability and peer review, where others who have ethics expertise evaluate whether the individual consultant is practicing in a way that is consistent with ethics expertise. Thus, at minimum, clinical ethics consultants should have some form of quality markers, peer review, and accountability.

Despite the difficulties and controversies surrounding certification, these considerations at minimum call for (a) some standardization or quality assurance in clinical ethics training, ensuring that consultants develop both sets of competencies required to exercise ethics expertise and (b) developing a process of regulation or standardization for the practice of ethics consultation, ensuring that consultants display the components and competencies of ethics expertise.

10.5 Conclusion

I've presented a theory of ethics expertise in clinical ethics based on an extended argument. In my view, ethics expertise consists of the ability to identify and analyze ethical conflicts and uncertainties, and to provide appropriate moral judgments as to the best way forward supported by objective reasons.The implications of this theory extend to the education of clinical ethics consultants and the practice of clinical ethics consultation, but also to other areas. Most notably, in my view it has implications for how we teach ethics to medical students and residents. I will not explore this in depth, but briefly – I would propose that the same considerations apply. Students need instruction in theoretical ethics method, and then a type of clinical ethics immersive experience that fosters the development of practical moral judgment. This could take various forms, and is in itself deserving of full consideration in an essay.

A further question that warrants deeper exploration is whether ethics experts themselves should also be moral exemplars; whether they should be the embodiment of moral expertise in how they conduct themselves (see Herdova, this volume, for an analysis of this issue). Related to this are questions over whether ethicists should develop professional traits and behaviors, and whether there is a set of virtues or character traits that make for a good clinical ethicist. These are questions worth exploring further, and which can provide expansion upon the theoretical base I've provided in this essay.

References

Acres, C.A., K. Prager, G.E. Hardart, and J.J. Fins. 2012. Credentialing the clinical ethics consultant: An academic medical center affirms professionalism and practice. *The Journal of Clinical Ethics* 23 (2): 156–164.

ASBH (American Society for Bioethics and Humanities) Core Competency Task Force. 2011. *Core competencies for healthcare ethics consultation*. Chicago: American Society for Bioethics and Humanities.

Beauchamp, T.L., and J.F. Childress. 2013. *Principles of biomedical ethics*. 7th ed. New York: Oxford University Press.

Cholbi, M. 2007. Moral expertise and the credentials problem. *Ethical Theory and Moral Practice* 10: 323–334.

Crosthwaite, J. 1995. Moral expertise: A problem in the professional ethics of professional ethicists. *Bioethics* 9 (5): 361–379.

Dubler, N.N., and C.B. Liebman. 2011. *Bioethics mediation: A guide to shaping shared solutions; Revised and expanded edition*. Nashville, TN: Vanderbilt University Press.

Dubler, N.N., M.P. Webber, D.M. Swiderski, and Faculty and the National Working Group for the Clinical Ethics Credentialing Project. 2009. Charting the future: Credentialing, privileging, quality, and evaluation in clinical ethics consultation. *The Hastings Center Report* 39 (6): 23–33.

Fiester, A. 2012. Mediation and advocacy. *American Journal of Bioethics* 12 (8): 10–20.

Hunter, K.M. 1996. Narrative, literature, and the clinical exercise of practical reason. *The Journal of Medicine and Philosophy* 21: 303–320.

Hurst, S.A., A. Perrier, R. Pegoraro, S. Reiter-Theil, R. Forde, A.-M. Slowther, E. Garrett-Mayer, and M. Danis. 2007. Ethical difficulties in clinical practice: Experiences of European doctors. *Journal of Medical Ethics* 33: 51–57.

Hursthouse, R. 1999. *On virtue ethics*. Oxford: Oxford University Press.

Jotterand, F. 2005. The Hippocractic oath and contemporary medicine: Dialectic between past ideals and present reality? *Journal of Medicine and Philosophy* 30 (1): 107–128.

Kälvermark, S., A.T. Höglund, M.G. Hansson, P. Westerholm, and B. Arnetz. 2004. Living with conflicts – Ethical dilemmas and moral distress in the healthcare dystem. *Social Science & Medicine* 58: 1075–1084.

Kass, L.R. 1983. Professing ethically: On the place of ethics in defining medicine. *Journal of the American Medical Association* 249: 1305–1310.

Kodish, E., J.J. Fins, C. Braddock III, F. Cohn, N.N. Dubler, M. Danis, A.R. Derse, R.A. Pearlman, M. Smith, A. Tarzian, S. Youngner, and M. Kuczewski. 2013. Quality attestation for clinical ethics consultants: A two-step model from the American Society for Bioethics and Humanities. *The Hastings Center Report* 43 (5): 26–36.

Pellegrino, E.D. 1990. The medical profession as a moral community. *Bulletin of the New York Academy of Medicine* 66 (3): 221–232.

Ross, W.D. 1930. *The right and the good.* Oxford University Press. (Reprinted copy: Indianapolis, IN: Hackett).

Smith, M.L., R.R. Sharp, K. Weise, and E. Kodish. 2010. Toward competency-based certification of clinical ethics consultants: A four-step process. *The Journal of Clinical Ethics* 21 (1): 14–22.

Tarzian, A.J., and the ASBH (American Society for Bioethics and Humanities) Core Competencies Update Task Force. 2013. Healthcare ethics consultation: An update on the core competencies and emerging standards from the American Society for Bioethics and Humanities' Core competencies update task force. *The American Journal of Bioethics* 13 (2): 3–13.

Yoder, S.C. 1998. The nature of ethical expertise. *The Hastings Center Report* 28 (6): 11–19.

Chapter 11
Moral Expertise in the Context of Clinical Ethics Consultation

Geert Craenen and Jeffrey Byrnes

11.1 How to Understand Clinical Ethics Expertise

There is little disagreement that the practice of clinical ethics consultation has grown, changed, and, indeed, advanced considerably in the last two decades. Yet, in many ways the theoretical underpinning of clinical ethics is still in it's early stages, with central concepts not yet fully defined, and even existing disputes where the battle lines are not clearly drawn. Thus, clinical ethical practice, by necessity, pushes forward, while the theoreticians strive to carve-out the conceptual foundations. To some of the exceedingly practical among us, the fact of this parallel march— with one pragmatic plodding practical group that marches ever onward and one reflective theoretical group that works the same ground over and over—demands that we ask if the second, theoretical, team is really of any use at all. Still, there are times when questions and disputes arise among practicing clinical ethicists that require them to turn to the theorists for better understandings of that very clinical ethical work. One such example of a practical dispute that requires input from the theoretical sphere comes from the recent debates about the need for, or the possibility of, a certification process for clinical ethicists. In order to have an established certification process, the discipline at large would have to be able explain exactly what knowledge a clinical ethicist should have, and exactly what skills she should master. In other words, we as a discipline would have to be able to provide an answer to the question, 'What does expertise in clinical ethics look like?' As if matters were not murky enough, this question in clinical ethics has overflown into the broader debate about whether ethical expertise is even possible.

G. Craenen (✉)
West Texas Veterans Affairs Health Care System, Big Spring, TX, USA
e-mail: geert.craenen@va.gov

J. Byrnes
Grand Valley State University, Allendale, MI, USA
e-mail: byrnesj@gvsu.edu

J. C. Watson, L. K. Guidry-Grimes (eds.), *Moral Expertise*, Philosophy and Medicine 129, https://doi.org/10.1007/978-3-319-92759-6_11

One initial line of response to that debate is pretty clear. Some would say that ethical expertise is impossible simply because, in a pluralistic society, it is self-referentially incoherent. For, as this line of thinking goes, to be an expert in ethics would mean to have specialized knowledge about which is the best ethical system, but this very knowledge would undermine one's ability to take seriously the notion of a pluralistic society, where multiple conceptions of the good are understood as plausible. Even if that may not be the last word on the issue, such an argument has considerable force. Nonetheless, in the domain of clinical ethics, many have tried to take different tack in order to defend the notion an expertise in ethics. Such a motivation probably arises out of an attempt to account for the phenomenology of learning about ethics. For frequently, someone who can recall the experience of her first ethics class remembers that it sure *felt* as though she can do something better than you could before. Then, when introducing ethics to attentive students, it is difficult not to recognize analogous experiences in them. And once again, when dealing with a case in the clinical consulting setting, one may well feel that he or she is grateful for that underlying ethics knowledge, even though no explicit discussion of it ever surfaces in the case.

Yet, even in light of these phenomena just what an introduction to ethics makes one *better at* is not immediately obvious. If that same introduction to ethics ventured to give us a dose of existentialism, then we will also understand that our own personal choice underlies any ethical justification that we might articulate for our actions. For, as Sartre explains to his student, once we understand that there are competing ethical systems, then any person who claims to have received advice from one particular theory, "would have decided beforehand the kind of advice he was to receive" (2007:PAGE). So, it would seem that what an introduction to ethics makes us *better at* is not employing some particular tool to resolve moral dilemmas automatically, as it were. Even so, understanding Sartre's point does not entirely eliminate the feeling of increased clarity and sense of empowerment, be it ever so small, that comes from learning something about ethics. Here our own experience is consistent with the phenomenology the clinic. There we see that untrained doctors, nurses, and patients can struggle to identify the salient parts of an ethically fraught situation, and thus can be inclined to report feeling disoriented, and frustrated by the fact that they are unable to articulate why a particular situation is troubling. And indeed, research has shown that being introduced to even basic ethical concepts and theories caused healthcare workers to feel empowered and more resilient in facing ethically complex situations. (Monteverde 2014; Auvinen et al. 2004) It is likely the case that this experience of empowerment from ethics education, observable in students, and born-out by the data, that motivates the continued search for an account of ethics expertise.

Of course, no one experience of ethics education is universal. In a play which itself focuses on questions of professional ethics, George Bernard Shaw wrote that when one learns something it, "always feels at first as if you had lost something." This sentiment is not difficult to understand. Imagine an intelligent young medical student who was raised in a society with a highly prescriptive imperial religious cult, and now is openly engaging introductory ethics education for the first time. It is not hard to imagine this student would report feeling quite unmoored by her education, for situations which she could previously navigate unreflectively, now open up and engage her intellect and concern in new ways. Indeed, one not be raised in an imperial cult to have this response. Anyone who has been taught that all complex ethical situations can be resolved, without remainder, by a wave of the hand, is likely to find ethics education disruptive. This response is the antithetical to the empowerment we noted above as a feature of the phenomenology and reflected in research. Are we then to understand this second kind of response to ethics education as a counterexample to the intuition that such education is empowering? I think that we should not.

The disruption experienced by the rigorous person is not an example of ethics education pushing the person in a different direction, rather it is pushing in the same direction, but starting from an earlier point on the journey. Just as in the history of science, unseating a comprehensive theory is disruptive when equally robust theory is ready to take its place. Yet, even theoretical uncertainty and openness is progress, when one's starting attitude is to cling to a falsifiable theory. Moreover, the setting of the clinic quickly reminds us that, in spite of all the accompanying feelings of security, unwavering confidence in one's unreflective intuitions can be very dangerous. In other words, the assurance of the unreflective intuitive clinical ethicist is not the kind of confidence that we would want to promote anyway. Thus, the feeling of being unmoored by an introduction to ethics may, for some, be a necessary first step toward another kind of empowerment that could come from internalizing tools available in that education.

As we said above, this feeling of empowerment serves to preserve the notion of ethical expertise in clinical ethics. Yet from this prospective, it is clear that the sense of ethical expertise available to us, and just what that ethical expert can offer in the clinic, is somewhat reduced. Whatever the remaining conception of ethical expertise can offer, is not the rapid, effortless, and complete resolution of ethical dilemmas. This could lead us to suggest that ethical expertise in the clinic should be thought of as a kind of formal or procedural expertise. While this is a promising first step in preserving the possibility expertise in clinical ethics, a purely formal account of expertise has elsewhere already been described as "thin gruel," because, "if this is all clinical ethics expertise amounts to, then what many would think is the core sense of expertise is not being claimed" (Wear 2005: 243).

There are several possible responses to this sobering point. One possible response is to attempt to thicken the gruel by elaborating and specifying the necessary procedural skills to be acquired through education and training. This is sensible, if we define expertise as simply a high level of knowledge and skills in a limited subject area, typically in a professional field (Steinkamp et al. 2008). To this can be added the notion that clinical ethics consultation also has a substantially performativity quality, which is also teachable (Weinstein 1994). This is a more robust account of ethics expertise in the clinical setting, but more could be said still. Another possible response to that challenge is to build upon the the account expertise as externally acquired knowledge and skill to include the qualities already internal the person, such as the capacity for empathy, the ability to acquire and apply knowledge, and the capacity for compassion. To be clear, the purpose of what follows is not to provide arguments from first principles for the existence of expertise in ethics. At least as it regards professional practice in clinical ethics, we believe that such expertise is at least conceivable. For now, we aim to examine the intrinsic qualifications of a potential expert in the performance of that professional practice of clinical ethics consultation.

We hold that expertise of the clinical ethics consultant springs from both the intrinsic quality of the consultant as a professional and from the extrinsic properties of the role. Individually the intrinsic and extrinsic qualities are insufficient to achieve expertise in performing clinical ethics consultations, but together they make up the ideal version of formal ethics expertise in the clinic. In what follows, we will examine the existing calls for external standards of skill and knowledge, then we will reflect upon the relevant intrinsic qualities, with an eye to sketching an expanded account of ethical expertise in ethics consultation. But first, let's turn our attention to the concrete current situation in order to see if there is any need for more expertise in clinical ethics.

11.2 The Current State of Affairs

Despite steady improvement, the current state of the field in clinical ethics consultation is far from ideal. Note that the fact that the United States has approximately 5000 hospitals, yet the ethics flagship organization ASBH has only a little over 2000 members, and its Clinical Ethics Consultation Affinity Group only 450. In a survey of 600 American hospitals of various sizes and structures, 81% had working ethics consultation services. While this seems like a solid start, further examination reveals something of a gloomy picture. The median number of consults addressed in the year prior to the survey was three, with 22% reporting no consults at all in the previous year. Only 16% of hospitals surveyed had an ethics consultation service supported by a paid ethicist, but 83% of respondents maintained that financial support for ethics consultation was sufficient. When survey respondents were given the opportunity to explain why they found the funding sufficient, some respondents said things like, "We're doing a good job, not spending money", "often when a case comes up with an ethical dilemma, it can be handled appropriately by the risk manager", or "we are supported adequately by our salaries so we do not mind." (All survey data comes from Fox et al. 2007.)

One take away from this is that, although many the world's leading ethics centers are associated with American medical centers and universities, by and large, clinical ethics consultation in most hospitals is still a minor-league endeavor, often undertaken by well-intentioned amateurs, and perhaps not by someone who meets the definition of an expert in clinical ethics. This data reveals that the still emerging field of clinical ethics consultation still has a lot of positions to fill, and thus that the need for a better account of what clinical ethics expertise looks like practical and substantial. Fortunately, several professional organizations have attempted to formulate just what this clinical ethics expertise looks like. Let's begin our account by considering two of those accounts.

11.3 Expertise per the Various Professional Societies

In the last 10 years both the UK Clinical Ethics Network (henceforth The Network) and the American Society for Bioethics and Humanities (henceforth ASBH) have published detailed statements on the qualifications that they deem necessary in order to "practice" clinical ethics. There are many similarities between the two. In their statement for The Network, Larcher, Slowther, and Watson divide the essential attributes of a clinical ethics consultant into three domains: skills, knowledge, and personal characteristics (Larcher et al. 2010). For our purposes here, we can treat that as The Network's outline of what clinical ethics expertise looks like. They emphasize the importance of knowledge, in keeping with the traditional definition of an expert as one with specialized knowledge. Still, the signigicance of the role played by skills allows their account to be largely procedural, so that no ethics expert would have to claim special knowledge about the best ethical system. Finally, they also give a nod to the importance of professional qualities, but they go on to distinguish the first two as requirements for a competent ethicist, while they describe the last of these as merely aspirational, saying that they should be pursued "as a long-term project in an analogous way to continuing professional development". (Larcher et al. 2010).

The ASBH takes a similar approach in its second edition of their *Core Competencies for Healthcare Ethics Consultation.* The ASBH divides the relevant domains into knowledge and skills; skills are then subdivided into assessment skills, procedural skills, and interpersonal skills. At a glance, this division seems largely analogous to the Brisitsh divison. Both emphasize knowledge and skills, and one might see the ASBH's 'interpersonal skills' as a version of the British 'personal characteris'. A closer examination reveals that there is a substantial difference.

While the British version emphasizes what we might call maters of character, or even virtues, such as tolerance, honesty, courage, prudence, and integrity, the interpersonal skills in the American version are more recognizable as discriptions of management techniques, such as 'listens well,' 'communicates empathy,' and 'enables communication between other parties.' Even in a passing nod to 'professional obligations,' the ASBH endorses virtues only instrumentally saying that, "consultants who have developed certain attributes, such as tolerance, patience, compassion, courage, prudence, humility, and integrity, are more likely to be effective" (Tarzian 2013).

Some additional emphasis on skill and education has emerged in the second edition of the ASBH's *Core Competencies.* The first edition, if anything, put such emaphsis on the personal attributes that it resulted in a nearly heroic view of the ethics consultant. For instance the first edition stressed the need for Healthcare Ethics Consultants (HCECs) to be amateurs. Not amateurs in the sense of ill-informed, unskilled dabblers. Instead, this call for amateurs is probably best understood as an idealization of the ethics consultant, in the same way that insisting that Olympic athletes be amateurs is an idealization of those athletes. No one who makes such a call for the Olympics desires to see less skilled athletes compete, rather, he or she simply wishes the athletes to be untainted by outside interests or motivation, such as a search for money or status. The second edition, on the other hand, shifted the emphasis by greatly expanding the table of necessary skills, and even creating a separate volume on educational expectations.

11.4 A Personalist View of Professionalism

In a field that is largely still emerging, both the UK and US bodies provide invaluable information about the important knowledge and skills needed for a kind of expertise as a clinical ethicist. Yet, there are important ways in which being an expert in a profession, indeed ways in which beign a 'professional' at all, goes beyond a list of quantifiable accomplishments. One such version of professionalism, or expertise in a profession, comes from the personalist view of what is professional.

By 'personalist' we mean the view that all the broad and diverse values that exist, all the value that there is, is derived from persons, and thus persons are the center of our concern. In this account, the *profess*ional, is, in the literal sense of the word, is a person who has made a *profess*ion, a person who has professed a commitment or taken an oath. Anyone who performs a service for a contracted fee is in the *employ* of the payer, he or she is an *employ*ee. In this use of the term, an employee has not necessarily made a profession, and thus is not necessarily a professional.

Indeed, in this view of professing there are many honorable careers and occupations which are not professions. Plumbing is a valuable skill, an important occupation, an essential craft, but it is not a profession simply because it does not demand practitioners to profess anything. Here we follow Edmund and Alice Pellegrino who explain that, "[b]eing a member of the profession is more than a mastery of a *techne*, in the Greek sense of 'an art or craft.' Therefore, it is also a way of life to which one publicly and voluntarily commits himself" (Pellegrino and Pellegrino 1988). A plumber may be very good at her job; she may dress and communicate in a manner we have come to describe as professional, and possess excellent skills, but that still only makes her a capable practitioner of a craft. She can choose her business, make the contracts that she wants, on the terms that she wants. What binds her in doing her work are the terms of that contract. Professionals, on the other hand, are bound by the very oath which they profess. Contracts are enforced by law, and interpreted by lawyers, whereas oaths are subject to a moral code.

That need for that moral code comes not merely from the fact of the professional having taking an oath, but also from those for whom the oath is taken. A professional's oath aims to commit her to a relation to the other persons for whom her work is intended. The character of that relation to others is best understood as one of service. In this view, a professional is someone who has professed herself willing and able to serve. By this definition, the list of professions is brief, including soldiers, first responders, doctors, nurses and priests make up most of it. For a variety of reasons around sustainability, these professionals need to be paid for their work, but imagine what it would be like if the only thing binding them was a contract.

Consider the way in which our expectations concerning interactions with soldiers, police, or physicians would change if the only governing standard their work was adherence to a contract. One such change would revolve around the extent of commitment. Contractual relationships require us to anticipate as many possible contingencies as possible, because the contracted is only bound to address that which is covered by the contract. One fear about dealing with contracted soldiers, police, or physicians would be that, the contracted person would have met the minimal necessary commitment, and thus fulfilled the contract, leaving us in the lurch when we most needed the support provided by their work. On the other hand, to take an oath to service, is a much more open-ended commitment.

Furthermore, an oath of service is a commitment to strive for the deep wellbeing of those served, and not to instrumentalise them with an eye toward meeting the letter of a contract. For instance, imagine a case in which some physicians have, to the best of their ability and in good faith, diagnosed a patient's condition and recommended surgery. Then, while preforming the recommended surgery, one of those phsyicans, under closer inspection, comes to understand that the patient's problem was more complex, or even wholly different, than originally understood. If bound only by a contract, then the surgeon would be permitted to complete only the procedure for which she was originally contracted. But, if bound by an oath of service, the surgon would be commited to any additional actions through which she could to bring about the patient's wellbeing.

Service, in the personalist sense, can be illustraited both in the ancient story of St. Martin, and today in the palliative care. St Martin, when still a young soldier serving in the Roman cavalry, comes upon a beggar suffering from cold in the street. St. Martin reaches for his sword and cuts his cloak in half, giving half to the beggar. Though wealthy, he does not buy the beggar clothes without real inconvenience to himself, but he cuts his own cloak in half. The beggar is now warmer, and St. Martin is now colder. His commentent to serve has lead him to take some of the beggar's suffering on himself. In modern medicine, a similar example can be seen in the palliative—the term itself derived from the Latin for cloak— care physician, who is often called upon to share a patient's or family's grief. The dedication to serving something beyond oneself, to respond to the appeal of another with sacrifice then becomes one necessary condition to becoming a true professional. The personalist view of a profession calls the professional to this level of service, and the success or failure in this endeavor is not a matter of contractual obligation, but a matter of character.

It is in this way that the personalist view of professionalism informs an account of clinical ethics expertise, that is, it emphasizes the intrinsic qualities of a professional ethicist—much more so than either The Network or ASBH lists of core competencies do. In the personalist view, a service oriented professional character is a necessary part of expertise in a clinical ethicist. Necessary, but not sufficient. As in any other field in medicine, a well-intentioned amateur is a very dangerous thing. A commitment to serve, to share the burden of another, and to adhere to a moral code is merely a foundation from which to develop expertise. Upon this feature one can add the other qualities identified by the professional organizations knowledge and the development of skills. Here we come full circle upon some of the intuitions and phenomenology that initiated our inquiry. If expertise is typically understood as highly specialized knowledge, but highly specialized ethical knowledge would include knowledge about which ethically system was correct, and if a strictly procedural account of ethics expertise seemed would be kind of thin gruel, then this tripart account of expertise in a clinical ethicist.

Ethical expertise takes on a special significance when we apply the term in the context of clinical ethics consultation. After all, the question of moral expertise remains largely academic until clinical ethics consultation attempts to translate ethical theory into recommendations about care decisions. In one way it is not difficult to understand why both The Network and the ASBH have taken the approach that they have. It is much easier for universities and hosptials to teach knowledge and skills thatn it is for them to develop, or even determine, something about a person's character. Yet, the mere fact that the personal character of an ethicist is not amenable to existing structures of education is not sufficient reason think that there is nothing that be done about it. Thus it is worth considering whether a better model exists to both determine and develop *each* of the parts of a clinical ethicsist's expertise. In what remains we offer some guidance first on how such a model of clinical ethics expertise could be incorporated into our exisiting education and training models, and second some thoughts on the way in which assessment could be undertaken.

11.5 Apprenticeship

Clinical ethics consultation has been defined as "a set of services provided by an individual or group in response to patients, families, surrogates, health care professionals, or other involved parties who seek to resolve uncertainty or conflict regarding value-laden concerns that emerge in health care." (ASBH 2011. Pg. 2) Understood in this way, clinical consulting is structurally similar to consulting services provided by the medical specalites, such as cardiology, oncology, and psychology.

These specialties long ago instituted professional practice standards and methods for evaluating quality of care and are recognized as expert consultants. In western medicine, at least, the formative model for medical clinical consultants has been standardized for well over a century as an institutionalized apprenticeship. (Also see Evan DeRenzo, this volume (Chap. 17), for an exploration of the meaning and importance of apprentice in clinical ethics consultation.)

Trained and accomplished medical professionals (MD, DO, MBBS, et cetera) enter into an apprentice-mentor arrangement within the bounds of institutionalized programs called residencies or assistantships. There, they accumulate knowledge specific to the field, acquire the relevant skills, and are given increasingly more responsibility as their progress merits. An evaluative process called specialty boards caps the experience, frequently consisting of separate assessments of knowledge and skills. An ongoing practice evaluation and recertification process continues after graduation, ensuring ongoing expertise in the chosen field of clinical consultancy. This model which has proven so successful in medicine could also be applied in the case of clinical ethics preparation. Potential clinical ethics consultants could undergo residency programs in order to allow members of the professon to guide and evaluate them on the three parts of clinical ethics expertise: knowledge, skill, and character.

11.6 Quality Attestation as a Mark of Expertise

Health care ethics consulting would then resemble other medical disciplines in the formative process of the expert clinical ethics consultant, and their accepted method for achieving expertise, with one exception: unlike those other disciplines, health care ethics consulting does not yet established methods for establishing competency. Yet both the ASBH and the UK Network have called for such methods and have proposed mechanisms to do just this.

As clinical ethics consultations recorded in the medical record have the potential to significantly help or harm a patient, it is imperative that the consultant possesses expertise in addressing moral issues. Other disciplines in medicine have long ago instituted professional practice standards and methods for evaluating quality of care or the practitioner's expertise involving initial board certification upon completion of the apprenticeship and requirements for periodic recertification. Clinical ethics consultation lags behind significantly in this regard. Fox et al. amongst others have clearly shown a wide disparity in the qualifications of those acting as clinical ethics consultants in the USA (Fox et al. 2007).

The ASBH has initiated a process of quality attestation, in an attempt to govern quality in clinical ethics consultation and transition the practice of clinical ethics consulting from the amateur to the professional level. As Lisa Rasmussen has stated, "[clinical ethics] consultants aiming to professionalize must articulate the specific set of skills, knowledge, or expertise possessed by clinical ethics consultants" (Rasmussen 2011). The ASBH report is the premier standard by which CEC expertise should be judged. ASBH is the foremost umbrella organization in the field, the report went to several drafts, and is currently in its second iteration, it has been circulated widely and extensively commented on by the foremost recognized experts in the field, and has matured into a cogently reasoned, imminently practical standard for judging moral expertise in clinical consultative ethics (Kodish et al. 2013; Tarzian 2013).

As it stands today, the process would involve portfolio review of cases stresses practical, rather than theoretical knowledge, respecting the diverse backgrounds represented in the clinical ethics consultation community and avoid biases. This is combined with an evaluation of formal education values the diverse theoretical knowledge an ideal CEC possesses. The process concludes with an oral examination evaluates, and to some extent simulates, the skills required of an ideal CEC, including discursive and analytical abilities, communicative and interpersonal skills, and the service orientation required of one who professes a willingness and ability to serve.

11.7 Conclusion

Certainly, it appears that the expertise of clinical ethics consultation is limited to the practical. There are indeed many moral theories, none of which enjoy primacy or hegemony or even overwhelming acceptance on a theoretical level (Wear 2005). Beauchamp and Childress may claim that many theories lead to the same result in the clinic, those of us who have attempted to use the Georgetown mantra in clinical ethics consultations know that while the so-called principles may assist in gaining a better understanding of the nature of a case, and serve as useful heuristics during discussion, they do not help in resolving it (Beauchamp and Childress 2001). But this precisely is our point. Theories belong in classrooms and seminars. In the clinical setting, the picture changes dramatically. Time for debate and exploration of theoretical nuances may be endless in a graduate seminar, but clinical case deliberation often comes with a deadline or a practical time limit. This is acceptable, since the purpose of clinical ethics deliberation is not to differentiate right from wrong, or to arrive at the ultimate truth, but to generate a recommendation which is deemed best among the available options. Furthermore, in our many combined years of clinical case consultation, we have never seen a case "decided" by choosing one theoretical approach over another. In other words, clinical ethics consultation is not intended to occupy some privileged position from whence to proclaim a course of action, but it is, at a minimum, a facultative process thereto.

11.8 Caveats

Once clinical ethics consultants become established as experts based on training and accreditation analogous to other medical consultants, the danger arises that expectations placed on clinical ethics consultants will be similar to those placed on other medical consultations, that a clinical ethics consultation will resolve the precipitating conflict, deliver a prescriptive judgement, issue a practical recommendation. However, while medical consultations do sometimes deliver a prescriptive analysis or a textbook recipe, many also conclude by offering a variety of options or deliver situational recommendations tailored to the specific patient and her circumstances, similar to an ideal ethics consultation. It remains important in all consultation processes to distinguish between an expert and an authority Fig. A clinical ethics consultant should be an expert in this process. That does not mean she should be regarded as an authority figure. An infectious disease consultant may be expected to order whatever antibiotics they deem most fitting, and a cardiothoracic consult may result in an immediate barrage of orders, tests, and even interventions. Such actions are integral to the role of those specific medical specialties. An ethics consultant however fills an advisory, explanatory, mediating, or even an advocacy role. She should neither be expected nor allowed to act in an authoritative fashion, when the situation does not call for it. Humility is a key virtue in any clinical consultant, one of the many reasons why we conceive of this process as a professional practice in the terms outlined above.

Acknowledgement The authors are gratefull to Dr. Stephen Wear, Ph.D. for germinating the idea of this chapter and guiding its early development.

References

American Society for Bioethics and Humanities. 2011. *Core competencies for healthcare ethics consultation* (2nd ed.), 385–401. Glenview, IL.

Auvinen, J., T. Suominen, H. Leino-Kilpi, and K. Helkema. 2004. The development of moral judgement during nursing education in Finland. *Nurse Education Today* 24: 538–546.

Beauchamp, T.L., and J.F. Childress. 2001. *Principles of biomedical ethics*. New York, NY: Oxford University Press.

Fox, E., S. Myers, and R.A. Pearlman. 2007. Ethics consultation in United States hospitals: A national survey. *The American Journal of Bioethics* 7 (2): 13–25.

Kodish, E., J.J. Fins, C. Braddock, F. Cohn, N.N. Dubler, M. Danis, and A. Tarzian. 2013. Quality attestation for clinical ethics consultants: A two-step model from the American Society for Bioethics and Humanities. *Hastings Center Report* 43 (5): 26–36.

Larcher, V., A.-M. Slowther, and A.R. Watson. 2010. Core competencies for clinical ethics committees. *Clinical Medicine* 10 (1): 30–33.

Monteverde, S. 2014. Undergraduate healthcare ethics education, moral resilience, and the role of ethical theories. *Nursing Ethics* 21 (4): 385–401.

Pellegrino, E.D., and A.A. Pellegrino. 1988. Humanism and ethics in roman medicine: Translation and commentary on a text of Scribonius Largus. *Literature and Medicine* 7 (1): 22–38.

Rasmussen, L.M. 2011. An ethics expertise for clinical ethics consultation. *The Journal of Law, Medicine & Ethics* 39 (4): 649–661.

Sartre, Jean-Paul. 2007. *Exisitetntialm as a humanism*. New Haven, CT: Yale University Press.

Steinkamp, N.L., B. Gordijn, and H.A. Ten Have. 2008. Debating ethical expertise. *Kennedy Institute of Ethics Journal* 18 (2): 173–192.

Tarzian, A.J., and ASBH Core Competencies Update Task Force, 1. 2013. Health care ethics consultation: An update on Core competencies and emerging standards from the American Society for Bioethics and Humanities' Core competencies update task force. *American Journal of Bioethics* 13 (2): 3–13.

Wear, S. 2005. Ethical expertise in the clinical setting. In *Ethics expertise*, 243–258. Dordrecht: Springer.

Weinstein, B.D. 1994. The possibility of ethical expertise. *Theoretical Medicine and Bioethics* 15 (1): 61–75.

Chapter 12
Are Hospital Ethicists Experts? Taking Ethical Expertise Seriously

David M. Adams

12.1 Introduction

Clinical ethics consultants (CECs) are now a common presence in many hospitals in the U.S. Somewhat less common is agreement on why they are there, what methods they should follow, and what responsibilities they should have. While a uniform standard of practice for healthcare ethicists has yet to mature, basic competencies for the conduct of ethics consultation have been devised, educational guidelines published, and a code of professional conduct promulgated.[1] The professionalization of ethics consultation (in the U.S. at least) is on the horizon. These developments all spring from the key assumption that these consultants are *experts* in ethics.

Expertise is usually thought of as relative to a domain of inquiry.[2] A dentist is an authority on human teeth; a Shakespearean, a specialist on the works of the eponymous playwright; a hydrogeologist, an expert on the movement of groundwater in the earth's crust. In the case of a clinical ethicist, the relevant expert domain is picked out by the fundamental task assigned her: To resolve moral uncertainty, confusion, and disagreement.[3] CECs specialize in fulfilling this aim;

[1] See ASBH 2011; ASBH 2009; ASBH 2014, also reprinted in Tarzian and Wocial 2015.

[2] See Weinstein 1993.

[3] This conception of the goal of CEC is explicitly endorsed or assumed in numerous sources spanning several decades. See, e.g., Ackerman 1987a, p. 146; Ackerman 1987b, p. 313; Andre 2002, p. 17; ASBH 1998, p. 3; ASBH 2011, p. 2; ASBH 2014, p. 2; Aulisio 2003, p. 9; Aulisio et al. 2009, p. 422; Berkowitz and Dubler 2007; Berkowitz *et al.* 2015, p. vi; Drane 1994, pp. xv,3; Dubler et al. 2009, p. 25; Dubler and Liebman 2011, p. 9; Fletcher et al. 1989, p. 1; Fletcher and Siegler 1996, p. 125; Hester 2008, p. 22; Jonsen et al. 2002, p. 1; Kanoti and Younger 1995, p. 404; La et al. 1991, p. 141; Macklin 1987, pp. 8, 17; Miller et al. 1996, p.3, 28; Pearlman, *et al.* 2015, p. 3;

D. M. Adams (✉)
California State Polytechnic University, Pomona, CA, USA
e-mail: dmadams@cpp.edu

J. C. Watson, L. K. Guidry-Grimes (eds.), *Moral Expertise*, Philosophy and Medicine 129, https://doi.org/10.1007/978-3-319-92759-6_12

they are, so it is believed, experts in providing guidance on practical questions in normative medical ethics.

In this paper I argue that an increasingly well-received conception of ethics consultation, exemplified in standards endorsed by the American Society for Bioethics and Humanities (ASBH) and a growing number of healthcare institutions in the U.S., incorporates an understanding of ethics expertise that is untenable in light of what that conceptions says is the aim of ethics consultation. For reasons that will become apparent shortly, I shall refer to this increasingly accepted model of clinical ethics consultation as the *conventional norms* account (CN). As I show, arguments supporting this account betray a deep ambivalence over the idea of ethics expertise, seeming at once both to affirm and deny that clinical ethicists can possess it. I argue that this ambivalence is traceable to a central assumption about how CECs should operate: that is, by supplying answers to moral problems reflecting conventionally-accepted views of what may and may not be done in clinical medicine. I then show that this assumption is incompatible with the attribution to CECs of expertise in doing what they are supposed to do—resolve moral puzzlement and conflict. Having shown that CN undermines its own claims to expertise on behalf of its practitioners, I suggest that being expert at addressing moral problems does not consist in furnishing answers, but rather in the deployment of skill in moral *reasoning*. Expertise in guiding others to think clearly *about* a moral problem as opposed to giving them an answer *to* it returns us to an understanding of the ethicist's role espoused decades ago, though recently lost sight of in the field. But even if ethicists are (or can be) experts in practical moral reasoning, I conclude, it is not at all clear (contrary to what proponents of CN assume) that ethics expertise can play a useful role in the day-to-day struggle to settle moral issues arising in the care of particular patients. For, if I am right, making use of the reasoning expertise needed to resolve such problems may be difficult in the units of a modern hospital.

12.2 Equivocation Regarding Clinical Ethics Expertise

Recent literature on ethics consultation makes two distinct sorts of claims about clinical ethicists—claims that sit very uneasily with each other. On the one hand, various sources describe the ethicist's remit in a guarded series of caveats: Ethicists do not have special insight into moral truth[4]; they are not persons with "right" answers about morality.[5] Ethicists must be wary of making recommendations reflecting their own moral opinions as this risks imposing their own judgments or rankings of values upon others, thus usurping the advisee's decision making role and

Siegler and Singer 1988, p. 759; Shelton and Bjarnadottir 2008, p. 56; Tarzian Tarzian, Anita J. and the ASBH Core Competencies Update Task Force 2013, p. 4.

[4] DeRenzo 1994, p. 384.

[5] DeRenzo 1994, p. 384.

encouraging deference to the personal beliefs of the ethicist.[6] CECs should not tell others what to do; they should limit themselves to sharing information derived from a broad societal consensus regarding acceptable value priorities and moral views.[7]

On the other hand, much of the literature insisting upon these provisos and qualifications at the same time confidently asserts that ethicists are ethics experts who can and should give an "expert response" to those who wish their moral assistance.[8] The CEC is the purveyor of expert answers to moral problems. To see why these assertions of ethics expertise appear discordant with the preceding set of disclaimers, it will help to begin by comparing the following two cases.

My Lawyer I consult an attorney specializing in business torts to determine whether I am vicariously liable for conduct of an independent contractor I hired whose negligence has injured a third party. "Given the facts of your case," my lawyer advises, "the answer depends on whether your contractor engaged in an abnormally dangerous activity within the meaning of *The Restatement of Law, Third:* Torts, section 58.20 (b)." I ask my lawyer what she thinks I should do. "I have a thorough grasp of this part of the law," she replies, "and I'm confident how the courts will rule. I believe the best answer is to argue against liability because no reasonable person could have foreseen the risk here." Because she is so well-versed in this part of current tort doctrine, my attorney can speak with some authority on this matter and I acknowledge I have good reason to defer to her judgment on how best to proceed.

My legal troubles in good hands for the moment, I now confront a new problem, for which a different kind of professional help is needed.

My Surgeon I experience distressing neurological symptoms and am seen at a hospital where tests reveal I suffer from a rare brain tumor. I am referred to a specialist, a noted expert in surgical management of brain tumors. Unlike my attorney, my surgeon begins our discussion with some odd disclaimers. "I am not here," he

---

[6] ASBH 1998, p. 7; ASBH 2011, pp. 6–9; 8–9; Aulisio 2003a, pp. 10–11; Aulisio, Arnold, and Younger 2000, p 9; Fletcher & Moseley 2003, pp. 103–104; Thomasma 1991, p. 137; Tarzian 2013, p. 5, p. 9. See also Zoloth-Dorfman & Rubin 1997, p. 430: "It is not the ethicist's claim or privilege to impose a particular set of moral beliefs on those who seek out her counsel"; Cummins 2002, p.25: "[I]t is not the role of the healthcare ethics consultant to make moral decisions, rather, she must acquire the skills necessary to facilitate moral discussions, to interpret participants' positions to each other, to discern the values at stake, and to assist others in making medical moral choices."

[7] ASBH 2011, p. 6, 7; Aulisio 2003a, p.p. 13–14; Geppert & Shelton 2012, p. 385; Tarzian 2013, p. 10.

[8] See ASBH 2011, p. 8. Ethics consultants possess "expertise in ethical analysis" (ASBH 2011, p. 6); ethicists are "sharing expertise" with others when drawing upon "ethical assessment" skills (ASBH 2011, pp. 8; 22, 55). See also Dubler 2011, p. 182; Dubler, et al. (2009); Glover et al. 1986, p. 23; Tarzian 2013, p. 5. See also Crosthwaite 2005; Meyers 2007; Steinkamp et al. 2008; Rassmussen 2011; Varelius 2008; Weinstein 1994. Some authors argue for or against '*moral* expertise', seeming to use that phrase interchangeably with 'ethics expertise.' See David 2011; Baylis 1989; Cross, 2016' Caplan 1989; Yoder 1998. As I suggest below, these terms should be carefully distinguished.

says, "as someone who claims special insight into the truth about tumors. I should not give you my own professional recommendation as I might usurp your role in making decisions for yourself about this tumor and thereby impose upon you my own professional judgments or conclusions about it. I cannot therefore tell you what you should do about your tumor. All I can do is share with you the general standards broadly agreed upon in the profession about how to evaluate tumors like yours."

While these two cases share a key feature, plainly they are also importantly dissimilar. The common feature is the initial assumption I made in seeking professional help. I assumed in each case that I had consulted an *expert*. An expert is generally thought to be an individual possessing a distinctive body of knowledge and skill not typically attributable to non-experts, conferring an ability to issue reasoned judgments within a specialized domain, such that he or she can claim the authority to make recommendations or give advice to which non-experts have strong (though perhaps not decisive) reason to defer.[9] Taking this as a roughly accurate characterization of what it means to be an expert, it seems clear that my lawyer fits the bill. She has provided authoritative advice grounded in a specialized body of knowledge and experience. She has addressed my uncertainty and replied to my questions with answers based upon professional judgment—judgment I have good reason to follow.

Things are otherwise, however, with my surgeon. His claims are decidedly strange, coming as they do from someone who purports to be an expert and whose help I desperately seek. His statements are puzzling because of course my surgeon has failed to do what I want: Tell me what he thinks I should do about my tumor. I need professional advice and answers to my questions about how my medical issue is best handled. I assumed I had consulted someone who does have some special insight into what really going on with tumors, and I expected I would have good reason to defer to his professional judgment and be guided by his decision. All of which is obviously the reason I sought out my surgeon in the first place. Yet my presumed medical specialist appears to disavow his own expertise. He backs away from any suggestion his views connect with medical truth and is hesitant to share with me his own conclusions and reluctant to exercise his own professional judgment in giving me answers. It is not clear why I should defer to his authority since (given what he says) it is not obvious he has any.

The disclaimers set forth by my surgeon are peculiar. Yet they are, as we have seen, precisely the sorts of disclaimers issued about CECs in the influential ASBH *Core Competencies in Healthcare Ethics Consultation*, which form the backbone of what I'm calling the conventional norms view. The second of my imagined cases accordingly exposes a tension between claims made in the two kinds of assertions about ethicists I distinguished above, suggesting it is very odd to make both kinds of assertions about them; this gives us good grounds for thinking that at least one set of claims is false. In other words, either (1) it is false that the role and authority of

[9] Here I draw upon similar analyses of expertise found in David 2011, Cross 2016, Weinstein 1994, and Varelius 2007.

hospital ethicists should be peculiarly hedged and qualified in the ways I noted at the outset, or (2) ethicists aren't really ethics experts.

A number of writers of course have adopted the strategy of expanding upon and defending (2).[10] Various arguments have been given to show that, unlike my lawyer, a CEC cannot be an expert in her professional domain of inquiry. Here I will try to pursue the opposite strategy, offering reasons in support of (1) and suggesting that clinical ethicists can properly be called ethics "experts," though in a form sharply at odds with the kind of expertise attributed to them by CN.

12.3 A Normative Challenge to Ethics Expertise

The claim that ethicists can provide expert answers to moral questions and expert solutions to moral disagreement is challenged in several ways. Moral non-cognitivists insist that since moral judgments or propositions have no truth-value, the impossibility of *expert* moral knowledge is just a special case of the claim that *all* moral knowledge is inconceivable.[11] Others maintain that purported ethics experts cannot, consistent with a commitment to clarifying common-sense morality, claim the ability to make judgments ordinary non-specialists cannot make; therefore they cannot be genuine experts.[12] Still others argue that even if there are experts with respect to morality, non-experts could not properly identify them.[13] These (and related) problems with the idea of ethics expertise raise metaethical and epistemic worries, many of which are explored by other contributors to this volume. But such concerns do not seem uppermost in the mind of those who seek to defend the conventional norms model of clinical ethics consultation. They, by contrast, are most troubled by a *normative* objection to the possibility of ethics expertise. It is important to get clear on what this objection says, as preoccupation with it, I believe, both explains the peculiar ambivalence evident in the contrasting claims made about the CEC's expertise and ultimately reveals the central weakness of CN's account of that expertise.

The central idea behind the normative objection to ethics expertise is that deployment of and dependence upon the advice and guidance of such an expert in ethics cannot be reconciled with a proper respect for the advisee's autonomy. It fails to show equal concern and respect for the deliberative capacities of one's fellow moral agents it is alleged, to insist they defer to the testimony and guidance of another on substantive issues in practical normative ethics.[14] There are two general reasons offered for this conclusion. First, such deference inappropriately elevates or

[10] See Cowley 2005; Engelhardt 2003; Engelhardt 2009; Engelhardt 2011; Engelhardt 2012a; Engelhardt 2012b; Frey 1978; Noble 1982; Scofield 2008a; Scofield 2008b; Swales 1982.

[11] See, e.g., Scofield 1993; Cowley 2005.

[12] See, e.g., David 2011.

[13] See Cholbi 2007.

[14] See, e.g., Caplan 1989; Nussbaum 2002; Parker 2005.

privileges the moral conclusions of an individual living in a diverse community of persons, holding competing moral views and different rankings of values, and lacking a rich body of shared normative commitments, so that no one individual's value priorities can be shown to be superior to those of anyone else.[15] Second, reliance upon another for expert moral advice substitutes the expert's moral conclusions for one's own. The advisee thereby allows the moral question at issue to be decided by the expert rather than independently examining it for him- or herself. Persons should not accept the moral conclusions of others on trust,[16] for such deference to an expert's moral views can come to constitute a kind of moral servility, a surrender of one's own powers of judgment and an abdication of one's responsibility to exercise them.[17] The basic thought here seems to be that I don't act morally when I simply do what someone else has told me to do. I should be able to justify my moral conclusions to myself, not indirectly through the word of an expert.[18]

It is not my purpose in recounting this objection either to endorse or take exception to it[19]; rather, I shall simply assume it raises genuine worries about ethics expertise. For making that assumption allows us to frame a central question to which I contend the proponents of CN have no adequate response: How can a hospital ethicist provide *answers* to moral problems while at the same time circumventing the normative objection and retaining her status as an expert? I will argue she cannot.

It will help to back up at this point and ask: What sorts of answers to moral questions might we expect an ethics expert to provide? Here is one possibility: The specialist we are imagining is someone who gives *correct* answers. She is a *moral* expert, where this means an individual who can reliably arrive at correct answers to a variety of questions in practical normative ethics or who makes practical moral judgments that are dependably morally true—or at least someone who can do these things with a frequency exceeding that of the non-expert.[20] I take it to be plain that the idea of a moral expert so understood is controversial, freighted with conceptual and normative assumptions and requiring much work to make plausible. I will not pursue that line of investigation here, however; for with some possible exceptions,[21] almost no one in the literature defends the claim that CECs possess *moral* expertise in this sense—indeed, the attribution of such expertise to hospital ethicists is explicitly rejected in several sources.[22] The CEC is not someone who claims to deliver

---

[15] See, e.g., Aulisio 2003; DeRenzo 1994; Engelhardt 2003; Engelhardt 2009; Engelhardt 2012; Scofield 1993.

[16] Nussbaum 2002.

[17] See Nelson 2007.

[18] For other articulations of this argument see, e.g., Driver 2006; Parker 2005; Rassmussen 2011.

[19] For what seem to me insightful replies to this objection, see Parker 2005.

[20] Some would add: And can do so *for the right reasons*. See Gesang 2010. For useful accounts of moral expertise understood in this way see, e.g., Cholbi 2007; McGrath 2008. There is a noteworthy literature on whether *moral philosophers* qualify as moral experts in this sense. See, e.g., David 2011; Caplan 1989; Cross 2016; Frey 1978; Gesang 2010; Singer 1972; Weinstein 1994.

[21] See Meyers 2003.

[22] See, e.g., Aulisio 2003; DeRenzo 1994; Macklin 1987; Rassmussen 2011.

right or *true* answers to the puzzled and conflicted. As the backers of CN worry, such an "authoritarian" conception of ethics consultation would "create the impression that the consultant's expertise in ethical analysis amounts to moral 'hegemony'."[23] Consistent with this concern, the first edition of the *Core Competencies for Health Care Ethics Consultation* rejected efforts to certify ethics consultants because certification might suggest "that certified individuals have special standing in ethical decision-making."[24] Thus, according to CN, a CEC's expert solutions, whatever they may be, do not amount to the decrees of a moral expert.

An alternative way of understanding the contributions of an ethics expert might go like this: She gives not *right* answers to moral problems but *conventionally accepted* answers. Not surprisingly, this is just what proponents of CN want to say. The ethicist facilitates "the building of consensus among the involved parties" within "a range of morally acceptable options."[8] The CEC's role as an ethics expert is simply to ensure that any solution to moral uncertainty or disagreement is limited to "a range of ethically acceptable options,"[25] clearly falling "within accepted ethical principles, legal stipulations, and moral rules defined by ethical discourse, legislatures, and courts…"[26] Such a "principled resolution" is not dictated by the ethicist but is selected from among a set of "acceptable clinical, ethical, and legal outcomes."[27] As the influential *Core Competencies* standards have stressed from the beginning, ethics consultation should respect boundaries set by "societal values, law, and institutional policy,"[28] identifying accurately those points of settled agreement that delineate the boundaries of permissible choice.[29]

Viewed in this way, the ethicist's expertise consists in resolving moral uncertainty and disagreement by appeal to widely-accepted judgments regarding what is and isn't allowable in clinical medicine. She does not to give her own opinion of what ought to be allowed; rather, she simply reports on which options are in fact

---

[23] ASBH 2011, p. 9; see also Aulisio 2003, pp. 10–11.

[24] ASBH 1998, p. 31.

[25] ASBH 1998, p. 8. A revised version of the same report (ASBH 2011) also charges the ethicist with determining "whether ethics consultation resulted in decisions or actions that are consistent with established ethical standards" (ASBH 2011, p. 39). Summaries of ASBH 1998 also appeared in Aulisio, Arnold, and Younger 2000 and in Aulisio 2003. See also Kodish and Fins 2013, p. 29.

[26] Dubler, et al. 2009. See also Kon 2012, P. 15: "The primary goals of a clinical ethics consultation are to (1) help patients and providers understand which options are ethically required, which are ethically permissible, and which are ethically unsupportable…"

[27] Dubler 2011, p. 185; see also Dubler and Liebman 2011, p. 12; 14–15; ASBH 2011, p. 7. See also Lowey 1990, p. 357: Ethicists should support "reasonable" decisions falling "within the bounds of communal and institutional acceptability."

[28] ASBH 1998, p. 7. The second edition of the *Core Competencies* is more specific: Allowable options are to be ascertained by reviewing "the bioethics literature, medical literature, other relevant scholarly literature, current professional and practice standards in the field of [CEC], statutes, judicial opinions, and pertinent institutional policies." ASBH 2011, p. 6.

[29] The ethicist's recommendations "should comport with the bioethics literature…current professional and practice standards in the field…statutes, judicial opinions and pertinent institutional policies." ASBH 2011, p. 6. See also Tarzian 2013, p. 5.

judged allowable according to the established views to which she is bound.[30] This understanding of ethics expertise is attractive because it offers a way to sidestep the normative objection to ethics expertise previously discussed. An ethicist giving conventionally accepted answers both recognizes and displays proper regard for the multiplicity of values and conceptions of the good present in a diverse society by limiting her advice to only those norms supportable by a kind of Rawlsian "overlapping consensus," reflecting judgments that are in fact the subject of large-scale agreement.[31] She makes no claim of privileged access to moral truth or for the superiority of her moral convictions vis-à-vis those of others. Hence, she cannot be accused of doctrinaire demands that others acquiesce to her in her own moral conclusions about the cases on which she is called to consult. Nor does she displace her advisee's exercise of his deliberative capacities by imposing her own moral views upon him or usurping that person's decision making role—for she does not share her own moral conclusions or value priorities at all. Instead, she merely relates information derived from a societal consensus regarding acceptable moral choices. Thus, the advisee owes no deference to the opinions of the ethicist and therefore retains final authority to decide for himself what to do. Here then is a way to give what looks like expert answers to moral problems while avoiding the normative objection to ethics expertise.

In the next section I will argue that this conventional norms model is unworkable, as it can yield no account of why the clinical ethicist remains an *ethics* expert.

12.4 Conventional Norms and Ethics Expertise

The claim that hospital ethicists should seek to resolve moral uncertainty and disagreement through adherence to conventionally accepted norms undercuts the grounds for asserting that they are ethics experts. To see why this is so, consider a further, illustrative case.

My Cousin Imagine I have a cousin who is hospitalized with respiratory insufficiency and renal dysfunction secondary to lung cancer that has metastasized. He is unresponsive, placed on ventilator, and started on medication to cope with his worsening kidney function. His attending physician believes the only appropriate goal for my cousin at this point is comfort care and palliation of symptoms, and the doctor believes a "do not resuscitate" (DNR) order should be written. The members of my cousin's family, however, strongly disagree, insisting that he wishes to live as long as possible and does not want his loved ones to "give up" on his treatment. Several contentious discussions take place between the family and the doctor; on the afternoon of the last of these meetings, I discover that the physician has

[30] We should evaluate the outcomes of ethics consultations, in part, based on whether they "result in decisions or actions that are consistent with established ethical standards." ASBH 2011, p. 39.

[31] See Rawls 1993, pp. 133–172.

unilaterally (that is, without the consent of that individual or his surrogate) taken it upon himself to write a DNR order and place it in my cousin's chart. I request a meeting with the hospital's ethics consultant and I ask her if the doctor's order is ethically justifiable.

On the conventional norms view, my cousin's ethicist should put my queries to rest with answers expressing widely-accepted positions (in this case on unilateral DNR orders). And similarly for all other ethics questions brought to her attention. But mere recitation of such orthodox answers is insufficient to qualify the person giving them as either an *expert* or as an expert *in ethics*. Several grounds can be given to support this verdict.

First, what method or procedure is my cousin's ethicist supposed to utilize in order correctly to identify whether the conventional standards of practice or prevailing policies allow the doctor to give a unilateral order not to resuscitate? Proponents of CN say very little about how to identify conventionally sanctioned options in clinical medicine, presumably because there isn't much to be said. To find answers to questions of the sort I put to my cousin's ethicist she simply "looks them up," referencing a list of conventional sources. But if this is how an ethicist is to operate, where does her expertise come in? The claim that CECs are professional specialists seems to entail that they alone have the training and knowledge to give authoritative advice on moral issues in medicine. But if such knowledge simply reduces to familiarity with the answers conventional norms endorse, and those answers are readily apparent, then little basis remains for a claim to expertise—for asserting, in other words, that hospital ethicists have professional insight into what is and isn't ethical not available to anyone else, or to maintain that their professional judgment on moral matters carries a degree of authoritativeness such that advisees have reason to defer to the ethicist's answers.

A further worry arises from the fact that reciting orthodox answers cannot always meet the aim of resolving *moral* disagreement, even if it serves other purposes. Some contributors for example, most notably Lisa Rassmussen, argue that the task of identifying the conventionally sanctioned courses of action in any given clinical situation is a complex one, beyond the capabilities of the lay person and requiring specialized knowledge and training. According to Rassmussen, clinical ethics expertise consists in "superior familiarity" with a "pervasive ethos or practice," including (for example)

> command of the moral arguments behind applicable law; knowledge of academic consensus or dissensus on a particular question; knowledge of public sentiment on a particular issue when it has been compiled; and the historical moral foundations of clinical practice.[32]

As advocates of CN have argued, CECs not only must "understand the milieu in which they operate, including local and federal law, institutional policy"[33] and the

[32] Rassmussen 2011, p. 650.

[33] *Ibid.*, p. 650.

like, they must also have the ability to impart this information in a way that brings contending parties together in a constructive discussion.

We can safely set to one side further details of Rassumussen's account and just assume that something like her view is what advocates of CN have in mind. The problem with such an account in the context of our present inquiry, however, is that while it may show CECs have *some* kind of expertise, it is difficult to see how it amounts to *ethics* expertise, where (as I've observed) that necessarily means the ability to resolve moral uncertainty, confusion, and disagreement. To begin with, though the CEC must presumably draw upon significant interpersonal skills to facilitate a respectful and productive dialogue amongst the involved parties who have requested her help, such skills have no necessary connection with the resolution of *moral* problems, as they can be utilized in a whole variety of non-moral settings. To be sure, an individual working as a hospital ethicist may well possess certain competencies and be proficient in them to a degree that we could call her an expert. For example, if she is a trained mediator, she may well deploy specialized skill reflecting mastery of an important body of relevant knowledge, qualifying her as a specialist in dispute resolution.[34] But deference to her *qua* mediator is not, in itself, an assertion of ethics expertise. If she is an attorney, an ethicist can be expected to have expert knowledge of medical jurisprudence; and an ethicist who serves as a hospital risk manager may have an unusually deep understanding of institutional policies and procedures. But these types of skill or competence by themselves confer no obvious insight into or authority regarding the resolution of moral disagreement or puzzlement—the ethicist's distinctive responsibility.

Most importantly, the account of ethics expertise adopted by both Rassmussen and supporters of CN seems to be premised on the belief that moral disagreement and confusion will always and straightforwardly turn on *what the conventional norms say*, not on *whether those norms are right (or justifiable)*. But there is no basis for this supposition.

Imagine people disagreeing, say, over whether sedation to unconsciousness is equivalent to euthanasia, whether allowing Grandpa to "pull the plug" on himself is letting him commit suicide, or whether a patient diagnosed with "total brain failure"[35] is really dead. If the parties in such cases are in fact only disagreeing over how courts have ruled on these questions, what hospitals policies mandate, or what response the AMA *Code* gives, then of course sharing what those conventional sources say will resolve their problems—but such cases are surely rare. People in hospitals with serious moral concerns about what to do are not just quarreling over what the College of Physicians thinks is right or what institutional procedures allow; they disagree over—and want to know—what courses of action are *in fact* right, what decisions are *all things considered* ethically justifiable (or unsupportable). Recall the case of my cousin. As it turns out, there is in fact continuing controversy over whether physicians may unilaterally withhold from a critically ill patient CPR

[34] See, e.g., Fiester 2007; Fiester 2015.

[35] This is the expression preferred to "brain death" by the President's Council on Bioethics. See President's Council 2008, p. 12.

and other interventions intended to postpone death on the grounds that offering them is judged futile.[36] But let us imagine that this is not so: In talking with the ethicist about his case my cousin's loved ones are told that unilaterally imposed DNRs are sanctioned by conventional norms of practice, policy, and law. Why should informing my cousin's family of that fact resolve or even address their moral concerns? Surely they will not be convinced that unilateral DNRs must be justifiable simply because the medical standards and policies say so. For in that case their entire point would be to question why the established norms are correct and their own beliefs mistaken. More generally, what Rassmussen calls the prevailing "ethos" or legal, institutional, and professional context of clinical medicine incorporates any number of substantive moral views that are contestable. So, for example, the conventional context to which CN points maintains that a commitment to relief of pain is incompatible with physician-assisted suicide,[37] intending to end suffering is morally distinct from intending to end life,[38] and continuous sedation until death is permissible only as a last resort and not for "existential" suffering.[39] Such orthodox views certainly can become a source of conflict and confusion in a clinical setting, and when they do, it will not dispel questions that have been raised or resolve disagreements that have arisen to insist upon adherence to them.

Therefore, expertise of Rassmussen's sort—proficiency in navigating the intricacies of law, policy, and standards of practice to determine the orthodox answers to moral questions—is not always sufficient for resolving genuine moral uncertainty and disagreement, for those answers may themselves be the subject of controversy. Thus, mastery of such knowledge and skill cannot be sufficient to bestow upon ethicists possessing it bona fide ethics expertise.

12.5 Ethics Expertise Re-Considered: A Modest Proposal

If expert ethical guidance in the clinic consists neither in the ability of the *moral* expert to offer the correct answer, nor in the competence of the *conventional norms* expert to describe accurately the widely-accepted answers, then in what *can* ethical expertise consist? To find out, we need to get clear on the key error in CN's account of ethics expertise. Where CN goes wrong, I submit, is in its fundamental assumption that the way to resolve moral puzzlement and conflict is *with answers*. Expert help in dealing with moral questions, if we are to make sense of it, must have a different form.

---

[36] For a small sample of literature testifying to this ongoing debate, see the articles in *The American Journal of Bioethics* 10(1) (2010) and *The American Journal of Bioethics* 15(8) (2015). See also Pope 2011; Pope & Kemmerling 2016; American College of Physicians 2012, p. 84

[37] American College of Physicians 2012, p. 85.

[38] Berlinger et al. 2013, p. 59.

[39] Opinion 2.201, AMA 2016; American Academy of Pain Medicine 2003; Berlinger et al. 2013, p. 184; VHA 2006, pp. 5–7.

The first step in grasping what ethical expertise might actually look like is to recall that assertions of expertise are relative to a domain of inquiry, and the CEC's domain is, we are told, the resolution of moral uncertainty and disagreement. So we must start by asking what 'resolution' can mean in this context. Think again about the controversy over my cousin. The parties in that case disagree over whether physicians may unilaterally determine that attempts to resuscitate a patient will not be undertaken because the patient cannot derive a meaningful benefit from them. What would it take to *resolve* such a conflict? This particular dispute is of course ultimately a disagreement over who determines whether and when further medical treatment is futile—a significant source of heated bioethical controversy for decades. An ethicist consulting on my cousin's case could not reasonably be expected to settle, once and for all, *that* debate. She can only be called upon to assist my cousin's doctor and family in reaching, if possible, a level of rational moral agreement with which *they* are satisfied in *this* particular situation. But how is she even to accomplish this more limited goal?

If, as supporters of CN insist, the CEC is responsible to ensure that whatever decision is reached in a given case consultation is morally defensible,[40] then it seems the only way to bring about that result is to involve those who are morally confused or at odds with each other in a process of *moral deliberation*: Thinking together about what exactly the problem is and how one could articulate and defend various possible solutions to it. Such careful thinking would presumably include a set of related tasks: Surfacing assumptions and uncovering theoretical commitments, highlighting key distinctions, examining justifications for opposing positions, detecting inconsistencies and fallacious inferences—in short, engaging persons in sustained, reasoned moral reflection. So, in the case of my cousin, this would involve, *inter alia,* examining the notion of medical futility, asking whether it can be defined non-normatively, exploring what it means for a medical intervention not to be worth doing, discussing the reasons for thinking physicians can judge the appropriateness of a procedure like CPR apart from the patient's preferences, and so on. To resolve the dispute over my cousin, then, the ethicist's role requires her to help the members of the family and care team think through the moral issues confronting them, examining how they might weigh up reasons for pursuing one course of action rather than another.[41]

If the CEC's role is understood in this way, then in what sense, if any, is she functioning as an ethics expert? When I sought help from my attorney I did so because she was better able than a non-lawyer to assist me in arriving at a legally well-reasoned position or conclusion on the issue of tort liability for which to argue in court. In a similar way, clinical ethics consultants could be viewed as experts at

[40] See ASBH 2009, p. 12: "[C]linical ethics consultation is focused on 'achieving defensible solutions to clinical ethical problems'."

[41] Some of what I say in this section about the kind of facilitated deliberation required to resolve moral disagreement and uncertainty has interesting parallels with *moral case deliberation,* a model of ethics consultation developed in parts of Europe, especially in The Netherlands. For a sample of that work, see Molewijk, *et al.* 2008; Widdershoven & Molewijk 2010; Stolper, *et al.,* 2010.

imparting and deploying standards of reasoning used to evaluate factors bearing on moral problems in clinical medicine. They are experts in that they can assist people in formulating better reasoned justifications for such judgments or answers than non-experts are capable of doing.[42] Ethicists are experts, then, not because they invariably think correctly about moral issues or think about them only in orthodox ways, but because they think more *clearly* and *deeply* about them. They possess the knowledge and skills necessary to help those struggling with moral disagreement or uncertainty arrive at the most ethically well-reasoned decision of which they are capable, given constraints of limited time, medical uncertainty, the urgency to reach resolution, and similar practical limitations. An ethicist is therefore a specialist by virtue of being a kind of *reasoning* expert.[43]

Some may balk at the idea of a specialist in practical moral reasoning, but it is not obvious why we should reject it. It is of course by no means new—writers in the field of clinical ethics consulting going back to the 1970s have defended it.[44] But if I am right, it is an idea of which the field has unfortunately lost sight; and this is regrettable, as it affords a defensible conception of the ethicist's distinctive expertise. There certainly appears to be an expert body of skill and knowledge essential to thinking carefully about moral matters. An ethicist seeking to resolve moral confusion and disagreement must possess the skill needed to identify reasons bearing upon one or another point of view on a given moral question; the understanding required to connect reasons into a coherent argument, draw attention to important disanalogies, and recognize relevant distinctions; the knowledge necessary to explain how certain reasons might reflect problematic assumptions; the insight called for in order to clarify concepts embedded in larger theories of moral normativity. These capabilities are not ones typically had by those without specialized training and experience, and a person so equipped appears able to speak with some authority on what it means to reason well or poorly from an ethical standpoint.

It is not my aim here to set out a comprehensive account of what it means for conclusions arrived at or decisions reached on a moral issue to be ethically well-reasoned. Nor will I offer an account of the kind of professional background or training one must have to qualify as a specialist in practical moral reasoning, or how hospital ethicists could be certified or even licensed to conduct expert ethics consultations. But it is important to see that, whatever exactly the details, an ethically well-reasoned decision is *not* one that can always be arrived at by limiting moral deliberation to the confines of established conventions and norms—to what, in other words, is generally accepted in practice, policy, and law. When such norms become the subject of controversy in the units of a hospital, thorough moral deliberation must include the ability to challenge accepted ways of thinking and examine

---

[42] Compare Weinstein's claims that ethics experts can provide stronger justifications for particular moral positions than non-experts. Weinstein 1993.

[43] See the discussion in Driver 2006, to which I am indebted.

[44] See, e.g., Macklin 1987; Singer 1972. For more recent accounts, see Weinstein 1994; Yoder 1998.

arguments both for and against them, exploring reasons for resisting or rejecting established norms as well as those supporting them.

12.6 Conclusion

Proponents of CN attempt to avoid the normative objection to ethical expertise by so limiting the kinds of answers CECs can give to moral questions that, though their responses might still display expertise—a deep understanding of conventional views firmly rooted in prevailing legal and practice standards—it seems no longer to count as *ethics* expertise. This should not be a surprising result; it is difficult to see how, on any account of clinical ethics consultation, an ethicist could supply genuine expert answers to moral questions—answers to which non-experts should defer—without appearing to privilege her own moral views and undercut others' autonomy by telling them what they may or may not do. The solution to this conundrum is to realize that experts in ethics do not function as authorities by issuing recommendations or giving advice in the form of answers. On the view I am suggesting, the only kind of ethics expertise properly attributable to CECs charged with resolving moral uncertainty and disagreement is a distinctive mastery of moral reasoning. Ethicists are therefore what Bruce Weinstein calls "performative experts": They are proficient in knowing *how* to carry out a certain task with great skill.[45] The ethicist *shows* others how to think productively about ethics, but doesn't *tell* them what to think or instruct them on which courses of action are or are not allowable. Those she assists have reason to defer to the ethicist's skill and knowledge in demonstrating to them how to arrive at an ethically well-reasoned decision in a given case, whatever that decision turns out to be.

The clinical ethicist *qua* reasoning expert does not run afoul of the normative objection to ethics expertise I outlined earlier. CECs are not on this account moral experts, endowed with the competence to arrive at moral judgments or answers that other cannot, still less do they have some special insight into moral truth or conclusions about the correct resolution of moral puzzlement. And the ethics expertise I claim CECs possess is fully consistent with respect for the autonomous agency of the advisee, since on this view the CEC does not foist her own substantive moral answers upon those she assists nor expropriate their authority to make decisions for themselves.

It must be said, however, that if my argument in this paper is correct, then the distinctive nature of clinical ethics expertise—proficiency in helping people reason morally—may be of questionable value in an actual clinical setting. Crowded hospital units and the tense environment of the ICU are not places conducive to the sustained moral reflection necessary to genuinely promote reasoned discussion and deliberation. Even if the only expert ethical guidance that could help my cousin's

[45] Weinstein 1993. This does not mean I wish to rule out that clinical ethicists cannot also qualify, in Weinstein's terms, as "epistemic" experts.

doctor and family is something that a CEC with excellent moral reasoning abilities is fully able to supply, a room in an acute care hospital, caught in a web of competing institutional demands, miscommunication, frayed nerves, and pressures from discharge planners and insurance providers is not a promising venue for careful moral reflection of the sort that appears essential to the exercise of real clinical ethics expertise. Hospital ethicists may well be able to contribute usefully to their organizations in various ways—by, for example, giving educational presentations at grand rounds or by assisting in the formulation of institutional policy. But, ironically, they may find themselves in possession of an expertise ill-suited for the context in which it was primarily intended to be used.

It is not clear what all this means for people like my cousin's family members, who truly need moral guidance of some kind, not to mention for the larger issue of whether it makes sense to certify a group of professionals as expert moral problem solvers. But plainly these are important concerns that need to be taken up elsewhere.

References

Ackerman, Terrence. 1987a. Medical ethics in the clinical setting. In *Clinical medical ethics: Exploration and assessment*, ed. Terrence Ackerman et al., 141–197. Lanham: University Press of America.

———. 1987b. The role of an ethicist in health care. In *Health care ethics: A guide for decision makers*, ed. Gary R. Anderson and Valerie Glesnes-Anderson, 308–320. Gaithersburg: Aspen Publication.

American Academy of Pain Medicine 2003. Ethics Charter. http://www.painmed.org/practicemanagement/ethics-charter/. Accessed 7 July 2017

American College of Physicians. 2012. Ethics manual, 6. *Annals of Internal Medicine* 156: 73–104. https://www.acponline.org/clinical-information/ethics-and-professionalism/acp-ethics-manual-sixth-edition-a-comprehensive-medical-ethics-resource/acp-ethics-manual-sixth-edition. Accessed 12 Dec 2016

American Medical Association. *Code of Medical Ethics*. https://www.ama-assn.org/about-us/code-medical-ethics. Accessed 9 Dec 2016

Andre, Judith. 2002. *Bioethics as practice*. Chapel Hill: University of North Carolina Press.

ASBH—American Society for Bioethics and Humanities. 1998. *Core competencies for health care ethics consultation*. 1st ed. Glenview, IL: ASBH.

American Society for Bioethics and Humanities. 2009. *Improving competencies in clinical ethics consultation: An education guide*. http://www.asbh.org/publications/.

———. 2011. *Core competencies for health care ethics consultation, Second Edition*. http://asbh.org/publications/books

ASBH—American Society of Bioethics and Humanities. 2014. *Code of ethics and professional responsibilities for healthcare ethics consultants*. http://www.asbh.org/publications/.

Aulisio, Mark P. 2003. Meeting the need: Ethics consultation in health care today. In *Ethics consultation: From theory to practice*, ed. Mark P. Aulisio, Robert M. Arnold, and J.Younger Stuart, 3–22. Baltimore: The Johns Hopkins University Press.

———. 2003a. *Ethics consultation: From theory to practice*. Baltimore: The Johns Hopkins University Press.

Aulisio, M.P., R.M. Arnold, and S.J. Younger. 2000. Health care ethics consultation: Nature, goals, and competencies. *Annals of Internal Medicine* 133 (1): 59–70.

Aulisio, Mark P., Jessica Moore, May Blanchard, Marcia Bailey, and Dawn Smith. 2009. Clinical ethics consultation and ethics integration in an urban public hospital. *Cambridge Quarterly of Healthcare Ethics* 18: 371–383.

Baylis, Francoise. 1989. Persons with moral expertise and moral experts: Wherein lies the difference? In *Clinical ethics: Theory and practice*, ed. Barry Hoffmaster, Benjamin Freedman, and Gwen Fraser, 89–99. Clifton: Humana Press.

Berkowitz, Kenneth A., and Nancy Neveloff Dubler. 2007. Approaches to ethics consultation. In *Handbook for health care ethics committees*, ed. Linda Farber Post, Jeffrey Blustein, and Nancy Neveloff Dubler, 139–143. Baltimore: The Johns Hopkins University Press.

Berkowitz, Kenneth, et al. 2015. *Ethics consultation: Responding to ethics questions in health care, Second Ed.* Veterans health administration. http://www.ethics.va.gov/integratedethics/ecc.asp. Accessed 28 April 2016

Berlinger, Nancy, Bruce Jennings, and Susan M. Wolf. 2013. *Hastings center guidelines for decisions on life-sustaining treatment and care near the end of life*. New York: Oxford University Press.

Caplan, Arthur. 1989. Moral experts and moral expertise: Do either exist? In *Clinical ethics: Theory and practice*, ed. Barry Hoffmaster, Benjamin Freedman, and Gwen Fraser, 59–87. Clifton: Humana Press.

Cholbi, Michael. 2007. Moral expertise and the credentials problem. *Ethical Theory and Moral Practice* 10 (4): 323–334.

Cowley, Christopher. 2005. A new rejection of moral expertise. *Medicine, Health Care, and Philosophy* 8: 273–279.

Cross, Ben. 2016. Moral philosophy, moral expertise, and the argument from disagreement. *Bioethics* 30 (3): 188–194.

Crosthwaite, Jan. 2005. In Defence of ethicists: A commentary on Christopher Cowley's paper. *Medicine, Health Care, and Philosophy* 8: 281–283.

Cummins, Deborah. 2002. The professional status of bioethics consultation. *Theoretical Medicine* 23 (1): 19–43.

David Archard. (2011). Why moral philosophers are not and should not be moral experts. *Bioethics* 25 (3): 119–127.

DeRenzo, Evan. 1994. Providing clinical ethics consultation. *HEC Forum* 6 (6): 384–389.

Drane, James F. 1994. *Clinical bioethics: Theory and practice in medical-ethical decision making*. Kansas City: Sheed & Ward.

Driver, Julia. 2006. Autonomy and the asymmetry problem for moral expertise. *Philosophical Studies* 128: 619–644.

Dubler, Nancy N. 2011. A 'Principled Resolution': The fulcrum for bioethics mediation. *Law and Contemporary Problems* 74 (Summer 2011): 177–200.

Dubler, Nancy N., and Carol B. Liebman. 2011. *Bioethics meditation: A guide to shaping shared solutions*. 2nd ed. Nashville: Vanderbilt University Press.

Dubler, Nancy Neveloff, et al. 2009. Charting the future: Credentialing, privileging, quality, and evaluation in clinical ethics consultation. *Hastings Center Report* 39 (6): 23–33.

Engelhardt, H. Tristram, Jr. 2003. The bioethics consultant: Giving moral advice in the midst of moral controversy. *HEC Forum* 15 (4): 362–382.

———. 2009. Credentialing strategically ambiguous and heterogeneous social skills: The emperor without clothes. *HEC Forum* 21 (3): 293–306.

———. 2011. Core competencies for health care ethics consultation: In search of professional status in a post-modern world. *HEC Forum* 23: 129–145.

Engelhardt, Jr., H. Tristram (Eds.). 2012. *Bioethics critically reconsidered*. Dordrecht: Springer.

Engelhardt, Jr., H. Tristram. 2012a. A skeptical reassessment of bioethics. In Bioethics critically considered: Having second thoughts, ed. H. Tristram Engelhardt Jr, 1–28. Dordrecht: Springer.

Engelhardt, H. Tristram, Jr. 2012b. Why clinical bioethics so rarely gives morally normative guidance. In *Bioethics critically considered: Having second thoughts*, ed. H. Tristram Engelhardt Jr., 151–174. Dordrecht: Springer.

Fiester, Autumn. 2007. The failure of the consult model: Why 'Mediation' should replace 'Consultation'. *The American Journal of Bioethics* 7 (2): 31–32.

———. 2015. Contentious conversations: Using mediation techniques in difficult clinical ethics consultations. *The Journal of Clinical Ethics* 26 (4): 324–330.

Fletcher, John C., and Kathryn Moseley. 2003. The structure and process of ethics consultation services. In *Ethics consultation: From theory to practice*, ed. Mark P. Aulisio, Robert M. Arnold, and J.Younger Stuart, 96–120. Baltimore: The Johns Hopkins University Press.

Fletcher, John C., and Mark Siegler. 1996. What are the goals of ethics consultation? A consensus statement. *The Journal of Clinical Ethics* 7 (2): 122–126.

Fletcher, John C., Norman Quist, and Albert R. Jonsen. 1989. Ethics consultation in health care: Rationale and history. In *Ethics consultation in health care*, ed. John C. Fletcher, Norman Quist, Norman, and Albert R. Jonsen, 1–15. Ann Arbor: Health Administration Press.

Frey, R.G. 1978. Moral experts. *Pacific Philosophical Quarterly* 59 (1): 47–52.

Geppert, Cynthia M.A., and Wayne N. Shelton. 2012. A comparison of general medical and clinical ethics consultations: What can we learn from each other? *Mayo Clinic Proceedings* 87 (4): 381–389.

Gesang, Bernward. 2010. Are moral philosophers moral experts? *Bioethics* 24 (4): 153–159.

Glover, J.J., D.T. Ozar, and D.C. Thomasma. 1986. Teaching ethics on rounds: The ethicist as teacher, consultant, and decision-maker. *Theoretical Medicine* 7: 3–12.

Hester, D. Micah. 2008. The 'What' and 'Why' of ethics. In *Ethics by committee: A textbook on consultation, organization, and education for hospital ethics committees*, ed. D. Micah Hester, 21–26. Lanham: Rowman & Littlefield Publishers.

Jonsen, Albert, Mark Siegler, and William Winslade. 2002. *Clinical ethics*. 5th ed. New York: McGraw-Hill.

Kanoti, G.A., and S.J. Younger. 1995. *Clinical ethics consultation. Encyclopedia of Bioethics*. New York: Macmillan.

Kodish, Eric, and Joseph J. Fins. 2013. Quality attestation for clinical ethics consultants. *Hastings Center Report* 43 (5): 26–36.

Kon, Alexander A. 2012. Clinical ethics consultants: Advocates for both patients and clinicians. *The American Journal of Bioethics* 12 (8): 15–17.

La, Puma, John, Schiedermayer, and L. David. 1991. The clinical ethicist at the bedside. *Theoretical Medicine* 12: 141–149.

Loewy, Erich H. 1990. Ethics consultation and ethics committees. *HEC Forum* 2 (6): 351–359.

Macklin, Ruth. 1987. *Moral choices: Ethical dilemmas in modern medicine*. Boston: Houghton Mifflin Company.

McGrath, Sarah. 2008. Moral disagreement and moral expertise. In *Oxford studies in Metaethics*, ed. Russ Shafer-Landau, vol. 3, 87–108. Oxford: Oxford University Press.

Meyers, Christopher. 2003. A defense of the philosopher-ethicist as moral expert. *Journal of Clincial Ethics* 14 (4): 259–269.

Meyers, Christopher. 2007. *A practical guide to clinical ethics consulting: Expertise, ethos, and power*. Lanham: Rowman & Littlefield Publishers.

Miller, F.G., J.J. Fins, and M.D. Bachetta. 1996. Clinical pragmatism: John Dewey and clinical ethics. *Journal of Contemporary Health Law and Policy* 13 (1): 27–51.

Molewijk, Bert, and Guy Widdershoven. 2010. Philosophical foundations of clinical ethics: A hermeneutic perspective. In *Clinical ethics consultation: Theories, methods, implementation, evaluation*, ed. Jan Schildmann, John-Stewart Gordon, and Jochen Vollmann, 37–52. Surrey: Ashgate Publishing.

Molewijk, A.C., T. Abma, M. Stolper, and G. Widdershoven. 2008. Teaching ethics in the clinic: The theory and practice of moral case deliberation. *Journal of Medical Ethics* 34: 120–124.

Nelson, James Lindemann. 2007. Trusting bioethicists. In *The ethics of bioethics: Mapping the moral landscape*, ed. Lisa A. Eckenwiler and Felicia G. Cohn. Baltimore: The Johns Hopkins University Press.

Noble, Cheryl. 1982. Ethics and experts. *Hastings Center Report* 12 (3): 7–9.

Nussbaum, Martha C. 2002. Moral expertise? Constitutional narratives and philosophical argument. *Metaphilosophy* 33 (5): 502–520.

Parker, Lisa S. 2005. Ethical expertise, maternal thinking, and the work of clinical ethicists. In *Ethics expertise: History, contemporary perspectives, and applications*, ed. Lisa Rassmussen, 165–207. Dordrecht: Springer.

Pearlman, Robert A., et al. 2015. Ethics consultation quality assessment tool: A novel method for assessing the quality of ethics case consultations based on written records. *The American Journal of Bioethics* 16 (3): 3–14.

Pope, Thaddeus Mason. 2011. Legal briefing: Medically futile and non-beneficial treatment. *The Journal of Clinical Ethics* 22 (3): 277–296.

Pope, Thaddeus Mason, and Kristin Kemmerling. 2016. Legal briefing: Stopping nonbeneficial life-sustaining treatment without consent. *The Journal of Clinical Ethics* 27 (3): 254–264.

President's Council on Bioethics. 2008. *Controversies in the Determination of Death.* Available at https://bioethicsarchive.georgetown.edu/pcbe/reports/death/. Accessed 28 April 2016.

Rassmussen, Lisa M. 2011. An ethics expertise for clinical ethics consultation. *Journal of Law, Medicine, and Ethics* 39 (4): 649–661.

Rawls, John. 1993. *Political liberalism.* New York: Columbia University Press.

Scofield, Giles R. 1993. Ethics consultation: The least dangerous profession? *Cambridge Quarterly of Healthcare Ethics* 2: 417–448.

———. 2008a. What *is* medical ethics consultation? *Journal of Law, Medicine, and Ethics* 36 (1): 95–118.

———. 2008b. Speaking of ethics expertise.... *Kennedy Institute of Ethics Journal* 18 (4): 369–384.

Shelton, Wayne, and Dyrlief Bjarnadottir. 2008. Ethics consultation and the committee. In *Ethics by committee: A textbook on consultation, organization, and education for hospital ethics committees*, ed. D. Micah Hester, 49–77. Lanham, Rowman & Littlefield Publishers.

Siegler, Mark, and Peter A. Singer. 1988. Clinical ethics consultation: Godsend or 'God Squad'? *The American Journal of Medicine* 85: 759–760.

Singer, Peter. 1972. Moral experts. *Analysis* 32 (4): 115–117.

Steinkamp, Norbert L., Gordijn, Bert, and ten Have, Henk A.M.J. 2008. Debating ethical expertise. Kennedy Institute of Ethics Journal 18(2): 173–192.

Stolper, Margreet, Sandra van der Dam, Guy Widdershoven, and Bert Molewijk. 2010. Clinical ethics in the Netherlands: Moral case deliberation in health care organizations. In *Clinical ethics consultation: Theories, methods, implementation, evaluation*, ed. Jan Schildmann, John-Stewart Gordon, and Jochen Vollmann, 149–160. Surrey: Ashgate Publishing.

Swales, J.D. 1982. Medical ethics: Some reservations. *Journal of Medical Ethics* 8 (3): 117–119.

Tarzian, Anita J. and the ASBH Core Competencies Update Task Force. 2013. Health care ethics consultation: An update on Core competencies and emerging standards from the American Society for Bioethics and Humanities' Core competencies update task force. *The American Journal of Bioethics* 13 (2): 3–13.

Tarzian, Anita J., Wocial, Lucia D., and the ASBH Clinical Ethics Consultation Affairs Committee. 2015. A code of ethics for health care ethics consultants: Journey to the present and implications for the field. *The American Journal of Bioethics* 15 (5): 38–51.

Thomasma, David C. 1991. Why philosophers should offer ethics consultations. *Theoretical Medicine* 12: 129–140.

Varelius, Jukka. 2007. Ethics consultation and autonomy. *Science and Engineering Ethics* 14 (1): 65–76.

———. 2008. Is ethical expertise possible? *Medicine, Health Care, and Philosophy* 11 (2): 127–132.

Veterans Health Administration, National Ethics Committee. (2006). *The Ethics of Palliative Sedation*. http://www.ethics.va.gov/docs/necrpts/NEC_Report_20060301_The_Ethics_of_Palliative_Sedation.pdf

Weinstein, Bruce D. 1993. What is an expert? *Theoretical Medicine* 14: 57–73.

———. 1994. The possibility of ethical expertise. *Theoretical Medicine* 15: 61–75.

Yoder, Scot D. 1998. The nature of ethical expertise. *Hastings Center Report* 28 (6): 11–19.

Zoloth-Dorfman, Laurie, and Susan B. Rubin. 1997. Navigators and captains: Expertise in clinical ethics consultation. *Theoretical Medicine* 18 (4): 421–432.

Chapter 13
The Necessity of Clinical Experience in Medical Ethics Expertise

Matthew A. Butkus

The literature on moral expertise has raised objections salient to the role of theoretical versus clinical knowledge. The assumption that clinical ethics can be performed by simply applying a transcending ethical principle (the top-down or "engineering model" (Caplan 1983)) has been rightly critiqued as untenable – actual clinical cases are far more nuanced, and frequently what appears to be a philosophical conflict over principle A is actually a mish-mash of a number of competing values B and C, principle D, and concept E. Family conflicts, racial and gender issues, communication barriers, and a host of other unanticipated complicating factors make actual decision-making much more complex than simply "solving for ethical X."

Even the concept of theoretical approaches to applied ethics has been critiqued (Jonsen 1991; cf. Cribb 2011). I do not go so far as Jonsen to suggest that clinical ethicists can simply forgo reference to theoretical work, as there needs to be a larger understanding of the conceptual terrain that transcends the immediate issue or uncertainty. Likewise, theoretical ethicists need to set foot inside a hospital and encounter actual clinicians, staff, patients, and their families in order for their theories to be well-informed. More is needed than simple reflective equilibrium.[1]

This gives rise to my stronger claim – if expertise is possible, it cannot occur without practical knowledge, experience, or practice (see also Cribb 2011). To paraphrase philosopher Gilbert Ryle, one judges a good cook by the meal produced, not the knowledge of recipes (Ryle 1945–1946; see also Baylis 1999). Similarly, we

[1] Reflective equilibrium refers to the attempt to reconcile particular judgments with larger ethical frameworks. Our general concern is the extent to which our particular value judgments cohere with our broader understanding of moral and ethical principles (Rawls 1971). Our broader understanding of morality and ethics informs our decision-making, while individual cases may demonstrate deficiencies in our larger framework (e.g., via exceptions to rules, refinement of principles, additional nuance when judging cases that appear to be similar, etc.) – theoretical standards and practical judgments are mutually influential.

M. A. Butkus (✉)
McNeese State University, Lake Charles, LA, USA
e-mail: mbutkus@mcneese.edu

J. C. Watson, L. K. Guidry-Grimes (eds.), *Moral Expertise*, Philosophy and Medicine 129, https://doi.org/10.1007/978-3-319-92759-6_13

ought to evaluate judgments in clinical ethics not merely by how they cohere with established ethical principles and theory but by their utility in resolving family disputes, overcoming communication barriers between physicians and families, and recognizing and accounting for the myriad variables present in clinical contexts.[2] Clinical exposure sensitizes ethical agents to non-philosophical yet entirely relevant and important elements of patient care. Patients generally are not isolated entities, and the resulting psychosocial and support frameworks yield complex interactions that must be reconciled with the immediate philosophical dispute.

13.1 Assumptions

All of this occurs within a larger framework questioning the possibility of whether moral expertise is even possible (Archard 2011; Cholbi 2007; Cowley 2005; Cowley 2012; Davis 2015; Driver 2013; Gesang 2010; Gordon 2014; Hills 2009; Kovács 2010; McClimans and Slowther 2016; Powers 2005; Rasmussen 2016; Casarett et al. 1998; Singer 1972; Vogelstein 2015; Yoder 1998). The challenges to the concept of moral expertise are compelling and raise interesting epistemological questions that are, however, outside the scope of this chapter. Because the focus here is contrasting theoretical versus clinical expertise (and the necessity of clinical experience for *prescriptive* recommendations), I will not be devoting significant space to the possibility of expertise, whether one needs to be perceived as an expert, whether the ability of others to learn the requisite material undermines expertise, and so on.

Clinical ethics expertise is much more than a simple appeal to ethical theory – the practice of clinical ethics is not simply making an epistemic claim to a (non-verifiable) moral truth and rendering a simple judgment of how to proceed. Clinical ethicists must wrestle with complex psychosocial variables, legal requirements and mandates, cultural differences, and language and comprehension barriers, all within a complex and probabilistic diagnostic framework (i.e., what we *think* is actually going on in this particular patient's pathology). This complexity defies translation into easy or teachable conceptual algorithms. In short, book knowledge and theoretical grounding are the *starting points* of discipline expertise – they are necessary but not sufficient conditions for expertise. For the present argument, there are several assumptions that will drive what follows. Some of these assumptions have

[2] This is a source of conflict and personal amusement when discussing hypotheticals with my colleagues. When they present me with a sample case and ask for a resolution, I invariably ask for more information than they have anticipated. Less flippantly, I have been involved in a number of ethics consults concerning a hesitant or oppositional family that turned out to be problems with communication. It is common for a patient to be seen by a number of specialists or by multiple physicians on a treatment team, all of whom might interact with the family and provide conflicting information. The pulmonologist might tell them that their mother is doing well (because he observed good lung function) while the nephrologist tells them she's concerned (because of poor renal function), and a family without an understanding of the differences between specialties would rightfully be confused as to their mother's actual medical condition.

greater degrees of empirical validation, but even those that are more speculative should not be controversial in light of the discussion that follows.

13.1.1 Not all Opinions Are Equal

The first assumption is that not all opinions are of equal value, truth, or justification – they reflect different backgrounds and experiences to be sure (and in that sense some relativism is encouraged to keep us honest), but opinions can also reflect deficits in critical thinking.[3] We may feel passionately about the truth of our positions and arguments while simultaneously failing to realize that we have made critical errors (Hills 2009; McGrath 2011; Musschenga 2009; see also Hulsey and Hampson 2014). First, there are no perfect cognitive agents – no one is capable of stepping outside their own cognitive processing to judge the accuracy of the myriad conscious and unconscious influences on the arguments we make and the conclusions we reach (a concept that we'll revisit below). As such, we have to accept some epistemic uncertainty and make good faith efforts to correct our own biases and preconceptions (which can prove to be quite difficult). Second, we have an obligation to recognize that opinions and arguments can involve mistaken information, assumptions, assertions, formal and informal logical fallacies, and a host of related issues (such as the information we consume – confirmation bias is a pernicious problem[4]). It would seem, therefore, that opinions and positions that minimize these sources of error are more justified than those which do not, and as these are essentially correctable problems, it would seem that any attempt to provide authoritative judgment must begin with making a good faith effort to correct these deficits (see also Nussbaum 2002; Sliwa 2012; Vogelstein 2015). For the current argument, this translates into a recognition that the opinions we have towards complex and polarizing topics like forgoing treatment or if there is a right to health care might be mistaken and that our own opinion might be wrong. More pointedly, I would argue that those opinions which have empirical rather than theoretical validation are likely to be more justified, complex, and reflective of real world clinical applications.[5]

---

[3] The recognition of "bad" opinions is explicitly designed both to reject relativistic notions about ethical judgments as well as provide initial justification for singling out individuals as experts. If we recognize that some people know more and consistently make more informed decisions about a particular field, we are introducing a theoretical possibility of expertise in that field, whether it is physics, basket-weaving, or ethics.

[4] Confirmation bias refers to the tendency to pay more attention to evidence that confirms an existing opinion than evidence that refutes it. The flip side to it is *disconfirmation bias* which refers to the tendency to be more critical of evidence that refutes an existing opinion.

[5] And I would argue that this is generally true across disciplines – knowledge of the way something is *supposed* to work does not necessarily translate into how it *actually* works. *Theoretically*, patients are perfectly reasonable and unbiased deliberative agents just as their attending clinicians are dedicated and dispassionate partners in medical decision-making. Actual practice shows this to be an unjustified assumption, with actual human behaviors varying significantly in their applica-

A related facet of this assumption stems from apparent consensus about the possibility of academic moral expertise, i.e., that person A can know more about a given field than person B (Archard 2011; Hills 2009; Rasmussen 2011; Vogelstein 2015; cf. Cross 2016). My PhD gives me authority to lecture on philosophical topics to my students as well as justifying their faith that I am passing on good information to them. I am perfectly able to *describe* ethical conflicts, paradoxes, challenges, and so on as well as common arguments and fallacies. I can *describe* the current framework elucidated by law, theology, philosophy, communications theory, gender issues, and so on. Controversy emerges, however, when this *descriptive* claim extends into *prescriptive* territory – once I begin to tell people what they should do and how they should act, some argue that I have gone beyond the bounds of legitimate epistemic and moral justification (Archard 2011; Hills 2009; Rasmussen 2011; cf. Cross 2016). I might still be able to make this claim if I fit a specific set of criteria (e.g., a Catholic priest who has studied Catholic medical ethics extensively and is being asked by a Catholic patient in a Catholic hospital how he should act in light of the Church's position on vasectomy – see also Cross 2016), but many do not extend this expertise beyond such restrictions, arguing that pluralism,[6] the lack of a recognition of authority, and the correctable nature of the knowledge deficit (the ability of the patient to become educated about the relevant clinical and ethical issues) undermine the legitimacy of the expertise claim. This ties directly into the second assumption.

13.1.2 There is a Pragmatic Necessity for Expertise

In a clinical environment, difficult decisions have to be made, and while the recommendations of an ethics consultant or consult team are not necessarily binding, they very frequently heavily influence the final decision made.[7] Treatment teams are asking for guidance, and there is at least an implicit assumption that this specialist (or team) is capable of providing a means forward to resolve an apparent impasse or

tion of reasoning, evidence, personal experience, and prejudice – a multivariate issue that does not reduce down to a simple textbook case.

[6] There are a variety of understandings of pluralism. For my purposes here, I am content with a broad understanding of the term reflecting a wide variety of different religious and philosophical backgrounds, personal differences on what is meaningful in life, what constitutes an acceptable quality of life, and so on. I am not arguing for political pluralism, scientific pluralism, or moral relativism. Rather, I am merely noting the (hopefully uncontroversial) acknowledgement that it is entirely possible that the values of the patient will not parallel the values of the clinician or the consulting ethicist.

[7] And I would argue that this is a good thing – it is at least a tacit acknowledgement by the treatment team that there is some degree of moral ambiguity present and that the proper path forward is not clear to them. I much prefer an honest admission of uncertainty to a brash assumption that knowledge or expertise in the technical elements of a complicated field like medicine translates into knowledge or expertise into the non-technical and value-laden elements of that field. Bioethicist Robert M. Veatch calls that assumption the fallacy of the generalization of expertise (Veatch 1991).

problem (Adams 2013; Cowley 2012; see also Agich 1995; Reiter-Theil 2009). In a classroom, an impasse or problem may remain unresolved at the end of the discussion (or semester), but in the context of patient care, this is not a tenable position. There are no meaningful consequences of denying expertise in a theoretical environment – patient care doesn't afford that luxury.

As such, there is a compelling pragmatic reason to assume that moral expertise is possible – in order for real decisions to be made, there needs to be more than simple *descriptive* expertise.[8] Recommendations go beyond simple description of conflicts and positions, but make actual suggestions of how to proceed.[9] Whether the treatment team actually *acts* on these recommendations is a separate question – it is still possible for the treatment team to disagree with the consultant or consult team, but that doesn't in itself deny the possibility of expertise. Medicine is by its nature probabilistic and we accept good faith treatment decisions using the best available evidence and clinical judgment.[10] Pragmatically, similar reasoning applies to the epistemic uncertainty involved in ethics consults – we don't simply pull a

---

[8] This is not meant to suggest that simply because an ethicist *can* contribute to a decision that said ethicist (or *any* ethicist) is thereby an expert or that moral expertise is possible. I am not trying to create a quasi-ontological argument for the existence of expertise. I am merely suggesting that the practical utility of an ethics consult in resolving an apparent impasse suggests it has subjective value to that treatment team – they view the ethicist as an authority worth consulting, and would not do so if there was nothing to distinguish the ethicist from the average person on the street.

[9] Some ethicists suggest that this prescriptive role is inappropriate. I have concerns about this as it would seem to change the role of the ethicist from consultant to reference – i.e., someone who either serves to simply provide information or to provide general guidance about *decision-making*, rather than about this particular decision. I'm not sure that this is helpful – when I teach my students ethics, there are a number of different questions I encourage them to consider: who is directly affected, what principles or relationships might be involved, what their professional or legal obligations might be, reasonably foreseeable consequences, whether there is any intrinsic moral status to the action considered (like murder being an instance of "wrong killing"), and so on. I do not make the decision for them, but teach them how to make decisions, and there is certainly value to that, but there are also times when decisions must be made and recommendations provided – a urologist would not be consulted on a patient simply to teach the clinical staff about kidney function, but to provide recommendations for this particular patient's care that are in line with the best practices of that field. The ethics consults that I have answered have served both purposes – they provide *both* education about the ethical principles involved *as well as* provide recommendations for this particular case. The underlying argument being that having this "teachable moment" can both resolve the existing issue and increase the skill set for the clinicians involved to better manage similar cases in the future, akin to being a tutor. Sometimes a good tutor must work through the whole problem in order to demonstrate how to work similar problems in the future.

[10] Several authors have suggested that it is appropriate to draw comparisons between clinical ethics consultation and consulting a physicist or other scientist (Cowley 2012). Without taking a position on the specific claims being made, I would simply suggest that both applied ethics and science are essentially inductive and probabilistic – science is not simply deductive nor is the correspondence model of scientific truth tenable due to our implicit epistemic barriers and limitations. Rather, science involves a coherence model that, despite its strong justification and general utility, may or may not actually reflect objective universal truths (see also Tännsjö 2011). If we are willing to defer in good faith to educated scientists on *probabilistic* scientific matters, it stands to reason that we could plausibly be willing to defer in good faith to educated ethicists on *probabilistic* ethical matters (see also Gesang 2010). Martha C. Nussbaum (2002) argues, however, that while philosophers may in fact be moral experts, they ought not take on public roles that are "incompatible with the committed ethical searching of their fellow citizens," e.g., in court cases on controversial public topics.

random person off the street and ask them what to do, but rather we ask those who have studied the relevant law, theology, and philosophy, and have some faith that the resultant recommendations will reflect the same care and concern that the clinician applies in light of her own clinical studies (Rasmussen 2011). Some may take exception to these assumptions, but I believe that they will generally hold up in real world environments. The argument that follows thus assumes that *prescriptive* ethical expertise is possible and that it is reasonable for others to defer to it. This deference, however, is neither categorical nor blind.

13.1.3 Expertise Does Not Require Infallibility

My third assumption is that it is uncontroversial to note that human agents of all types are limited and imperfect beings, and as such, even those who make good faith efforts to transmit accurate information, relay the professional standards of their fields, minimize personal biases, and ensure the logical consistency of their positions are still capable of making mistakes. This does not undermine expertise or render the concept incoherent. Rather, it suggests that we can and ought to maintain a healthy critical attitude towards what we hear from experts. This does not require us to descend into the depths of epistemic relativism – this is not a conflict of "mere opinions" as noted above (see also Cowley (2005)). Rather, it is a recognition that even otherwise intelligent and reasonable people can make mistakes in good faith – even philosophers (Schwitzgebel and Cushman 2012).[11]

Additionally, it places a burden on those who would claim expertise to ensure – to the best of their ability – that the information that they convey is accurate and that the descriptive and prescriptive elements cohere with the relevant law, theology, philosophy, and medicine. Claiming expertise is a serious consideration, and it must be able to be demonstrated via verifiable knowledge claims (e.g. do the standards of the references fields actually support the claims being made?). We would rightly require our experts to be reliable and to provide quality information.[12] This, however, might prove contentious, and these claims warrant further exploration elsewhere,

[11] When I discuss this with my students, I reference a patient who was admitted to a major university research hospital with ankle pain and an apparent skin reaction on her lower leg. She was seen by multiple specialists up to and including infectious disease, and she was treated with a variety of antibiotics and other medications to which she had a negative reaction. Each specialist suggested a different course of action, viewing her condition through the lens of their own experience. Ultimately the case was resolved by an orthopedist who immediately recognized it as a sprain, as the patient was a ballerina.

[12] Additionally, there is significant debate about what constitutes verifiable knowledge claims, reliability, and quality information. For my purposes, I understand these to refer to the ability to analyze an ethical dilemma in light of existing medicine, case law, philosophy, and other fields, i.e., is what this person saying supported by these fields, and are the conclusions reached (descriptive and prescriptive) routinely supported by evidence drawing from these fields or another strong evidentiary or logical basis. Whether this would satisfy Plato is a separate question.

such as the current discussion of proper accreditation for clinical ethicists (Kipnis 2009; Kodish et al. 2013).

In light of these assumptions, it is useful to turn now to the central thesis – clinical experience is necessary to justify a claim of ethical expertise. This may seem, at first, to be a mild point to make, but the claim becomes stronger once it is unpacked. More explicitly, while philosophers can study ethics and applications in medicine, I argue that they cannot claim expertise unless this theoretical knowledge is complemented by practical experience. Theoretical knowledge is a *necessary but not sufficient* condition for clinical ethics expertise.[13] The potential applications of clinical ethics are not limited to case consultation. Rather, ethicists are consulted in committee work, policy development, staff education, and other *ad hoc* responsibilities that would require a much longer work, but it *is* possible to elucidate this central thesis through a shorter analysis by exploring several cases.[14]

13.2 Cases

Case-based reasoning (both formal and informal) are routine elements of medical ethics classes. Anonymous cases are presented in order to highlight salient lessons or critical thinking elements, and the class spends hours (or days) deliberating over different possible courses of action. They are generally excellent methods of teaching critical concepts in applied ethics – showing how abstract theories can translate into practical situations. Students are capable of weighing multiple options and making decisions that are essentially consequence-free – no one is actually helped or harmed in the classroom as a result of the decisions made. The same is not true once the classroom becomes the clinic, and three cases should help to demonstrate this.

First, consider Martha, who is in the ICU after being found nonresponsive. She is in her mid-80s and has a medical history significant for diabetes mellitus (which resulted in her right leg being amputated below the knee) and partial paralysis (she

[13] A recurring frustration during my graduate education, clinical placement, and subsequent career was reading articles making theoretical assertions about patient care that were immediately and directly refuted by the patients I was working with. If, for instance, autonomy is an absolute and inviolable principle, how could we justify forcibly hospitalizing patients and holding them against their will? What principles of compassionate care would apply to malingerers who were actively abusing the system? How well would principles apparently derived from house medicine translate into behavioral health contexts?

[14] The central thesis of the necessity of actual experience translates to issues at multiple levels of a healthcare organization. For instance, a classroom discussion of institutional resource allocation as a concept is a radically different experience than serving on a dialysis committee that has to actually decide which patients will be served and which will not. Discussing whether healthcare is a right is a very different question than doing the policy work of crafting a national single-payer policy. Practical and pragmatic concerns don't disappear simply because one is no longer dealing with the issues facing individual patients.

is unable to move her left side). Her treatment has plateaued – she is not getting better or worse with the standard of care – and the treatment team isn't sure whether it would be appropriate to use more aggressive measures with her. In an attempt to resolve this impasse, the team calls an ethics consult. Because the hospital uses a team approach (rather than a single individual responding to provide a consultation), a doctor and nurse in the treatment team meet with the hospital ethicist, who is accompanied by two observing graduate students. The consult team meets to review Martha's chart and put them all on the same page, and then they meet with the patient's family – two daughters and a son. The first daughter is a registered nurse, and she was the one who spotted the gangrene on Martha's foot which necessitated the amputation. She believes that her mother has had enough and is trying to let go. The second daughter has been her mother's caregiver for the past 10 years but did not appreciate the severity of her mother's circulation problems. She believes that her mother would want to live. The daughters disagreed vehemently about the necessity of the amputation and did not speak to each other for 3 years after the procedure was performed. The son says he will go along with whatever is decided, but he is also visibly intoxicated. What should be done?

As presented, the case suggests a number of recurring concerns that are common in medical ethics: the patient's previously expressed autonomy interests, the potential utility of advance directives and the appointment of a proxy decision-maker to resolve treatment and value questions, and so on. The case could easily be used to support arguments for requiring individuals in frail health to inform their families and caregivers about their preferences. The case could easily demonstrate that expressions of personal autonomy aren't limited to decisions to forgo treatment – a patient can just as easily exercise their autonomy by telling their caregivers or proxies that they "want everything" or that they are willing to consent to more invasive procedures on a trial or temporary basis (as many patients are not necessarily afraid of interventions like ventilators but rather afraid of dying on them). The case could easily stymie both undergraduates and graduate students alike – both sisters genuinely believe that their mother would want what they say she would. There is a real conflict, too, as to what would we expect to happen to their relationship if the treatment team were to pick one over the other. If they didn't speak for 3 years after the amputation, what are the odds that they would speak to each other again at all if a decision were made that resulted in their mother's death? We will return to Martha's case below.

Second, consider William, a man in his mid-40s who is dependent on dialysis three times per week due to kidney failure. He has informed his treatment team that he no longer wants to be dialyzed and has said that he understands that this will result in his death. He also told his treatment team that if they did not honor his request, he would take steps to hasten his death. This last statement worried his treatment team, which first called a psychiatric consultation to rule out depression. Following the psychiatric consult, which did not find evidence supporting a diagnosis of depression, the treatment team still felt that the request was worrisome and called an ethics consult. As above, the consult is answered by a team of a variety of disciplines, which meets with the William to discuss his wishes. He feels that dialysis is too burdensome, as it impacts his ability to socialize with his friends, eat out, and so on. He has friends who were on dialysis who also found it interfered too much with their lives. He does not disclose anything in the course of the conversation that suggests that his decision-making process is compromised by a physical or mental condition.

As in Martha's case, this would seem to clearly involve a conflict between patient autonomy versus beneficence – what the patient wants is in direct conflict with what would seem to be in his best medical interest.[15] The principle of autonomy is rightly held as essential to responsible and ethical conduct in medical treatment. It is not an absolute, in that there are circumstances when other philosophical or practical considerations can outweigh it, but these are relatively rare: medically inappropriate treatment, suicidal or homicidal ideation, a gross inability to care for oneself, delirium, other disease states that compromise cognition, adjudicated incompetence, and so on.[16] The default assumption is that patients have the ability to decide for themselves what treatments are value-congruent and tolerable and what treatments are intolerable and too burdensome.

Third, consider Steven, a 70-year-old man with non-Alzheimer's dementia in a geriatric psychiatry unit. He was admitted for anxiety, which is exacerbated by an inability to urinate, as he had been previously diagnosed with bladder cancer. His dementia has advanced to such a degree that he is no longer oriented to place and time, and he experiences memory lapses so severe that he forgets conversations within a few minutes of them taking place. His family has asked that his cancer diagnosis not be revealed to him, as they are worried that it would increase his anxiety. The treatment team did not call an official ethics consult, but rather agreed with the family's preferences and did not disclose the diagnosis to him. He was treated for anxiety and discharged back to his care facility.

Was it right to withhold this information from him? In general, principles of autonomy and informed consent require physicians to be honest and forthright with their patients, disclosing to them information relevant to their medical decision-making. However, issues get murky once psychiatric disorders and cognitive impairment enter into the situation – it is unclear that our obligations to a patient's autonomy have the same force. While opinions are divided on what to do

[15] As a point of clarification, I understand beneficence to refer primarily to the patient's physical and mental best interests (physical and mental health). I do not adhere to arguments that would include a patient's value structure or preferences, as I would argue that this blurs the line between beneficence and autonomy. As such, when I refer to beneficence throughout this chapter, it should be understood in this light.

[16] Delirium, compromised cognition, and adjudicated incompetence may prove controversial in that a legitimate question can be asked as to whether autonomy is being overridden or is simply inapplicable in these circumstances. It is right to note that respecting autonomy is not simply respecting whatever the patient says. However, this raises additional questions about capacity as we learn more about how we make decisions even in "normal" cognition, which is a fascinating discussion and well outside the purview of this chapter. For the moment, it suffices to say that autonomy as a principle of self-direction and governance is not an absolute, but it is one that *prima facie* requires us to be mindful of whether our decision not to apply it in a particular case is reasonable/unreasonable or justifiable/unjustifiable.

in this *particular* case, the underlying philosophical commitment to truth-telling, even in the face of unpleasant news, remains an essential element of responsible clinical practice.

Taken at face value, these three cases represent standard hypothetical cases in clinical ethics that highlight essential and recurring principles of biomedical ethics. It would not be unexpected to see them appearing in a textbook for the chapter on forgoing medical treatment. Cases and thought experiments like these are quite useful for teaching, and it is interesting to see undergraduates wrestle with them. Thought experiments are essential in both theoretical and applied ethics – they can elucidate paradoxes and contradictions, cast light on hidden logical fallacies and unstated assumptions or assertions, and explore potential avenues of action.[17] Thought experiments alone, however, are not sufficient for working in a clinical setting (Baylis 1999).

Textbook cases and thought experiments do have a role in terms of *introducing* ethical concepts, but they do not suffice to generate expertise. The theoretical principles brought forth by thought experiments may be useful in introducing scarcity (e.g. asking students how they would allocate resources for a hypothetical medical population) or whether we ought to allow active euthanasia, but it isn't clear that this is sufficient ground to claim justification for a particular judgment about a particular patient.

Thought experiments tend not to have the nuance or concreteness of case studies – thought experiments are designed to raised specific issues or concepts. They may test the ability of a particular proposition to stand up to scrutiny (e.g. does it produce a paradox), may show the incompatibility of two principles, or may question a "meta" element of a given discussion by raising a *conceptual* issue informing the current problem. While these are all useful activities, they are not necessarily helpful in resolving *particular* clinical problems. Additionally, much like conducting clinical research, there are concerns about internal and external validity (Wilson 2014). These essential concerns challenge "expertise" derived from considering thought experiments rather than case studies.

To extend the above research analogy briefly, the first concern when designing a research study is whether the methodology proposed is capable of actually measuring or testing the behavior or intervention in question. If, for instance, I am looking to study rates of depression in my clinical patients, I must make sure that I am using a metric that (1) is capable of detecting depression and (2) is capable of distinguishing symptoms stemming from physical pathology from those stemming from psycho-

[17] It is worth noting that thought experiments are distinct from case studies. Thought experiments are generally abstract 'questions to consider' like whether it is preferable to act and cause one death or refrain from acting and allow five deaths to occur (the standard trolley problem in philosophy). Case studies tend to be more concrete and frequently contain examples drawn from actual events and individuals, necessitating anonymity and demonstrating the concrete reality of the problem.

pathology. If these criteria are not met, I am engaging in a futile exercise – I cannot actually study what I want to study (a failure of internal validity).[18]

Similarly, it is possible that a given thought experiment does not actually raise the ethical issues it is meant to raise. A proposed thought experiment may try to draw out a particular value conflict when there are deeper issues or confounding ethical principles present, or the case is not developed enough to elicit the specific conflict desired. Questions of patient autonomy might actually bring forth cultural issues, underlying biases of the designer, or may not actually reflect the conflict in question (e.g., a discussion of autonomy may actually require a prior discussion of the obligation to disclose particular information to a patient, so a question of autonomy becomes a question of truth-telling). If a thought experiment isn't actually exploring the problem at hand or is more indicative of a different type of value conflict, it is not well-designed for that particular issue. As a consequence, there are significant concerns about expertise predicated solely on thought experiments and controlled cases.

13.2.1 External Validity (Ecological Validity)

The larger concern is translating research from controlled environments into the wild. What happens when we step outside the lab and begin observing behaviors and interventions when there are variables outside of our control? If a rigid experiment yields promising results, but the behaviors observed in the lab do not correlate well with behaviors in freer environments, it fails a test of external (or ecological) validity. It fails to measure the behavior in question because it has created an artificial construct of behavior, and as such, is not a reliable guide for real environments. This is a recurring concern in the social sciences – the behaviors modeled in psychology labs, for instance, might not translate well into the real world.

In thought experiments, the designer is omnipotent – she is free to place whatever constraints she desires on the agents in question. She is also omniscient, and is able to assert motivations, contexts, beliefs, and so on in order to control the environment as much as possible. Neither omnipotence nor omniscience translate well into the real world of clinical environments. We frequently are unaware of patient value structures, personal beliefs, familial interactions, professional relationships, education levels, or any of a wide variety of factors that impact their beliefs, behaviors, and decision-making abilities. The interventions we propose are also probabilistic – this is what we think *could* or *should* work, but it isn't something we can guarantee. What seems like a simple case in a textbook can become nightmarish in

[18] Depression, physical illness, and medical interventions can all produce similar physical diagnostic symptoms (e.g., sleep changes, loss of energy, appetite, sexual activity, and interest in hobbies, etc.), all of which may be present in a patient in the ICU. As such, both false positives and false negatives are possible if a metric is employed that suggests a psychiatric condition is present based on physical criteria, making the results unreliable.

clinical reality. Further, there is a temptation in a textbook case to try to find *the* right answer, when the clinical reality might be more about finding *a* right answer or the *least wrong* answer. As I tell my students, ethical deliberation in the real world is messy, and does not lend itself to controlled environments and studies. If professionals are simply focusing on theoretical examples, "meta" issues of language or philosophical frameworks, or other concepts not derived from or immediately applicable to *actual patients*, *actual institutions*, and *actual conflicts*, then there is legitimate ground to question whether they are *actual* experts.

The three cases presented above are not simply thought experiments – I was one of the students in attendance as part of the consult team for both Martha and William as I cut my clinical ethics teeth, and Steven was one of my patients while I worked as an aide in behavioral health during graduate school.[19] As with many thought experiments, the clinical realities of the cases are not completely reflected in the text. They were complicated by additional factors that generally are not presented in textbooks, simply because many (if not most) clinical ethics consultation cases involve contextual elements that are impossible to anticipate – the casuistry of the clinic is different than the casuistry of the classroom. We will return to this issue later.

When we analyze Martha's case, for instance, there are factors not present in the case scenario. The son's intoxication clearly precludes him from making a life or death decision for his mother at this point, leaving the team with the two conflicting daughters. In the course of the conversation, it was revealed that there had been a fourth child – a daughter who had had a stroke 20 years prior and who had been ventilator-dependent in critical care. Although she had ultimately died, Martha had told her daughter's treatment team "Don't you turn off that machine!" which clearly expressed her wishes towards her daughter's care. This invites conflicting interpretations. On the one hand, Martha may have been expressing her feelings about continuing treatment in critical situations. If that was the case, then it would seem that she would want ventilation maintained. On the other hand, she may have felt differently towards maintaining her daughter's care than she would have felt towards her own – it shouldn't be controversial to note that generally parents don't want to see their children die. Even this split, however, doesn't capture the complexity entirely. People change and their preferences change with them.[20] Martha was younger (mid-60s) when her daughter had the stroke, making her daughter even younger (somewhere in her 30s or 40s). Given her daughter's youth there are compelling reasons to believe that Martha might have made a different decision for her daughter in the past than she would have made about herself in the present, as Martha may have felt

[19] Needless to say, none of these are the patients' real names nor are the details exactly as presented in order to preserve their anonymity.

[20] In fact, it is entirely possible for people's preferences to change so radically that adhering to an older document may yield an autonomy violation. The patient at T_1 may not have the same value preferences at T_2, so holding T_2 patient to the terms of T_1 patient's advance directive may violate their contemporaneous preferences if they have not been adequately expressed. We may end up undermining autonomy in our good-faith attempt to uphold it.

that her daughter could recover and live a long and happy life. Because she hadn't articulated her contemporaneous values, the consult team was left with information that was both helpful and confounding.

Ultimately, the treatment team found a path forward by discussing the motivations of the sisters. Neither of them wanted their mother to die, and both agreed that it would make sense to think about how Martha had reacted to their sister's death, as it was the only insight available into how she might feel. Since she was currently stable, they agreed that they would revisit the issue if her condition worsened, as it would a week after the consultation. Martha ultimately died, but her family wasn't fractured as a result. The case illustrates a recurring critique of top-down "engineering" approaches to ethics, in which it is assumed that problems can be resolved by simply applying an abstract philosophical principle. Nuance and context matter, and essential information isn't always immediately evident. You have to ask *people* questions, not books or articles.

William's case presents other difficulties. He is still a young man, and it is entirely understandable that the treatment team would be worried about an apparent desire to die. This is not to say that it is impossible for young people to justifiably or rationally conclude that their lives are not worth living, and it is legitimate to ask whether it is appropriate for person A to tell person B whether B's life is or is not worth living. But this seems to assume idealized cognitive agents, not how we are really wired – cognitive psychology and neuroscience have demonstrated that we have both emotional and rational elements to our cognition,[21] and it is entirely reasonable for someone to have an emotional response that is then overridden by further reflection. For instance, it might not make visceral sense to a clinician if her Jehovah's Witness patient elects to forgo a lifesaving therapy because it involves accepting blood products, but her legal and ethical training may remind her that patients have the right to refuse treatment and are not necessarily cognitively impaired because they make decisions with which we don't agree. She may still feel strongly that a patient should do X, despite recognizing that she cannot force the patient to do X.

In William's case, it is understandable that the treatment team may feel consternation and perceive a need for psychiatric and ethics consults. Patients can withhold information just as well as clinicians can, which can also include their motivations for seeking treatment or refusing it.[22] It is entirely possible that William was actually

[21] See, for instance, George Lakoff and Mark Johnson's *Philosophy in the Flesh* (1999) for a fascinating discussion of what cognitive neuroscience tells us about essential elements of human cognition.

[22] I spent several years working in behavioral health units and the psychiatric emergency room. Malingering was, is, and will continue to be a problem so long as diagnostic criteria are poorly defined, psychiatric admissions criteria remain soft, and institutional, state, and federal policies remain punitive. I am not suggesting that hospitals simply turn patients away, but I have met a remarkable number of patients who knew what to say in order to secure themselves admission and who would walk down the street to the next behavioral health facility once they were discharged from ours. The take home message for the present discussion is that the truth of the matter might not always be evident or may be intentionally obscured on *both* sides of the clinical interview.

suicidal and depressed, but he answered questions in a way that prevented psychiatric detection. The case is complicated by theoretical issues as well – he did not disclose information at the time which would have undermined his decision-making capacity, but it is entirely possible that there was an issue or problem with his thought process. Again, cognitive psychology and neuroscience provide significant insights into actual decision-making processes, as well as providing a picture of all of the elements that can influence that process at both a conscious, unconscious, and preconscious level, which I discuss elsewhere (Butkus 2015). Our thought processes are replete with unconscious associations and biases that can alter how we experience and process information. These biases and automatic processes are frequently useful – they provide us with an easy means of navigating our everyday lives and novel situations. However, they can also be sources of error – we place too much emphasis on a particular memory or experience, which can alter our perception of a current event or situation.

In short (and hopefully without causing controversy), our memories and past experiences can cause us to make incorrect, inappropriate, or uninformed decisions. While we can normally use these "bad" decisions to guide decision-making in the future, this can be next to impossible when deciding to terminate treatment – our own death is not a teaching experience.[23] As such, there was a chance that William's decision was an honest but mistaken processing of his current situation and consequently there is a chance that he might have made a different decision if different questions were posed to him. This does not mean that the consult team arrived at the wrong recommendation in supporting his decision to terminate dialysis, but it does mean that there are a number of factors that must be considered beyond rote application of philosophical principles or uncritical application of standards of decision-making capacity.

Finally, Steven's case illustrates nonstandard conflicts that can occur, specifically between our obligation to tell the truth to patients versus delivering compassionate care. Steven's decisional capacity is profoundly challenged by his dementia, and the necessity of reorienting him and reminding him of previous conversations was much more severe than presented above. At one point, we were reminding him of basic elements of his care on a minute-by-minute basis. The family was concerned that if he were made aware of his cancer diagnosis, it would exacerbate his anxiety – the effort to be honest and truthful with him would likely have directly worsened his quality of life. To help manage his symptoms, he was provided with a Texas catheter and advised to urinate when he could and he was told that sometimes people who felt the urge to urinate had a hard time going. There is a body of literature on both positive patient attitudes towards compassionate deception as well as other philosophical justifications, which I discuss elsewhere (Butkus 2014).

[23] I am aware that people have been revived after being declared biologically and clinically dead and who could potentially apply life lessons learned (like not walking on thin ice), but these situations are *extremely* infrequent and generally not reflective of the type of decisions exemplified in William's case.

For the present work, it is important to note that in a clinical setting, our obligations to individuals with profound cognitive impairments may be different than those we have towards patients who are capable of making their own treatment decisions, and the right thing to do might not be to uphold autonomy over all other values. At the same time, it is extremely difficult to establish circumstances when it is morally permissible to withhold information or lie to a patient – it would seem that discussions of morally permissible duplicity open a door to potentially undermining the trust essential in physician-patient relationships. Such a contextual nuance defies easy categorization, and because of the variety of interpersonal variables (physician-patient, physician-family, patient-family, physician-institution, etc.), they may simply have to be decided on a case-by-case basis, which affirms the need for casuistry without providing clear guidance.[24] It can suggest questions to ask, but only within a clinical and situationally-appropriate context, and it points to the need to have clinical cases to serve as bases for evaluation or comparison.

13.3 Clinical Ethics Expertise Requires On-Going Experience

These cases illustrate the essential thesis of this chapter – practical experience is an essential element of any claims to medical ethics expertise. It is entirely possible to familiarize oneself with the relevant literature in law, philosophy, theology, etc., but without the appreciation of case complexity, such knowledge does not translate into wisdom, just as a theoretical physicist is not by default also an engineer. For instance, the moral prohibition of administering blood products to a patient who is Jehovah's Witness is well-understood. But this prohibition isn't an absolute – it is entirely possible that *this particular* Jehovah's Witness is willing to receive blood products or refuses when his spouse is present but consents when she visits the cafeteria. Similarly, Catholicism carries with it suppositions about morally ordinary versus morally extraordinary care, as well as morally permissible versus impermissible procedures (e.g., those failing to meet the criteria for double effect, those violating the principle of totality, etc.). But individual Catholic patients may or may not endorse these beliefs (and, in fact, there is a significant role of individual conscience in following Catholic medical ethics), and without knowing a particular patient's degree of adherence (or awareness of doctrine), it is difficult to assume what this *particular* patient may want. As such, there is a critical need to have experiences that develop contextual understanding and awareness of factors that may influence decisions outside of the parameters of a simple hypothetical case or top-down

[24] Casuistry is case-based reasoning in which the current case is compared to previous cases. If, for instance, a case resembling the current one facing clinicians was considered in the past, it would be reasonable to see if there are sufficient similarities between the two that would allow it to serve as a guide. Alternatively, the case may be judged only a continuum of cases in which a particular action was considered to see where the current case would fall.

analysis (Driver 2006; Molewijk et al. 2008; cf. Ryberg 2013) – in short, in order for us to develop the necessary skills and abilities to become experts, we need practice (Rossano 2008).

At this point, I want to extend my premise – not only does clinical ethics expertise require experiential learning and contextual awareness, it requires *on-going* learning. It is entirely possible that a once-skilled ethicist is no longer an expert as the field has advanced.[25] While John Stuart Mill, Immanuel Kant, and Aristotle are not writing any new books, the same cannot be said about politicians writing new health policy, regulatory or accrediting bodies changing the standard of care, the evolving state of medical technology, or the new and unique risks faced by patients and practitioners. The fields defining the framework of a coherent proposition in clinical ethics are themselves defined by dynamism – my lectures on health care policy must change with each shift in federal, state, or institutional policy, each shift in technology, and each new meta-analysis of effective medical practice (see also Vogelstein 2015). Working in clinical ethics is like exercise – there is a skill set that must be used if it is to be maintained. The longer one goes without clinical exposure and context, the less sensitive and aware one is to the actual ethical issues as well as subtle nuances in otherwise straightforward contexts. Just as we expect our doctors and lawyers to engage in lifelong learning to stay abreast of the changes in their fields, so too should we be rightly concerned about ethicists staying up to date.

As mentioned at the outset of this chapter, I have assumed throughout that expertise is possible, for both theoretical and pragmatic purposes – we have to be able to make clinical recommendations, as patient care routinely raises complex and confounding ethical dilemmas. The realities of patient care do not allow us to simply throw up our hands and declare impossibility - we must act as if moral expertise in clinical ethics is possible. That being said, however, expertise in theory is not the same as expertise in application, and it would be inappropriate to assume that one can hold forth with authority simply by doing a lot of reading. Ethical theory and research are essential to moral expertise but are insufficient when we enter into the clinic. Despite controversy, theory and principle can provide us with a compass and map of the ethical terrain, but alone they are not reliable guides. They must be informed by direct experience – the principle must be reconciled with practice.

References

Adams, D.M. 2013. Ethics expertise and moral authority: Is there a difference? *The American Journal of Bioethics* 13 (2): 27–28.

Agich, G.J. 1995. Authority in ethics consultation. *Journal of Law, Medicine & Ethics* 23: 273–283.

[25] And I do not hold myself exempt from this concern – my career path has taken me far afield from direct patient care and interaction, and I can feel the justification for my own claims of expertise weakening as I spend more time on theoretical issues in cognitive psychology and neuroscience. This expertise is a muscle that must be exercised like any other, and it runs the risk of atrophy without use.

Archard, D. 2011. Why moral philosophers are not and should not be moral experts. *Bioethics* 25 (3): 119–127.

Baylis, F. 1999. Health care ethics consultation: 'Training in virtue'. *Human Studies* 22: 25–41.

Butkus, M.A. 2014. Compassionate deception: Lying to patients with dementia. *Philosophical Practice* 9 (2): 1388–1396.

———. 2015. Free will and autonomous medical decision-making. *Journal of Cognition and Neuroethics* 3 (1): 75–119.

Caplan, A.L. 1983. Can applied ethics be effective in health care and should it strive to be? *Ethics* 93: 311–319.

Casarett, D.J., F. Daskal, and J. Lantos. 1998. The authority of the clinical ethicist. *The Hastings Center Report* 28 (6): 6–11.

Cholbi, M. 2007. Moral expertise and the credentials problem. *Ethical Theory and Moral Practice* 10: 323–334. https://doi.org/10.1007/s10677-007-9071-9.

Cowley, C. 2005. A new rejection of moral expertise. *Medicine, Health Care and Philosophy* 8: 273–279. https://doi.org/10.1007/s11019-005-1588-x.

———. 2012. Expertise, wisdom and moral philosophers: A response to gesang. *Bioethics* 26 (6): 337–342. https://doi.org/10.1111/j.1467-8519.2010.01860.x.

Cribb, A. 2011. Beyond the classroom wall: Theorist-practitioner relationships and extra-mural ethics. *Ethical Theory and Moral Practice* 14: 383–396. https://doi.org/10.1007/s10677-011-9289-4.

Cross, B. 2016. Moral philosophy, moral expertise, and the argument from disagreement. *Bioethics* 30 (3): 188–194.

Davis, M. 2015. On the possibility of ethical expertise. *International Journal of Applied Philosophy* 29 (1): 71–84. https://doi.org/10.5840/ijap201561738.

Driver, J. 2006. Autonomy and the assymmetry problem for moral expertise. *Philosophical Studies* 128: 619–644. https://doi.org/10.1007/s11098-004-7825-y.

———. 2013. Moral expertise: Judgment, practice, and analysis. *Social Philosophy & Policy* 30 (1–2): 280–296. https://doi.org/10.1017/S0265052513000137.

Gesang, B. 2010. Are moral philosophers moral experts? *Bioethics* 24 (4): 153–159.

Gordon, J.-S. 2014. Moral philosophers are moral experts! a reply to david archard. *Bioethics* 28 (4): 203–206.

Hills, A. 2009. Moral testimony and moral epistemology. *Ethics* 120: 94–127.

Hulsey, T.L., and P.J. Hampson. 2014. Moral expertise. *New Ideas in Psychology* 34: 1–11.

Jonsen, A.R. 1991. Of balloons and bicycles: Or, the relationship between ethical theory and practical judgment. *The Hastings Center Report* 21 (5): 14–16.

Kipnis, K. 2009. The certified clinical ethics consultant. *HEC Forum* 21 (3): 249–261. https://doi.org/10.1007/s10730-009-9104-y.

Kodish, E., Fins, J. J., Braddock III, C., Cohn, F., Dubler, N. N., Danis, M.,. .. Kuczewski, M. G. (2013). Quality attestation for clinical ethics consultants: A two-step model from the american society for bioethics and humanities. *Hastings Center Report,* 43(5), 26–36. doi:https://doi.org/10.1002/hast.198

Kovács, J. 2010. The transformation of (bio)ethics expertise in a world of ethical pluralism. *Journal of Medical Ethics* 36: 767–770. https://doi.org/10.1136/jme.2010.036319.

Lakoff, G., and M. Johnson. 1999. *Philosophy in the flesh: The embodied mind and its challenge to western thought*. New York: Basic Books.

McClimans, L., and A. Slowther. 2016. Moral expertise in the clinic: Lessons learned from medicine and science. *Journal of Medicine and Philosophy* 41: 401–415. https://doi.org/10.1093/jmp/jhw011.

McGrath, S. 2011. Skepticism about moral expertise as a puzzle for moral realism. *The Journal of Philosophy* 108 (3): 111–137.

Molewijk, A.C., T. Abma, M. Stolper, and G. Widdershoven. 2008. Teaching ethics in the clinic: The theory and practice of moral case deliberation. *Journal of Medical Ethics* 34: 120–124. https://doi.org/10.1136/jme.2006.018580.

Musschenga, A.W. 2009. Moral intuitions, moral expertise and moral reasoning. *Journal of Philosophy of Education* 43 (4): 597–613.

Nussbaum, M.C. 2002. Moral expertise? constitutional narratives and Philosophical Argument. *Metaphilosophy* 33 (5): 502–520.

Powers, M. 2005. Bioethics as politics: The limits of moral expertise. *Kennedy Institute of Ethics Journal* 15 (3): 305–322.

Rasmussen, L.M. 2011, Winter. An ethics expertise for clinical ethics consultation. *Journal of Law, Medicine, and Ethics* 39: 649–661.

Rasmussen, L. 2016. Clinical ethics consultants are not 'ethics' experts - but they do have expertise. *Journal of Medicine and Philosophy* 41 (4): 384–400.

Rawls, J. 1971. *A theory of justice*. Cambridge: Harvard University Press.

Reiter-Theil, S. 2009. Dealing with the normative dimension in clinical ethics consultation. *Cambridge Quarterly of Healthcare Ethics* 18: 347–359. https://doi.org/10.1017/S0963180109090550.

Rossano, M.J. 2008. The moral faculty: Does religion promote "moral expertise"? *The International Journal for the Psychology of Religion* 18: 169–194. https://doi.org/10.1080/10508610802115727.

Ryberg, J. 2013. Moral intuitions and the expertise defence. *Analysis* 73 (1): 3–9.

Ryle, G. 1945–1946. Knowing how and knowing that. *Proceedings of the Aristotelian Society*, 46, 1–16.

Schwitzgebel, E., and F. Cushman. 2012. Expertise in moral reasoning? order effects on moral judgment in professional philosophers and non-philosophers. *Mind & Language* 27 (2): 135–153.

Singer, P. 1972. Moral experts. *Analysis* 32 (4): 115–117.

Sliwa, P. 2012. In defense of moral testimony. *Philosophical Studies* 158: 175–195.

Tännsjö, T. 2011. Applied ethics. A defence. *Ethical Theory and Moral Practice* 14 (4): 397–406.

Veatch, R.M. 1991. *The patient-physician relation: The patient as partner, part 2*. Indianapolis: Indiana University Press.

Vogelstein, E. 2015. The nature and value of bioethics expertise. *Bioethics* 29 (5): 324–333.

Wilson, J. 2014. Embracing complexity: theory, cases and the future of bioethics. *Monash Bioethics Review* 3: 3–21. https://doi.org/10.1007/s40592-014-0001-z.

Yoder, S.D. 1998. The nature of ethical expertise. *The Hastings Center Report* 28 (6): 11–19.

Chapter 14
Clinical Ethics Expertise & the Antidote to Provider Values-Imposition

Autumn Fiester

Despite divergent philosophies and methods of clinical ethics consultation (CEC), there is universal agreement in the field that consultants may not impose their own moral views on stakeholders in a clinical ethics conflict. Setting the standards for the field, the American Society for Bioethics and Humanities' *Core Competencies for Healthcare Ethics Consultation* clearly states, "Ethics consultants need to be sensitive to their personal moral values and should take care not to impose their own values on other parties" (American Society for Bioethics and Humanities, 2011, p. 9). Such imposition, they write, amounts to a kind of "moral 'hegemony'" in which the consultant "usurps the authority of the primary decision makers" (American Society for Bioethics and Humanities 2011, p. 7). The recent Quality Attestation Presidential Task Force (QAPTF) (Kodish and Fins 2013), a group charged with creating a credentialing process for CEC, echoes these warnings that consultants "must be trained to avoid the risk of imposing their values and judgments" (Kodish and Fins 2013, p. 27). Part of clinical ethics expertise, therefore, is knowing how to avoid values-imposition on stakeholders in a value-laden conflict, which demonstrates recognition of the profound values pluralism that exists in the US.

This concern about values-imposition is part of an on-going dialogue about the best practice standards for CEC and the type of moral expertise ethics consultants possess. But there is another potential source of "moral hegemony" in bedside ethical conflicts that largely goes unnoticed and unpoliced, and there is little dialogue in those circles about what constitutes moral expertise and its limitations. Those circles are the providers who care for patients and interact with their families. Unlike bioethics' mandate for CEC training that "requires that consultants be able to identify and articulate their own views and develop self-awareness regarding how their views affect consultation" (American Society for Bioethics and Humanities 2011),

A. Fiester (✉)
University of Pennsylvania, Philadelphia, PA, USA
e-mail: fiester@pennmedicine.upenn.edu

J. C. Watson, L. K. Guidry-Grimes (eds.), *Moral Expertise*, Philosophy and Medicine 129, https://doi.org/10.1007/978-3-319-92759-6_14

the training of nurses and physicians rarely demands such introspection and caution. But if it is a Herculean task for a trained clinical ethics consultant to avoid values-imposition on stakeholders in a conflict (Fiester, 'Quality Attestation' and the Problem of the False Positive 2014a), the risk of that problem is magnified many-fold in professionals who are not trained to reflect on the limits of moral expertise or scrutinize their interactions in search of values-imposition. Physicians and nurses do not include the goal of avoiding values imposition in their criteria of professional expertise. They are unlikely to have ever been asked to answer questions like the QAPTF's "How do you recognize and handle your personal beliefs…with others who may or may not share those beliefs" (Kodish and Fins 2013)? Values-imposition by providers on patients and their surrogates exacerbates the already looming power imbalances that make patients and families morally vulnerable (Fiester, Weaponizing Principles: Clinical Ethics Consultations & the Plight of the Morally Vulnerable, 2015b). Given the high stakes involved, I will argue that part of exercising the moral expertise of clinical ethics consultation is detecting values-imposition by one stakeholder on another, their own and anyone else's. The national bioethics organization holds clinical ethicists accountable for being experts in "process," that is, in the facilitation of dialogue about stakeholders that "helps to elucidate issues, aid effective communication, and integrate perspectives of the relevant stakeholders (American Society for Bioethics and Humanities 2011, p. 7). As experts in process, clinical ethicists have a two-fold obligation in their facilitative role (American Society for Bioethics and Humanities 2011, p. 9) in ethics consultation: while consultants must certainly avoid imposing their own values on the other parties in the conflict, they must also guard against the values-imposition of the treating team on their patients and surrogates. The moral expertise of the clinical ethicist includes knowing when there are deeply held competing values at stake in a conflict about which there is no societal consensus or governing law.

One locus of bedside disputes that carries a high risk of values-imposition is conflict between providers and surrogates over care at the end of life. Values-imposition by providers in end-of-life decisions can be blatant and overt, or they can be inadvertent and subtle, cloaked in language that appears to be merely value-neutral patient advocacy. In this essay, I will offer a description and anecdotal evidence for both categories of values-imposition – the overt imposition of provider values and an inadvertent values-imposition. I will argue that, as part of the obligation of clinical ethicists to discharge their specific species of moral expertise – which includes safeguarding against values-imposition on patients and surrogates – ethics consultants must be on the lookout for instances of a care-team's insertion of their own values and life priorities, particularly in end-of-life treatment disputes. And, as most ethics services are also committed to a mission of ethics education, I will argue that the CEC service can play an important role in educating clinical teams about the risk of values-imposition and the harms it can cause. In this way, they use this aspect of their moral expertise for capacity-building in best clinical practice.

14.1 Values-Imposition & the Provider Conception of the Good

Political philosopher John Rawls coined the term "conception of the good" (Rawls 1971, p. 11) to capture the constellation of "core values, religious and moral beliefs, and life-priorities that give our lives normative structure and meaning" (Fiester, Teaching Non-Authoritarian Clinical Ethics: Using a Positions-Inventory in Bioethics Education, 2015a, p. 21). One's conception of the good defines what makes life worth living, what constitutes "quality of life" and the necessary conditions for its continuation. But different people hold widely different and sometimes incompatible conceptions of the good, a feature of American pluralism that we deeply value. But when a provider advocates for a particular treatment choice, this may not come from the patient's first-person account of her conception of the good, but might be sourced in the provider's *own* value system (Cook and Guyatt 1995) (Sugarman 1994) (Cassell 2003) (Danis and Gerrity 1988) (Schneiderman 1993). Providers' own normative views about resource allocation, meaningful existence, withholding or withdrawing treatment, death-prolonging measures, dignity, obligations to treat, futility, and the good death have been empirically shown to undergird their clinical recommendations (Garland and Connors 2007).

As an example of values-imposition sourced in a provider's own conception of the good, consider the use of the concept "suffering" in clinical care. Suffering was identified as an important, neglected part of the patient experience decades ago in a now-famous piece by Eric Cassell (1982). The concept was highlighted by Cassell to draw attention to the very real, multi-faceted ways that patients can and do suffer, and that suffering clearly provokes genuine moral anguish on the part of the care-team that witnesses it. But like so many well-intended bioethical concerns, "suffering" has a less noble application: providers sometimes invoke patient-suffering in cases where the care-team disagrees with the choices of the surrogates, augmenting medical judgment with a normative one. For example, physicians in one study reported witnessing providers "using guilt in an overt way" (Brush et al. 2012, p. 1083) to bring about their desired outcome. One intensivist admitted, "I tell people, 'If this was somebody I loved, I would not do that to them because it's a horrible way to die" (Brush et al. 2012, p. 1082). Others said they knew physicians who tell surrogates, "We are torturing this person, you must stop" (Brush et al. 2012, p. 1083). One clinician acknowledged only emphasizing the patient's suffering when driving an agenda of withdrawal: "I never use that approach on people that I think we should press ahead on" (Brush et al. 2012). The problem with invoking patient suffering in treatment disputes with surrogates is that it may reveal more about the providers' preferences and priorities than the patient's own. Providers are not trained in a process of critical reflection to be able to analyze the exact language they use to describe suffering, the way they frame conversations around suffering, the motivations they have for bringing up questions of suffering when they do (and when they decide not to). They are not taught to lay out evidence for the conclusion that the patient is in pain, or to give indications about the level of pain the patient is in.

One explanation for invoking suffering is that a provider has reached the conclusion that the patient has a "value-negative existence," that the patient's "existence, on balance, entails far more pain and suffering than any amount of good that can come from continuing it" (Fiester, When It Hurts to Ask: Avoiding Moral Assault in Requests to Forgo Care, 2014b). But the assessment of a value-negative life is, in all but the rarest cases, purely subjective and perspectival. Unless the provider is merely conveying the patient's *own* judgement about the benefits and burdens of continued existence, the assessment of the provider cannot stand as objectively true. Take the following example: imagine a situation in which the providers have been caring for a patient in an ICU setting for a sufficiently long time to have strong evidence of what her conditions are and her overall trajectory and prognosis are. They have good evidence, based on patient responses to painful stimuli and monitor readings, that the patient is highly distressed and in a great deal of pain. Because of the patient's condition and because she is already maxed out on pressors, the physicians (including those from palliative care) cannot fully manage her pain without risking a life-threatening drop in blood pressure. The patient is dying, and there is no evidence that the situation is reversible or that the patient could regain capacity or even leave the ICU. A provider might argue that the patient's experiential interests have been whittled down to only one: to not suffer since pain is the only possible experience left to this patient in her final days or weeks. One might even add to the case that there is no evidence that the patient valued suffering from some religious belief, for example. So the providers might argue that they would be *remiss* if they didn't explain to the surrogate that the patient's life involves pain and suffering that will perhaps only get worse over time, and continuing aggressive measures will not provide any benefit beyond mere sustenance of biological existence. Would they then be justified in naming this a "value-negative existence" and pressuring the family to withdraw life support? My answer is "no." All the providers in this case have established is that there is objective evidence that the patient experiences discomfort and that there is no documented evidence that the patient values pain. But what the providers don't know is what else the patient does and does not value. The patient may be opposed to having life-support withdrawn but not opposed to risking the deadly drop in blood pressure from increasing the sedation. The patient's surrogates may be correct that the patient would choose the extra time of being alive for the sake of her family and their needs. I often joke – but also mean it quite literally – that if my children needed me to stay alive in such a state to use my tuition benefits for their college education, they should absolutely do it (and without guilt). It is a presumptuous arrogance for providers to believe they know when a life is no longer of value to the patient, and that assessment cannot be made on any "objective" facts about the patient's condition.

The failure to acknowledge the influence of a provider's own conception of the good can result in a profound imposition of values on patients and their families. These impositions can occur as either a deliberate, overt attempt to achieve the provider's treatment choices, or through subtle and unwitting interactions in which the provider's conception of the good plays a stealth, unconscious role in steering the direction of values-based patient care. Let's look at examples of each in turn.

14.2 Overt Imposition of Provider Values

Summarizing the account of physician approaches to negotiation with surrogates about end-of-life care plans, a study in *Critical Care Medicine* concluded that "[m] any physicians had witnessed colleagues negotiate in ways they found objectionable such as providing misleading information, injecting their own values into the negotiation or behaving unprofessionally toward surrogates" (Brush et al. 2012, p. 1083). Anchoring the conclusion that physicians provide misleading information to drive patient choice, a recent study of cardiology interventions published in *JAMA Internal Medicine* found that physicians "contribute to patients' misperceptions of benefit through explicit or implicit overstatement of benefits, understatement of risks, and communication styles that may hinder patient understanding" (Goff et al. 2014). A different study in *JAMA Internal Medicine* a year earlier found similar results about ICD implantation, concluding that physicians engage in "unclear representations and omission of information to patients" (Hauptman et al. 2013). A 2012 study in *Health Affairs* probed physicians about these practices quite directly, asking: "In the past year, how often have you (d)escribed a patient's prognosis in a more positive manner than warranted?" Although 44.8% reported that they had never misrepresented a prognosis to a patient in the previous year, 55.2% admitted that they had misled patients at least some of the time (Lezzoni et al. 2012). Nurse testimonials in a qualitative study in the *Journal of Palliative Medicine* backs these data up, with nurse comments such as, "Updates from the primary are often unrealistic and only portray the 'small victories' instead of the overall prognosis," "False hope is given," and patients "are not told the truth about being in an end-of-life scenario" (Aslakson et al. 2012). Physicians sometimes mislead patients or families in the opposite direction as well, portraying the clinical situation more gravely than the clinical facts warrant. In the qualitative study of physicians' interactions with surrogates mentioned above, colleagues reported having witnessed situations in which "only one side of the coin is presented and the physician paints the picture in very dark colors" when the provider desired a transition to comfort care (Brush et al. 2012, p. 1083). Another clinician in that study commented, "I have witnessed physicians misrepresenting a patient's status…I've seen people representing that a cancer was back or growing when we had no objective evidence that that was the case…to generate an outcome that they wanted" (Brush et al. 2012, p. 1083).

There is evidence that some physicians are quite aware of their attempts to bring treatment plans in line with their own values and preferences. One physician in the *Critical Care Medicine* study referred to this as his efforts "to 'take the reigns' … and to drive the direction of care of the patient" (Brush et al. 2012, p. 1082). Physicians unabashedly revealed strategies to secure surrogate-compliance with the option the clinician preferred. One intensivist remarked on his technique for persuading surrogates, labeling it: "[t]he best thing I've found to make them make the right decision" (Brush et al. 2012, p. 1082). Referring to family meetings, one physician admitted that "with few exceptions the preponderance of these meetings are called to persuade the family to go along with a decision. And every word that is uttered by the physician in these discussions…is intended to produce – persuade them to accede to the recommendation" (Brush et al. 2012, p. 1084). One study found that surrogates perceive these overt attempts, with some reporting that they had "felt pressure from the staff to hasten their loved one's death" (Abbott 2001). In fact, some physicians were concerned, in retrospect, about the way they had imposed their values on patients and surrogates, like one who reported, "[S]ome of my greatest regrets as a practitioner have arisen from talking people into things that they were very reluctant to do, and really did not want to do, and only did because I was twisting their arm" (Brush et al. 2012, p. 1084).

My claim that the direction of care endorsed by physicians is closely correlated to their own conception of the good is not mere conjecture: there is sound empirical evidence to support this assertion. For example, one study in *Intensive Care Medicine* concluded, "Physicians who more strongly believed that the appropriate goal of care was life prolongation were less likely to inform surrogates about the option of comfort care" (Schenker et al. 2012). A different research team reported parallel findings, concluding that increased treatment withdrawal or hastening of death for patients was closely associated with the physicians' *own* lower preference for life-sustaining treatment (Neede et al. 2012).

Surely, the physicians in these cases are acting on what they perceive to be the patient's best interests. They press surrogates to choose a particular course of treatment because they are convinced of the benefits of that course of treatment. There is no malice here – only a desire to do right by their patients. The problem is that "best interest" is values-dependent, and that subjective assessment can lead the clinical staff into over-stepping their boundaries. One might object that a claim can be values-dependent and yet not subjective given that there can be objective moral values. But even if there are objective moral values, there is no obviously objective way to hierarchize those values. Returning to the ICU example above, avoiding suffering that one does not value in and of itself may be an objective value, but that doesn't mean that this value ranks highest among other, competing values.

14.3 Inadvertent Imposition of Provider Values

While some providers quite intentionally impose their values on patients by steering families towards a particular treatment decision and have a level of self-awareness about their actions, other providers do not intend to press their personal values on surrogates and are unlikely to perceive value-imposition when it occurs. Appeals to suffering, quality of life, and risk-benefit ratio in these cases are believed to be objective assessments of the clinical situation, much more akin to medical fact than to subjective values. But how the provider hierarchizes and ranks these values is perspectival, even if these are objective goods. This form of values-imposition has the semblance of neutrality, appearing as support for patient-centered care, a defense of a patient's autonomy, or recognition of the best interest standard. Because of its benevolent, neutral framing, this form of values-imposition will likely be denied, making it more challenging for ethics consultants to highlight or guard against. Two cases illustrate this more insidious form of values-imposition.

(1) The first case comes from a recent first-person testimonial published in the bioethics literature (Peace 2012). Anthropologist William Peace, paralyzed from a long-ago accident and hospitalized for a stage-four wound, writes about the way his suffering was used by his physician to suggest that he discontinue life-sustaining therapy. With no evidence from Peace that he wanted to stop curative treatment, the clinician pressed the comfort-care only option under the guise of patient-centered care. To bolster the case for withholding aggressive treatment, Peace's physician laid out a barrage of axes on which Peace would suffer during his long convalescence: "bedbound for at least six months," "good chance the wound [will] never heal," "never be able to work again," "life of complete and utter dependence," "kidneys or liver could fail at any time," "bankruptcy…likely" (Peace 2012, p. 15). Peace writes of the physician's perspective, "Clearly death was preferable to nursing home care, unemployment, bankruptcy, and a lifetime in bed" (Peace 2012, p. 15). But that view of Peace's existence was not shared by Peace. He writes poignantly: "I wanted to live" (Peace 2012, p. 15). He added, "Other people with a disability have been offered the same permanent solution to their perceived suffering that I was" (Peace 2012, p. 16).

It is very unlikely that the physician would have registered this exchange as an instance of values-imposition because it likely seemed to him to be a mere application of the universal principle of a patient's right to forego treatment. Peace writes, "His next words were unforgettable. The choice…was my decision and mine alone," emphasizing, "I had the right to forego any medication…I could be made comfortable…I would feel no pain…The message was loud and clear. I can help you die peacefully" (Peace 2012, p. 16). But the values-imposition in this case makes a mockery of "*patient*-centered care." Again, Peace: "It made me wonder, how do physicians perceive 'patient-centered' care"? Is it possible that patient-centered health care would allow, justify, and encourage paralyzed people to die" (Peace 2012, p. 16)? He adds, "When hospitalized, not once did I feel well cared for" (Peace 2012, p. 16). In fact, Peace calls his healthcare experience "a denial of personhood" (Peace 2012, p. 14) – a denial of his conception of the good, his view of what makes existence meaningful and justified.

The objection to the accusation of values-imposition by the physician will be that the clinician was merely acknowledging the patient's right to choose his own treatment. It will likely be said that the physician's action was motivated by a commitment to patient autonomy and that it does not constitute a commentary by the physician on the worthiness of the patient's life. From the editorial review of an earlier essay calling attention to Peace's experience, the editor argues that "the contention that asking a question about limiting treatment in non-terminal cases equals a negative assessment of one's quality of life is not established" (Fiester, email message to Journal of General Internal Medicine, 2013a). But an easy rejoinder to this objection is to point out the counter-case: how do physicians act when they believe that a life is unequivocally and unambiguously worth saving? The empirical literature cited show that they simply won't bring up the option of withholding treatment in the first place. They will continue to provide aggressive life-sustaining treatment because they see no need to secure any justification for its continuation. Asking Peace to make an active choice in favor of continuing his life is a clear indication that his life's current value is not self-evident from the physician's perspective. But it was self-evident to Peace, who never wavered in his commitment to curative treatment. Therefore, this line of questioning demonstrates the clinician's values-bias. Rehearsing the litany of a patient's future suffering to raise the possibility of withholding care – completely unsolicited – is the clear imposition of one person's conception of the good on another.

A second line of defense of the physician's exchange with Peace comes from another physician-reviewer of the earlier essay: "[I]t is assumed from that start that the physician did wrong. There is no discussion of possible beneficial motives of the physician, or whether the previous experiences of the patient misinterpreted the situation" (Fiester, email message to Journal of Pain and Symptom, 2013b). This physician-reviewer misses an essential point about values-imposition: it can be both unintentional and nevertheless pernicious. The imposition of a provider's values on a patient constitutes a wrong-doing, even if the provider had non-malevolent motives. Ethics consultants safeguarding patients and surrogates from values-imposition by a care-team should not be on the lookout for providers who harbor malevolent motives. Values-imposition is insidious precisely because the care-team defines their actions as benevolent.

(2) A second example of the unwitting imposition of provider values comes from the case of "Mr. J," published in *Chest* as a "Point/Counter Point" commentary on the Texas Advance Directives Act (TADA) (Fine 2009). Mr. J is an elderly man with renal failure, congestive heart failure, and advanced dementia, who has lived in a nursing home for the previous two years. There he received artificial food and hydration, as well as dialysis at an outpatient facility several times a week. His current hospital course is his third in six months, but he was able to return to the nursing home after each of the prior two acute crises. At the time of the case, the patient is extubated and in a regular room, having been successfully weaned off the ventilator after a week in the ICU. The nursing home is refusing to readmit Mr. J in his current state because he is too unstable, and the pulmonologist and nephrologist agree that he is too ill for that setting. The commentary's case is framed by the care-team's perception and concern about the patient's suffering: "The nursing staff, social workers, and physicians are very concerned about the patient's suffering and recommend a 'comfort care only' level of treatment" (Fine 2009, p. 963). All of the providers involved in this case and case commentary see themselves as altruistically motivated, and they are unanimously in favor of withholding future life-sustaining treatment for Mr. J against the family's clearly expressed wishes. The family's stance is that the providers "do everything," which includes "reintubation if needed,

and full cardiopulmonary resuscitation" (Fine 2009, p. 963). The providers want Mr. J to have a DNR order and to cease dialysis.

The case hinges on the level of suffering Mr. J experiences. The evidence that the patient is suffering comes from the palliative care team who observed "multiple nonverbal criteria for pain, including moaning and grimacing" (Fine 2009, p. 963). In contrast, "[t]he family expresses a strong belief that the patient is not in pain and that the patient 'wants to live'" (Fine 2009, p. 963). The care-team's perception of the patient's suffering is treated in the case as an overriding clinical fact that makes withholding (and maybe even withdrawing) life-sustaining care the obvious and incontrovertible clinical indication. In the view of the treating team and case commentators, death is undoubtedly preferable to living like this, and there is no room for divergent opinion in a situation that admits of such clinical and moral certainty.

But rather than an objective conclusion, the providers' subjective conceptions of the good are thoroughly infused in their interpretation of what is best for Mr. J. They unwittingly impose their own values on Mr. J and his family about meaningful existence, quality of life, the good death, the appropriate conduct of families and surrogates, and the purpose of medical care. Let's start with the assessment of the family. The case states that the "children are increasingly perceived as 'in denial,' meddlesome, and potentially litigious" (Fine 2009, p. 963). The case commentary describes the family as "acting in bad faith" (Fine 2009, p. 969) because "the demands of Mr. J's family were inappropriate and harmful to the patient" (Fine 2009, p. 971). Each of those accusations is inherently perspectival and subjective. Take the claim that the family is "in denial." There are two separate issues about which the family could be accused of denying: their father's prognosis and their father's discomfort. Regarding Mr. J's prognosis, it is easy to understand why they resist the team's assessment: their father has been able to be discharged back to the nursing home on two separate occasions in the past six months, has already been able to be weaned from the ventilator after just one week in the ICU, and the team has been advocating the withholding of aggressive interventions for two solid years, yet Mr. J has survived ("Mr. J's nephrologist notes that he has recommended this for the past 2 years, but the family refused" (Fine 2009, p. 963)). As for their perception of Mr. J's discomfort, we are left to wonder what evidence the physicians have amassed to show his family that the patient is in pain. Much is made of their refusal to recognize or allow treatment for his pain ("they reject suggestions for any pain medication" (Fine 2009, p. 963)), yet "[h]is children take turns staying in his hospital room around the clock to 'supervise' his treatment" and they consistently tell the staff to "do everything" (Fine 2009, p. 963). This hardly seems like the behavior of a family "acting in bad faith" or unconcerned about their father's well-being. In fact, the case raises the possibility that the family earned the reputation of being "meddlesome" because they perceive a need to protect their father from a staff that would clearly prefer to see him dead. This fear explains their resistance to the suggestion of hospice (they state: "hospice just kills people" (Fine 2009, p. 963)). It appears that in these providers' conception of the good, "good" surrogates are those who accept the recommendations of the medical staff, do not interfere with the team's care plan, do not threaten law suits when they feel powerless, and make decisions that conform to what the providers believe is in the patient's best interest. The family surely sees it differently. Might the family be imposing their own values on their father, who might have made different choices were he decisionally capacitated? It's surely possible, and that possibility is one reason why the physicians characterize the family in the way they do. But would Mr. J depict his children as "in denial, meddlesome, and potentially litigious," even if he would choose a different treatment course? That is a very particular narrative frame. Isn't it more likely, even in the case of disagreement with his children, that he would describe them as "struggling, devoted, and willing to protect him at all costs"?

But perhaps the axis on which the providers' values-imposition is most pronounced is their unanimous verdict on Mr. J's value-negative existence. Unlike the family, which "is clearly not making choices that are in the patient's best interests," the providers believe that they know that what is obviously best for Mr. J is death. That is why the case report states that the clinicians should immediately "notify the family that further requests for life-sustaining therapies would be refused" (Fine 2009, p. 970). The first commentator explains the rationale: "Our patient cannot recover enough to live free from the hospital, is either suffering or unable to appreciate any joy in being alive, and cannot die easily or peacefully without permission. We conclude that further aggressive treatment is futile" (Fine 2009, p. 963). He reiterates this justification, saying, "Mr. J displays no signs of joy in life and clearly shows signs of suffering. How can we justify aggressive treatment…" (Fine 2009, p. 964)? Twenty-five years ago, bioethicist Felicia Ackerman wrote the perfect rebuttal: "It is as presumptuous and ethically inappropriate for doctors to suppose that their professional expertise qualifies them to know what kind of life is worth prolonging as it would be for meteorologists to suppose their professional expertise qualifies them to know what kind of destination is worth a long drive in the rain" (Ackerman 1991). On the conception of the good of the two case authors – likely shared by the treating team – it is a necessary condition for Mr. J's continuation of life that he has the ability to experience joy, live outside the hospital, or interact meaningfully with his loved ones, but that is not a universal criterion for life being worth extending, and it is not one shared by Mr. J's family (and – for all we know – Mr. J himself). As the Schiavo case made painfully clear, Americans disagree fundamentally on this question. Even when the medical facts were given in polls at the time of this case, a full 35% of Americans polled disagreed with the decision to remove Terri Schiavo's feeding tube (Gallup 2005). While bioethicists and providers attributed the views of that 35% to ignorance of the medical facts, that is simply not true. Many Americans simply oppose the removal of food and hydration regardless of the prognosis.

One of the most worrisome features of cases like this is that the providers not only impose their conception of the good on Mr. J's family, but they also project their values onto Mr. J – as if they know him better than his children do and are his true protectors from his family's "bad faith" actions. There is a trope now in clinical ethics conflicts of the "bad family" who wants to keep the relative alive for the social security check, has some other conflict of interest, or is blinded by their own self-interest or greed. Disagreement with the treating team is all it can take to land a family into this bogeyman category. But data show significant skepticism on the part of surrogates that ICU providers could know patients well enough to make values-based recommendations. A recent study in the *American Journal of Respiratory and Critical Care Medicine* found that 42% of surrogates preferred to receive no recommendations at all regarding the limiting of life support (White 2009). This percentage is noteworthy because the fictional study-vignette used in the research explicitly tries to frame the discussion around the surrogate's perception of the patient's own values and preferences, and still almost half of the surrogates wanted to hear no opinion from the physician regarding withholding or withdrawing life-sustaining therapy. One study participant summarized the reasoning of many on the anti-recommendation side: "The simple fact is a doctor cannot give you that kind of information…in no way do I believe that he can, at any point in time, assume in any way that he knows…he knows anything about the person, except for the medical condition and outcome" (White 2009, p. 322). Another participant in the study said that the surrogate "would be the one who would really have the information towards what the patient's desires would be, not the doctor" (White 2009). Maybe a family that has organized shifts for a 24/7 bedside vigil doesn't know or care about the preferences of their father, but it is presumptuous for the treating team to assume they know more about Mr. J's values, or hold them in higher regard. They never even met Mr. J pre-dementia. The reason they believe that Mr. J would rather be dead than alive like this is that, given the providers' conception of the good, *they* would rather be dead than alive in his condition. That is the very essence of values-imposition.

14.4 Clinical Ethics Expertise, Ethics Consultation & the Obligation to Avoid Values-Imposition

The foregoing argument claimed that values-imposition by providers on patients and surrogates occurs in both overt and inadvertent ways. But why does it take a moral expert to guard against values-imposition? What can a CEC contribute expertly and distinctively in these sorts of cases that a sensitive professional couldn't do instinctively or with minimal training? First, clinical ethicists have long recognized an obligation to avoid values-imposition in the recommendations they offer during a consultation. This respect for the diversity of values is an essential part of clinical ethics expertise. And if this obligation is morally anchored by a respect for patient autonomy and values, then their obligation to protect patients extends beyond their own conduct. Clinical ethicists have both an obligation to avoid values imposition in their own actions *and* an obligation to ensure that values are not imposed that are held by other powerful stakeholders in the conflict. Ethics consultants must employ a high-level of scrutiny of what providers say in treatment disputes, being on the lookout for assessments based on the providers' own conception of the good.

But the second reason why this task falls to the moral expert, rather than merely to a sensitive professional, or someone with minimal training, is that it takes ethics expertise to recognize values – those that are latent, blatant, implicit, unstated, or articulated. One's deeply held values do not come with signage. Only a moral expert can mine for them, bring them out into relief, and name them. And that role extends to the providers involved in a case. As the national organization states in depicting the role of the clinical ethicist in working with "an administrator, nurse, or physician," they need to "help to identify their values and commitments" (American Society for Bioethics and Humanities 2011, p. 7). There is no other party with the moral expertise to assume this watchdog role because it requires moral expertise to identify values.

This obligation to guard against values-imposition in difficult ethics disputes is essential to the role of facilitative consultation that is endorsed by the national organization (American Society for Bioethics and Humanities 2011, p. 9). The duty to "facilitate" is not reduced to mere tasks of "keeping order" or "making sure everyone gets to speak." Rather, the facilitation approach requires the elucidation of "hidden or unnamed values," helping stakeholders – providers included – to "identify their values or commitments" (American Society for Bioethics and Humanities 2011, p. 7). This is a process I have elsewhere called "moral archaeology:" "a systematic uncovering of the moral values, interests, principles, and laws at play in an ethics dispute" (Fiester, Weaponizing Principles: Clinical Ethics Consultations & the Plight of the Morally Vulnerable, 2015b). The moral archaeology required of a consultant is not limited to mining for moral principles, but encompasses a responsibility to uncover the content, perspective, bias, or agenda of the individual stakeholders, including those of the clinical team. Until ethics consultants are on the lookout for the imposition of provider values onto patients and surrogates – especially in end-of-life disputes – moral hegemony in patient care will be impossible to avoid.

But what about the objection established by the Kantian adage that "ought implies can"? (Kant, Religion Within the Boundaries of Mere Reason, 6:50, p. 94). It can only be a moral obligation of ethics consultants to guard against values imposition if it is possible for them to achieve this type of monitoring. The objection continues: surely the ethics service isn't present in many (maybe most) instances of physician values-imposition. This is a powerful objection, and it demonstrates the clear limitations of CEC to address this problem as it occurs. My rejoinder is twofold. First, many of the most contentious values disputes between surrogates and the clinical team *do* involve the ethics service, and in those cases, the consultants can guard against values-imposition. But my second rejoinder is that an institution's ethics service usually has a secondary mission of promoting ethics training, and in this function, an ethics service can educate providers about the risks of values imposition and the harms it can cause. While this is clearly not a panacea for values imposition in the clinical context, it does encourage ethics consultants to make an important contribution towards reducing its occurrence. It is one way in which clinical ethics expertise can serve as the antidote to provider values-imposition.

References

Abbott, K.E. 2001. Families looking back: One year after discussions of withdrawal or withholding of life-sustaining support. *Critical Care Medicine* 29: 197–201.

Ackerman, F. 1991. The significance of a wish. *Hastings Center Report* 21: 27–29.

American Society for Bioethics and Humanities. 2011. The nature and goals of healthcare ethics consultation. In *Core competencies for healthcare ethics consultation*. IL: Glenview.

Aslakson, R., R. Wyskiel, and I.E. Thorton. 2012. Nurse-perceived barriers to effective communication regarding prognosis and optimal End-of-Life care for surgical ICU patients: A qualitative exploration. *Journal of Palliative Medicine* 15 (8): 910–915.

Brush, D., C. Brown, and C. Alexander. 2012. Critical care physicians' approaches to negotiating with surrogate decision makers. *Critical Care Medicine* 40: 1080–1087.

Cassel, E. 1982. The nature of suffering and the goals of medicine. *NEJM* 306: 639–645.

Cassell, J.E. 2003. Surgeons, intensivists, and the covenant of care: Administrative models and values affecting care at the end of life. *Critical Cae Medicine* 31: 1263–1270.

Cook, D., and G.E. Guyatt. 1995. Determinants in Canadian health care workers of the decision to withdraw life support from the critically Ill. *JAMA* 273: 703–708.

Danis, M., and M.E. Gerrity. 1988. A comparison of patient, family, and physician assesments of the value of medical intensive care. *Critical Care Medicine* 16: 594–600.

Fiester, A. 2013a, March 7. email message to Journal of General Internal Medicine.

———. 2013b, May 2. email message to Journal of Pain and Symptom.

———. 2014a. 'Quality Attestation' and the problem of the false positive. *Hastings Center Report* 44.

———. 2014b. When it hurts to ask: Avoiding moral assault in requests to Forgo care. *American Journal of Physical Medicine and Rehabilitiation* 93: 260–262.

———. 2015a. Teaching non-authoritarian clinical ethics: Using a positions-inventory in bioethics education. *Hastings Center Report* 45: 20–26.

———. 2015b. Weaponizing principles: Clinical ethics consultations & the plight of the morally vulnerable. *Bioethics* 29: 309–315. https://doi.org/10.1111/bioe.12115.

Fine, R.L. 2009. Point: the Texas advance directive act effectively and ethically resolves disputes about medical futility; The concept of futility. *Chest*: 963–967.

Gallup (2005). The Terri Schiavo case in review. http://news.gallup.com/poll/15475/terri-schiavo-case-review.aspx. Acccessed 29 Oct 2017.

Garland, A., and A. Connors. 2007. Physicians' influence over decisions to Forego life support. *Journal of Palliative Medicine* 10: 1298–1305. https://doi.org/10.1089/jpm.2007.0061.

Goff, S., K. Mazor, H. Ting, R. Kleppel, and M. Rothberg. 2014. How cardiologists present the benefits of Percutaneous Coronary interventions to patients with stable angine. *JAMA Internal Medicine* 174 (10): 1614–1621.

Hauptman, P., J. Chibnall, C. Guild, and E. Armbrecht. 2013. Patient perceptions, physician communication, and the implantable cardioverter-defibrillator. *JAMA Internal Medicine* 173 (7): 571–577.

Kodish, E., and J.E. Fins. 2013. Quality attestation for clinical ethics consultants: A two-step model from the American Society for Bioethics and Humanities. *Hastings Center Report* 43: 26–36. https://doi.org/10.1002/hast.198.

Lezzoni, L., S. Rao, C. DesRoches, C. Vogeli, and E. Campbell. 2012. Survey shows that at least some physicians are not always open or honest with patients. *Health Affairs*: 383–391.

Neede, J., R. Mularski, T. Nguyen, and E. Fromme. 2012. Influence of personal preferences for life-sustaining treatment on medical decision making among pediatric intensivists. *Critical Care Medicine* 40 (8): 2464–2469.

Peace, W. 2012. Comfort care as Denial of personhood. *Hastings Center Report 42*: 14–17.

Rawls, J. 1971. The man idea of the theory of justice. In *A theory of justice*, ed. J. Rawls. Cambridge, MA: Harvard University Press.

Schenker, Y., G. Tiver, S. Hong, and D. White. 2012. Association between Physicians' beliefs and the option of comfort care for critically Ill patients. *Intensive Care Medicine 38*: 1607–1615.

Schneiderman, L.E. 1993. Do Physicians' own prefences for life-sustaining treatment influence their perceptions of Patients' preferences? *Journal of Clinical Ethics 4*: 28–33.

Sugarman, J.E. 1994. Catalysts for conversations about advance directives: The influence of physician and patient characteristics. *Journal of Law, Medicine, and Ethics 22*: 29–35. https://doi.org/10.1111/j.1748-720X.1994.tb01272x.

White, D.E. 2009. Are Physicians' recommendations to limit life support beneficial or burdensome? *American Journal of Respiratory and Critical Care Medicine* 180: 320–325.

Chapter 15
Clinical Ethics Consultation: Moralism and Moral Expertise

Jennifer Flynn

15.1 Introduction

What kind of criticism does the term 'moralism' make? Usually, to call a moral judgment or attitude moralistic is to disparage it, by way of suggesting its presumption, sanctimony, or even its illegitimacy. In this paper, I examine whether clinical ethics consultation is ever moralistic, and the relevance to this question of the moral expertise of the clinical ethicist. More specifically, I pose two questions: why might clinical ethics consultation sometimes be perceived as moralistic, and does clinical ethics consultation ever require a certain kind of moralism? Though the first question references how clinical ethics consultation is in fact sometimes perceived, each question ultimately probes the inherent nature of clinical ethics consultation. The notion of moral expertise is invoked in my responses to these two questions. In the first instance, moral expertise can reduce the perception of moralism, which is desirable. My response to the second question is yes, clinical ethics consultation does sometimes require a certain kind of moralism, and this requirement relates to the moral expertise required of the clinical ethicist.

I begin by looking at the term moralism, providing an exposition of some relevant philosophical work. Moving to clinical ethics, I examine two concerns about clinical ethics, concerns at least partly underwritten by worries about moralism. I then address the issue of the perception of moralism in clinical ethics consultation, and suggest that the clinical ethics consultant's moral expertise can (and ought to) reduce that perception. Next I examine the requirement of clinical ethics consultation of a certain kind of moralism, this in fact as a result of the clinical ethics consultant's moral expertise.

J. Flynn (✉)
Faculty of Medicine, Memorial University, 300 Prince Philip Drive, St. John's, NL A1B 3V6, Canada
e-mail: jflynn@mun.ca

J. C. Watson, L. K. Guidry-Grimes (eds.), *Moral Expertise*, Philosophy and Medicine 129, https://doi.org/10.1007/978-3-319-92759-6_15

15.2 Moralism

What kind of criticism do the terms 'moralism' and 'moralistic' make? We should recognize that the term is sometimes used in non-disparaging ways. For instance, the late South African author Nadine Gordimer has been described in several obituaries as an "unwavering moralist" (Sithole 2014); Cora Diamond provides as an example of a non-pejorative use of moralism a comment of Henry James' on Henry Fielding, with James describing our seeing Fielding's character Tom Jones through Fielding's "fine old moralism" (Diamond 1997, 199); and Robert Fullinwider points out its more innocent use: "A moralist is one who teaches morality or gives counsel on moral matters. A moralizing account or explanation is one framed in moral terms" (Fullinwider 2005, 106).

Generally speaking, though, to call a moral judgment or attitude moralistic is to disparage it, by way of suggesting its presumption, sanctimony, or its being out of touch with the realities of the moral issue or question on the table. Connections are also drawn between moralism and hypocrisy (Fullinwider 2005, 106; Diamond 1997, 201). As Fullinwider has it, morality encourages one to scrutinize one's own moral character while being charitable towards others (Fullinwider 2005, 110): "our primary moral task is to improve our own character and think generously of others" (Fullinwider 2005, 112). Where morality directs our attention toward our own moral failings and endorses the silent judging of the faults of others only so that we can learn from them and improve upon our own moral dispositions, moralism tends not only toward criticizing the failings of others, but toward the public airing of that criticism.

There is a sense in which we take moralists to overstep certain boundaries in issuing their moral judgments or criticisms—in speaking of the moralistic Seth Pecksniff's (Pecksniff is the central character in Charles Dickens' *Martin Chuzzlewit*), Fullinwider describes Pecksniff as being so free in dispensing moral judgments that he "assumes prerogatives that aren't properly his" (Fullinwider 2005, 109). And, in returning to the distinction between moralism and morality, he reminds us that morality requires "that we be strict toward ourselves and generous toward others" (Fullinwider 2005, 117), and suggests that this can help us adjudicate when "people's judgments toward others exceed proper bounds." Fullinwider points out our general tendency to want to withhold (public) judgment of others, and notes this tendency's entrenchment in both social mores and religious and philosophical argument (Fullinwider 2005, 118). At moralism's core is the tendency to judge others uncharitably and publicly; such judging is underwritten by a sense of one's own righteousness (Fullinwider 2005, 109), and the moralist finds "easy the path to conclusive judgment" (Fullinwider 2005, 106).

Self-certainty is another trait associated with the moralist. Pecksniff's path to the conclusive judgments he so freely dispenses to others is an *easy* one (Fullinwider 2005,106).[1] Fullinwider finds objectionable the lack of hesitation with which the

[1] Fullinwider argues that in John Caputo's *Against Ethics*, what Caputo is actually 'against' is moralism, not morality (Fullinwider 2005, 115), with Caputo in fact resisting an 'ethics' that involves "easy, incautious, complacent, self-certain judgments" (Fullinwider 2005, 117).

moralist shares her moral judgments. On top of that, Pecksniff's morality "is never going to exact real sacrifice from him" (Fullinwider 2005, 109), in that his own life is not governed by the moral standards he applies to others. Another idea associated with the moralist is a lack of standing,[2] or a lack of authority to issue public moral judgments of the sort that violate the moral norms discussed (the norm of judging oneself and not others, and of refraining from the public judging of others). Fullinwider admits that there are exceptions to the general rule to refrain from public judging—and these exceptions come in the form of special licenses to hold the office of moralist.[3] According to Fullinwider, certain people hold an "office," or role, which comes along with such authority (he offers clergy, intellectuals, and public authorities as examples of such roles). On Fullinwider's characterization of moralism, there is a distinction between appropriate moral judgment and a moralizing judgmentalism. Building upon what Fullinwider actually writes, it seems that the content of a judgment could be completely appropriate, with the making of that judgment being moralistic. In such cases, its moralism stems from a certain orientation to the world such that one sees it as one's place to be concerned with judging others (before oneself), to do so quickly and publicly, without holding the proper office. So a moral judgment need not be uncharitable in its substance in order for it to be moralistic. Though Fullinwider discusses the virtues of an appropriate non-judgmentalism, he does warn against a "promiscuous nonjudgmentalism," which he amounts to mindlessness.

In a 1986 article, Larry Churchill and Alan Cross examine some of the possible roles of the clinical ethicist, such as technician,[4] sophist, and teacher/learner. In examining the clinical ethicist's role as moralist, these authors introduce the idea of crossing a boundary—"crossing the threshold," they say—in describing when it can be that a clinical ethicist is taken to be moralizing. Moralistic clinical ethicists cross this boundary when prescribing "moral formulae" to their students and colleagues (Churchchill and Cross 1986, 4). There is also the idea of trying to change another's conduct: the role of moralist gets associated with the clinical ethicist through the (largely misguided) notion that teachers (including here clinical ethicists) are "trying to teach people how to be good" (Churchchill and Cross 1986, 4). Albert Jonsen, in speaking of what he calls "American moralism," argues that the development of bioethics was stimulated by "the moralistic mentality." This mentality involves an

[2] Politicians tend to think about the notion of standing. Michael Ignatieff, former Canadian Prime Ministerial hopeful, writes about the notion of standing in connecting it to "forms of personal authority" (Ignatieff 2013, 126). He writes about what he takes to be the necessity of having standing to being elected to political office. Ignatieff speaks about how his main political opponent "denied him standing in his own country" (Ignatieff 2013, 126); Ignatieff's campaign was plagued by charges that having been away from Canada for so long, he was overstepping certain boundaries by running for Prime Minister. In the eyes of the electorate, Ignatieff lacked the authority to speak on Canadian political issues, and thereby lacked standing with the Canadian electorate.

[3] Fullinwider 2005, 111.

[4] In the technician role, the ethicist might clarify the topic of the consultation, or elucidate relevant concepts, stopping short of telling people "what is right or wrong" (Churchchill and Cross 1986, 5).

insistence upon "clear, unambiguous moral principles" (Jonsen 1991, 119), it seeks "patterns of control that will affirm order against disorder" (Jonsen 1991, 121), and current remnants of this moralism can be seen in the view that morality can only consist in absolutes (Jonsen 1991, 128).[5]

We can extract certain themes from this review of the literature, for instance the connection between moralism and the judging of others' moral characters and actions before judging one's own. Thinkers have associated being public in those judgments of others, as well as being quick in making those judgments, with moralism. Being uncharitable in those judgments is a habit of the moralist. Along similar lines, the moralist tends toward applying moral standards to others that she does not apply to herself. We have also seen that moralism is associated with an overstepping of boundaries, itself linked to a lack of authority to make a judgment on a topic over which one takes oneself to have jurisdiction.

15.3 Clinical Ethics Consultation

How might moralism play itself out in clinical ethics consultation, and how might the expertise of the clinical ethicist impact that playing out? We can see how the tendencies and attitudes associated with moralism could become associated with clinical ethics in general, and with the clinical ethicist in particular. After all, the clinical ethics consultant could be quite easily viewed as being in the business of making moral judgments.

As stated, I shall examine the perception of clinical ethics consultation as moralistic, and suggest that perception can in large part be mitigated by the clinical ethics consultant. I shall also urge, though, that clinical ethics consultation requires a certain kind of moralism. But first, there is concern about the current state of clinical ethics consultation, concern that seems to be motivated by concerns about moralism. Partly to motivate the examination that follows, we shall examine two such concerns.

15.3.1 Concerns about the Purview of the Clinical Ethicist

First, in the academic literature on clinical ethics, one can find skepticism toward clinical ethics, a skepticism grounded partly in a hesitancy about whether training in moral philosophy (an entry path into clinical ethics for many clinical ethicists) qualifies one as an expert in how to live and act. The concern is that expertise in an academic subject is conflated with an expertise in moral conduct, and that as a result, within the context of clinical ethics consultation, knowledge is being

[5]Three further interesting treatments of moralism are found in Diamond 1997, Baier 1993, and Crary 2009.

mistaken for virtue.[6] An overstepping of boundaries is the concern here: it is one thing for the clinical ethicist to have expertise in an area of philosophy (say), quite another for her to assume this background qualifies her to be in the position of directing others on how to behave.

Maria Hedlund is worried about the conflation of ethical expertise and moral expertise.[7] Hedlund's main concern is the threat to democracy posed by political decision-makers asking ethics experts for advice. The threat involves the transformation of "ethics expertise into alleged moral expertise," when ethics experts (such as those appointed to bioethics councils) are instructed to "provide consensus decisions on value questions" (Hedlund 2014, 282). According to Hedlund, we ought to make a distinction between ethics expertise, which involves the systematic analysis of moral problems, and moral expertise, which is expertise in "(providing the right) morality" (Hedlund 2014, 282, 285). Her concern is over the possibility of "ethics experts" offering advice on value questions, and overstepping their boundaries. For Hedlund, value questions should be addressed "in democratic process," not by expertise (Hedlund 2014, 282).

A more general concern in the neighbourhood is discussed by David Casarett, Frona Daskal, and John Lantos, who describe the debate over the moral authority of the clinical ethicist. As they set matters out, on one side of the debate are those who "are skeptical of the value of moral theory in general and of moral experts and 'right' answers in particular" (Casarett et al. 1998, 6). On the other, there are those who are comfortable with the idea of clinical ethics as a field of "expert knowledge." The authors describe the debate as one underwritten by a question about the clinical ethicist's moral authority. The debate is about which sort of topic the clinical ethicist can rightly provide guidance. And this is a question that relates to moralism, in that it is a question about when the clinical ethicist oversteps boundaries. David Adams ends his article on clinical ethics and physician assisted suicide with questions about the nature of clinical ethics, questions similar in spirit to what we see above. He asks whether a clinical ethicist is an expert, or a "moral authority," in the same way a neurosurgeon is "an authority in operating on the brain" (Adams 2015, 143).[8]

Finally, even on the bioethics blog *Impact Ethics*,[9] we find concerns about a moralism associated with medical ethics (though not with clinical ethics as such). In

---

[6]This is how it is put by Churchill and Cross (Churchchill and Cross 1986, 4), who here draw on Clouser 1972. More recent expressions of this concern can be found in Crosthwaite (1995), Gesang (2010), and Archard (2011).

[7]Thanks to Giles Scofield for directing me to the Hedlund piece.

[8]Adams (forthcoming) does not offer answers to these questions, though he takes himself to have initiated the relevant conversation. In his paper in this volume, he argues that the clinical ethics consultant's expertise lies in "the deployment of skill in moral reasoning," which suggests that his answer to such questions is no.

[9]Impact Ethics is an on-line forum for discussion of bioethical issues, which prioritizes the Canadian context. Its mission statement is: "(1) to promote critical discussion of bioethical/health policy issues relevant to Canadians; (2) to disseminate opinions on these issues from diverse perspectives to a wide academic and non-academic audience; and (3) to promote public engagement with and education on bioethical issues." See https://impactethics.ca/about/

"How We Learned Not to Drop the e-Bomb," Jeremy Snyder and Valorie A. Crooks describe their research on the ethical dimensions of decision-making surrounding medical tourism, for which 32 Canadian research participants were interviewed.[10] Snyder and Crooks found that in responding to questions about the influence of 'ethical issues' on decision-making, participants responded as though "they were being accused of unethical conduct" (by defending their choices to go abroad for medical care) (Snyder and Crooks 2014). Since having dropped the word "ethics" (or "the e-bomb") from most of their interview questions (but still asking "ethical questions"), Snyder and Crooks found that participants were less defensive, and that they were able to probe the relevant ethical issues in more depth. (Snyder also shares that a news article about Snyder and Crooks' website referred to Snyder as "'ethics' professor" and [thereby] "professional guilt tripper"). This does not address clinical ethics as such. However, it is still important for our purposes as evidence of an association between medical ethics and accusation and sanctimony, and also between one "doing ethics" professionally, and one being taken to be not only in the business of passing judgment on others, but of tending to arrive at negative judgments of others.

15.3.2 *Motivation for the Certification and Professionalization of Clinical Ethicists*

The second concern is found in the current conversation about whether clinical ethicists should professionalize, how they should be trained, the accreditation of training programs, and the certification of clinical ethicists themselves. For instance, Eric et al. (2013) discuss the accreditation of clinical ethics training programs (including the discussion within their historical overview) within the context of their discussion of a method of assessing clinical ethicists. The main point of the article is to set out a system for assessing clinical ethicists, their knowledge, skills, and practice. The authors justify the endeavor by pointing to the fact that data reveals unacceptable variance in the practice of clinical ethics consultation,[11] and to the unevenness of clinical ethicist qualifications.

Interestingly, a paper published 13 years earlier in 2000, which summarized the conclusions of the report of the Society for Health and Human Values-Society for Bioethics Consultation Task Force on Standards for Bioethics Consultation, concluded that "certification of individuals and accreditation of programs are rejected" (Aulisio et al. 2000, 66).[12] One reason given for this position is that certification would suggest that the certified ethics consultant has "special standing," which

[10] Thanks to Lynette Reid for directing me to this blog post.

[11] The authors cite E. Fox, S. Myers, and R.A. Pearlman, "Ethics Consultation in United States Hospitals: A National Survey," *American Journal of Bioethics* 7, no. 2 (2007): 13–25.

[12] Aulisio et al., "Health Care Ethics Consultation: Nature, Goals, and Competencies," *Ann Intern Med* 2000; 133; 59–69.

would displace health care providers and patients as moral decision makers, thereby encouraging the authoritarian approach to ethics consultation which the Task Force rejects. It would seem that a shift has occurred, one that has led to the view that without a standardization of practice and training, clinical ethicists may well overstep their boundaries and make decisions or judgments that are in some way or other beyond the bounds of their professional bounds, from the view that certification of clinical ethicists may mistakenly accord them special standing.

Further, we have the American Society for Bioethics and Humanities (ASBH) *Core Competencies for Healthcare Ethics Consultation* document itself, which sets out the expected competencies for clinical ethicists. There are three basic categories of competencies: competencies of skill (such as the ability to "clearly articulate the ethical concern(s) and the central ethics question(s)" (ASBH 2011, 22), competencies of knowledge (such as a knowledge of "moral reasoning and ethical theory") (ASBH 2011, 26), and (though not presented as a category of competencies as such) attributes, attitudes, and behaviors to which the clinical ethicist should aspire, such as honesty and prudence (ASBH 2011, 32). The setting out of such competencies provides structure to training and certification programs. It also clarifies the skills, knowledge, and attributes the clinical ethicist is to possess in order to be qualified to do her work. Such clarification is a clarification of the clinical ethicist's requisite expertise. It is also, ostensibly or not, a way to mitigate the possibility that the clinical ethics consultant assumes a prerogative that is not properly hers to assume, a possibility connected to moralism. In many professional contexts, one is taken to overstep boundaries in this way when one performs a task one is unqualified to perform. The *Core Competencies* document is an attempt to clarify what those qualifications are for the clinical ethicist.

Just recently, the Healthcare Ethics Consultation Certification Task Force of the ASBH circulated the results of its 'Needs Assessment Survey,' a survey completed by 787 respondents (working in some capacity in clinical ethics). 74.8% of those respondents were in favor of the development of a certification program for clinical ethicists. When asked about how certification would be a benefit, 75.3% of respondents cited a "validation of professional knowledge" as such a benefit, and 74.0% offered "professional legitimacy" as a benefit of certification. (Other reason options included "personal growth," cited by 57.5% by respondents, and "enhanced professional practice," cited by 59.0% of respondents.) In announcing its decision to support ("in concept") a certification program, the ASBH Board of Directors described such a program as one that will "validate HCEC [healthcare ethics consultant] professional knowledge and professional legitimacy" (ASBH 2017, 1). Implicit in the survey results and in the ASBH's description of what a certification program for clinical ethicist will do is the notion that 'knowledge' of the clinical ethicist is somehow sometimes called into question, that it is on some level in need of 'validation' (in need of support in some way). Even more directly in connection with our earlier discussion of moralism, the survey results and the description of the certification endeavor directly address the issue of legitimacy, suggesting that there is something in need of attention surrounding the professional legitimacy of the work of the clinical ethicist. The concerns just reviewed are underwritten by concerns about moralism; an overarching concern here is that the clinical ethicist will be asked to offer expertise on matters outside of her jurisdiction.

I will now address the issue of the perception of moralism in clinical ethics consultation, and will suggest ways that the clinical ethics consultant's expertise can reduce that perception. Next I shall address the question of whether clinical ethics consultation requires a certain kind of moralism, and how the answer is related to the required moral expertise of the clinical ethicist.

15.4 Perceptions of Moralism and the Importance of Moral Expertise

15.4.1 Models of Clinical Ethics Consultation

We should begin by reviewing three models of clinical ethics consultation, in order to remind ourselves of what, according to the endorsed model of clinical ethics consultation, is expected of the clinical ethicist. To review the three models discussed in the ASBH's *Core Competencies for Health Care Ethics Consultation* document, the goal of the first approach to clinical ethics consultation, the "pure consensus" approach, is "to forge agreement among involved parties" (ASBH 2011, 7). The ASBH task force deems this approach to be inadequate on the ground that any consensus reached could, for instance, fall outside widely accepted moral norms or standards (not to mention more theoretical standards). There seems to be little room on this model for charges of moralism directed toward either the clinical ethicist or the consultative endeavor in general. Second, the authoritarian approach involves the clinical ethicist as the primary decision maker, which comes along with the suggestion that the clinical ethicist's values or perspectives are "more correct or important than the moral perspectives of other participants in the consultation" (ASBH 2011, 6). The tendencies toward a moralistic approach to ethics consultation on this model are evident, with the emphasis not only upon the clinical ethicist making a decision or judgment, but upon that decision or judgment being made unilaterally and imposed upon others. The ASBH task force does not support this approach.

On the model endorsed by the ASBH, the ethics facilitation approach, the clinical ethicist both helps fashion a resolution that incorporates the concerns and feedback from those around the table, and makes sure that that plan is "a principled ethical resolution," by (we can imagine) subjecting that plan to the tenets of a particular moral theory, or through some other means of ensuring that the resolution is "ethically principled" or otherwise morally acceptable. David Adams describes the agreed upon model (which he refers to as the "received view" of ethics consultation) as involving the situating of the proposed resolution within normative constraints. The clinical ethicist is responsible for ensuring that the agreed upon resolution is ethically justified, partly by verifying that the proposed resolution fits within such constraints (constraints indicated by the relevant current literature, institutional policies, and laws). Adams points out that on this model it is the clinical ethicist's responsibility to "ensure that identified options comport with" the bioethics and scholarly literature, among other standards (Adams 2015, 130).

15.4.2 Perceptions of Moralism in Clinical Ethics Consultation

Ethics consultants are often called upon because an individual or team finds themselves in a moral dilemma, facing a moral question, or feeling moral distress, and help is sought in solving the dilemma, answering the question, or identifying the source of distress. Whether or not a clinical ethics service or a clinical ethicist in general purports to be able to solve an ethical dilemma or answer a particular moral question, the expectation is often that the ethics consult can solve the dilemma or answer the question.[13] The stage is set for at least the perception of moralism, in that the clinical ethicist is called in for the specific purpose of making a moral judgment. A more modest expectation of the clinical ethics consultant is that she will offer a recommendation that will help inform medical decisions. (Moralism enters as a possibility here, because the making of such a recommendation allows the clinical ethics consultant to impose her personal values, or her theoretical commitments, in a way that is seemingly neutral. We might think of this as covert moralism; I do not discuss this further in this paper, but find it an important issue in its own right.[14]).

I shall set out three ways in which there is fertile ground for the perception of moralism. First, the clinical ethicist could well appear to assume a perspective that is not hers to assume. It can be the case that for some involved in an ethics consult, knowledge of moral theory, for example, is not enough to give the clinical ethicist an understanding of the problem or issue from the inside, as it were.[15] Such reservations could be linked to the thought that unless one is working on the front lines in health care, one's understanding of the struggle at hand will necessarily be wanting, no matter how comprehensive one's knowledge of theoretical ethics.[16] This perception may be exacerbated in situations in which the clinical ethicist is not embedded in and employed by the health care organization requesting the consultations she facilitates. In cases where the clinical ethicist comes from outside of the

[13] This is often my own experience in doing clinical ethics consultation, and that of many clinical ethics consultants with whom I have worked. It is also my experience that those involved with a consult are often grateful for the consult even when it does not match these expectations.

[14] Autumn Fiester emphasizes this concern. See Fiester, this volume (Chap. 14).

[15] It is true that some argue that no one is in a position to say what others ought to do, and that therefore there are no moral experts. See, for example, the discussion of the views of C.D. Broad (1952) and Bernard Williams (1993 and 1995), in Watson and Guidry-Grimes (2018) introduction to this volume.

[16] They might also involve worries about the current lack of standard training or professional status of clinical ethicists, though such a concern is perhaps more likely to come from within clinical ethics than outside of it. For an argument that clinical ethicists need clinical experience (and not merely academic training) for robust expertise, see Butkus, this volume (Chap. 13).

hospital or health care organization—if she is a university professor, say, whose university has an arrangement with the health care organization[17]—the clinical ethicist stands a greater chance of being taken to overstep her boundaries.

Second, the circumstances of some clinical ethics consultations can lead to the perception of the consultant as sanctimonious and self-certain. This is not just tied to the fact that the personality of clinical ethics consultants will vary widely, and that we are all human and liable to imperfect behavior. It is tied to the use of specialized knowledge to which the clinical ethicist quite rightly refers in the facilitating the development of a resolution and subjecting that resolution to critical scrutiny. Her explaining her reasoning to those involved in highly specialized terms could well create the impression of sanctimony. Her drawing on specialized knowledge can render her explanation of her recommendation inaccessible to others, and her theoretical take on morality puts matters in terms that are removed from ordinary moral thought. In cases where that knowledge leads to a straightforward and uncomplicated conclusion, the impression can be created that the ethics consultant is making swift moral judgments about matters that are in fact extremely complex.

Those not trained in moral theory do not tend to think about morality in theoretical terms. Certain approaches to moral problems, and even certain words, will be unfamiliar to many involved with a consult. A lack of familiarity with phrases and frameworks used by another can often be enough to make one feel as though the other is being spoken down to. This feeling is likely to be enhanced when the situation is one about which one is personally invested and very concerned. Such an impression of moralism is closely tied to expertise in that the specialized knowledge that contributes to that expertise is taken to remove the clinical ethicist from others.

Third, the clinical ethicist will in fact find herself doing something very close to applying moral standards to others she does not apply to herself, at least in that she is unlikely to be implementing the resolution issuing from the consult she facilitated. One might wonder whether to find this aspect of ethics consultation moralistic is wrong-headed. After all, in other areas of life experts often offer advice on a course of action the expert has not herself undergone, and we find this totally acceptable. An example would be the orthodontist who is not herself wearing braces, but recommends them to a patient with crooked teeth, on the basis of a specialized body of knowledge.[18] The difference between the two, between the clinical ethicist and the orthodontist, I think, is two-fold: when we take advice from others about morality, we are uncomfortable if we suspect that the person offering the advice has never lived up to such advice (especially if we suspect he would be unable to do so). On

[17] This is the case for my colleagues and me. The upside of such arrangements is that not being employed by the organization for whom we facilitate ethics consultations, the perception (or reality) of a conflict of interest, one whereby there would be pressure to make a certain kind of recommendation on certain topics, for instance, is minimized. A possible downside is that in not being embedded in the health care organization, familiarity and trust relationships with those involved in a consult might be lacking.

[18] Thank you to volume co-editor Laura Guidry-Grimes for the suggestion of this example.

top of that, often the resolutions offered in a consult are difficult to follow. They are not difficult to follow in the sense that bearing the pain that braces can cause is difficult. To follow them is often *existentially* difficult.

15.4.3 The Importance of Moral Expertise in Reducing Perceptions of Moralism

Such perceptions ought to be avoided, and the expertise of the clinical ethics consultant can at least reduce them. The possession of good interpersonal skills—the ability to "listen well and to communicate interest, respect, support, and empathy to involved parties" (ASBH 2011, 24)—and attitudes of tolerance and prudence, for instance, can mitigate the perception of moralism. For instance, the self-certainty associated with a moralistic approach is rarely appropriate; even in cases where the ethically appropriate response is relatively straightforward, an attitude of humility and tolerance will be important to avoiding a moralistic consultation. Much of what is required of the clinical ethicist here—much of what is required to avoid moralistic tendencies—falls outside of the core competencies of knowledge and skill set out by the ASBH. The *Core Competencies* document, which implies a picture of the moral expertise of the clinical ethics consultant, includes attributes, attitudes, and behaviors the clinical ethicist should strive to exhibit and possess.

Starting with the perception that the clinical ethicist assumes a perspective not hers to assume, we can look to the core attributes, attitudes, and behaviors all clinical ethics consultants should, according to the ASBH, strive to embody (ASBH 2011, 32). Honesty about one's limitations, and an articulation of any gaps in knowledge about the workings of relevant institutions, or gaps in one's medical knowledge, will help avoid the impression of presumption. Being aware of and forthright about such limitations—forthrightness and self-knowledge are also listed as ideal attributes—can serve to reduce the perception of moralism. They will fuel an acknowledgement on the part of the clinical ethics consultant of the limitation of her position, which can be disarming and reduce the perception of moralism as well.

The impression of sanctimony can be reduced by avoiding the use of highly specialized language. Basic interpersonal skills (ASBH 2011 24), such as recognizing and attending to barriers to communication, incline a clinical ethicist away from supporting a recommendation via reference to highly specialized knowledge. It is important that the possession of such knowledge not alienate the clinical ethicist from those with whom she interacts during ethics consultations, those likely to be better versed in relevant details of working on the front line, or in the relevant medical details, but less well versed in moral theory and the current state of bioethical discussion. Good interpersonal skills (such as listening well and showing respect and support to others) will be important to bridging various knowledge differentials.

Though there is not much to be done about the fact that the clinical ethicist will not implement her own recommendations, that the recommendations she offers will not exact sacrifice from her, humility can help her acknowledge that this is true. A clinical ethicist can approach one's tasks in a way that is sensitive to this unbalance of burden. Where the *Core Competencies* notes how humility can help possible areas of conflict between the clinical ethicist's own moral views and her role in conducting a consultation, it can also reduce the impression of moralism that can be associated with one's making moral demands of others that one does not (or at least need not) live up to oneself. These are examples of how a clinical ethics consultant's expertise can reduce perceptions of moralism.

15.5 The Requirement of Moralism in Clinical Ethics Consultation

15.5.1 The Requirements of Clinical Ethics Consultation: Moral Expertise at Work

Autumn Fiester has argued that moralism is inherent to clinical ethics consultation, and that clinical ethics consultants must try to protect themselves from it through rigorous training.[19] The possible perceptions of moralism I have just set out are harmful to the ethics consultation process. They can, and should, be mitigated by certain core competencies, competencies that partly comprise the clinical ethicist's expertise, and that (it is hoped) can be obtained through training.[20] I want to suggest that at the same time, a certain kind of moralism is required by the clinical ethics consultant. This is partly due to the requisite expertise of the clinical ethics consultant and what our working model of clinical ethics consultation expects of clinical ethicists. On the ethics facilitation model, it is the clinical ethicist who is responsible for deciding whether the resolution is ethically justified. It is the clinical ethicist who is responsible for making the call as to whether that plan lines up with positions "adopted in the bioethical literature" (Adams 2015, 130), and to ensure that any resolution "falls clearly within accepted ethics principles, legal stipulations, and moral rules defined by ethical discourse, legislatures, and courts . . ." (Adams 2015, 129, quoting Dubler et al. 2009, 28). Insofar as a way of establishing whether a particular plan is ethically justified is by figuring out whether it can be verified by one moral theory or other, the clinical ethicist is the one choosing the relevant theory.

The authoritarian approach to clinical ethics consultation (rejected, recall, by the ASBH) involves the clinical ethicist as the primary decision maker, which comes

[19] Fiester (2015).

[20] For questions about the training of future clinical ethics consultants in the attitudes, attributes, and behaviors set out by the ASBH, see Flynn (2017).

along with the suggestion that the clinical ethicist's values or perspectives are "more correct or important than the moral perspectives of other participants in the consultation" (ASBH 2011, 6). On the ethics facilitation model, the clinical ethicist might not be the primary decision maker, but once a possible resolution is decided upon, it certainly is up to the clinical ethicist to decide whether that resolution is ethically justified.

It is required (by the ethics facilitation model) of the clinical ethicist that she appeal to the literature, and that she choose a high-level theory (e.g. utilitarianism), or a mid-level theory (e.g. principlism), or that she eschew theoretical approaches and opt for a casuistic approach (say), and that she make some assessment of the relevant normative constraints.[21] Which literature is relevant to the issue at hand? What is the state of the bioethical debate on the topic? Which philosophical approach should be employed in determining whether a proposed resolution is ethically principled? It is the clinical ethicist who makes such determinations.

15.5.2 Requirements of Moralism

Many other judgments are within the clinical ethicist's purview, and are in fact required by the accepted approach to clinical ethics consultation. Even if the clinical ethicist's expertise involves the full realization of the attributes, attitudes, and behaviors set out in the ASBH document—and this would go quite a long way in mitigating many tendencies associated with the perceptions of moralism—a structural connection to moralism remains. The clinical ethicist's very role as expert requires a certain kind of moralism. That the clinical ethicist independently makes the assessment regarding whether a proposed resolution is ethically justified, while often appropriate, can foster the perception of presumption and sanctimony and self-certainty.[22] It will sometimes be the case that the clinical ethicist will be very certain in her determination that a proposed resolution is, or is not, ethically justified. Such a determination could be straightforward and swiftly arrived upon. A clinical ethicist's recommendation may not only be implemented by others, but its implementation may involve details of health care administration and delivery with which only someone working in health care could be familiar. And scrutinizing a proposed resolution might involve if not sanctimony, engaging in reasoning that is not accessible to others involved in the consult. Self-certainty, subjecting others to

[21] For a discussion of the virtues and drawbacks of various theoretical (and non-theoretical) approaches to bioethics, see Arras 2010.

[22] The question of when self-certainty is appropriate in moral philosophy and clinical ethics is complex. Speaking to clinical ethics, I would hold that it is a mistake for a clinical ethicist to act as though the 'correct' resolution is obvious. Based on my experience doing clinical ethics consultation, the resolution to a situation is often not obvious, and it is not always clear that one course of action is clearly morally superior to other options.

standards one need not live up to oneself, and speaking in ways that can remove one from others are associated with moralism.

That these aspects of moralism could be at work, and are often likely to be at work during an ethics consultation, is related to the clinical ethicist's moral expertise. They are related not just to her experience with moral reasoning, but to her command over a specialized body of knowledge. Choosing a theoretical framework, deciding whether to employ a theoretical framework in the first place, deciding which literature to consult in an effort to assess the pulse of the current relevant bioethical debate—all are her tasks as per the structure of our accepted consultative model. Her ability to carry out these tasks is made possible by her expertise.

A certain kind of moralism will sometimes be required by clinical ethics consultation. A purpose of an ethics consult—and sometimes, its main purpose—is to formulate a resolution to a moral problem or puzzle, a resolution that falls within moral and otherwise normative constraints. This cannot be accomplished without the presence of the clinical ethicist, with her in-depth knowledge of moral theory and bioethics. The presence of someone with that knowledge, in combination with the expectations of the ethics facilitation model, will sometimes require from the clinical ethics consultant tendencies associated with moralism.

15.6 Conclusion

As long as we reject the pure consensus model—and I think we are right to reject it—and endorse a model that positions the clinical ethicist in the way the ethics facilitation model does, clinical ethics consultation will, in virtue of its very structure and purpose, sometimes involve and require a certain kind of moralism. Clinical ethics consultation from the perspective of the accepted ethics facilitation model also sets the stage for the perception of moralism, a perception which ought to be mitigated. The clinical ethicist's expertise, understood as involving the core competencies of skill, knowledge, and the attributes, attitudes, and behaviors identified by the ASBH, can do much to reduce these perceptions.

We have not yet worked out a model of clinical ethics consultation that occupies a space between the pure consensus model, and one that is oriented toward recommendations or other forms of prescription, such as the ethics facilitation approach. I am not sure such a model is possible, given the purposes and goals of clinical ethics consultation. As matters stand, we will do well to recognize the connections between our current model and moralism; the areas where perceptions of moralism can and should be reduced, and areas where moralism can be required, or at least acceptable, and the connection of each of these to the clinical ethics consultant's moral expertise.[23]

[23] Earlier versions of this paper were presented at the 2014 meeting of the Canadian Bioethics Society in Vancouver, and as part of the University of Toronto's Joint Centre for Bioethics Seminar Series in October 2016. My thanks to all in attendance. Thank you also to the co-editors of this collection, Jamie Watson and Laura Guidry-Grimes, for their detailed and helpful comments.

References

Adams, David M. 2015. Clinical ethics consultation and physician assisted suicide. In *New directions in the ethics of assisted suicide and euthanasia*, ed. M. Cholbi and J. Varelius, 125–147. Switzerland: Springer.

———. Forthcoming. Are hospital ethicists experts? Taking ethical expertise seriously. Forthcoming, this volume.

American Society for Bioethics and Humanities (ASBH). 2011. *Core competencies for health care ethics consultation*. 2nd ed.

American Society for Bioethics and Humanities (ASBH), and PSI Services. 2017. *Certification for health care ethics consultants: An update.*

Archard, David. 2011. Why moral philosophers are not and should not be moral experts. *Bioethics* 25 (3): 119–127.

Arras, John. 2010. Theory and bioethics. In *The Stanford encyclopedia of philosophy*, ed. E.N. Zalta https://plato.stanford.edu/entries/theory-bioethics/.

Aulisio, Mark P., Robert M. Arnold, and Stuart J. Younger. 2000. Health care ethics consultation: Nature, goals, and competencies. *Annals of Internal Medicine* 133: 59–69.

Baier, Annette. 1993. Moralism and cruelty: Reflections on Hume and Kant. *Ethics* 103: 436–457.

Broad, C.B. 1952. *Ethics and the history of philosophy*. London: Routledge.

Casarett, David J., Frona Daskal, and John Lantos. 1998. Experts in ethics? The authority of the clinical ethicist. *Hastings Center Report* 28: 6–11.

Churchchill, Larry R., and Alan W. Cross. 1986. Moralist, technician, sophist, teacher/learner: Reflections on the ethicist in the clinical setting. *Theoretical Medicine* 7: 3–12.

Clouser, K. 1972. Philosophy and medicine: The clinical management of mixed marriage. In *Proceedings, first session, institute on human values in medicine*, 47–80. Philadelphia: Society for Health and Human Values.

Crary, Alice. 2009. *Beyond moral judgment*. Cambridge: Harvard University Press.

Crosthwaite, Jan. 1995. Moral expertise: A problem in the professional ethics of professional ethicists. *Bioethics* 9 (4): 361–379.

Diamond, Cora. 1997. Moral differences and distances: some questions. In *Commonality and particularity in ethics*, ed. Lilli Alanen, Sara Heinämaa, and Thomas Wallgren, 197–234. Great Britain: MacMillan Press.

Dubler, Nancy N., Mayris P. Webber, and Deborah M. Swiderski. 2009. Charting the future: Credentialing, privileging, quality, and evaluation in clinical ethics consultation. *Hastings Centre Report* 39: 23–33.

Fiester, Autumn. 2015. Teaching nonauthoritarian clinical ethics: Using an inventory of bioethical positions. *Hastings Center Report* 45 (2): 20–26.

Flynn, Jennifer. 2017. Clinical bioethics and core competencies of attributes, attitudes, and behaviors: Foundations in philosophy and literature. *Ethics, Medicine, and Public Health* 3: 335–342.

Fox, E., S. Myers, and R.A. Pearlman. 2007. Ethics consultation in United States hospitals: A national survey. *American Journal of Bioethics* 7 (2): 13–25.

Fullinwider, Robert K. 2005. On moralism. *Journal of Applied Philosophy* 22: 105–120.

Gesang, B. 2010. Are moral philosophers moral experts? *Bioethics* 24 (4): 153–159.

Hedlund, Maria. 2014. Ethics expertise in political regulation of biomedicine: The need of democratic justification. *Critical Policy Studies* 8: 282–299.

Ignatieff, Michael. 2013. *Fire and ashes: Success and failure in politics*. Toronto: Random House Canada.

Jonsen, Albert R. 1991. American moralism and the origin of bioethics in the United States. *The Journal of Medicine and Philosophy* 16: 113–130.

Kodish, Eric, Joseph J. Fins, Clarence Braddock, et al. 2013. Quality attestation for clinical ethics consultants: A two-step model from the American society for bioethics and humanities. *Hastings Center Report* 43: 26–36.

Sithole, Ndundu. 2014. South African anti-apartheid author, Nobel winner Gordimer dies. http://www.reuters.com/article/us-safrica-gordimer-idUSKBN0FJ1F320140714. Accessed 11 Aug 2017.

Snyder, Jeremy, and Valorie A. Crooks. 2014. How we learned not to drop the e-bomb. https://impactethics.ca/2014/09/15/how-we-learned-not-to-drop-the-e-bomb/. Accessed 20 Oct 2016.

Watson, Jamie, and Laura Guidry-Grimes. 2018. Introduction to this volume.

Williams, Bernard. 1993. Who needs ethical knowledge? *Royal Institute of Philosophy* 35: 213–222.

———. 1995. Truth in ethics. *Ratio* 8 (3): 227–236.

Chapter 16
To Stretch toward without Reaching: Moral Expertise as a Paradox in Clinical Ethics Consultation

Salla Saxén

"The opposite of a correct statement is a false statement. But the opposite of a profound truth may be another profound truth."

Niels Bohr

The debate around the professional status and credibility of clinical ethics consultants has long centered on the problem of moral expertise. In a nutshell, the struggle centers on whether it is possible to identify unique and essential characteristic of moral knowledge that the clinical ethicists possess and other healthcare professionals—or laymen for that matter—do not. How to distinguish clinical ethicists' expertise in moral issues from everyday moral understandings? How to legitimize a professional claim of moral expertise if, at the same time, it is acknowledged that morality refers to subjective values rather than objective knowledge?[1] Consequently, the question of whether, even in a theoretical sense, expertise about values can exist—in other words, moral expertise—has been continuously and heatedly debated for decades. (See for example Noble 1982; Crosthwaite 1995; Shalit 1997; Yoder 1998; Archad 2011; Gordon 2014; Cross 2015; Iltis and Rasmussen 2016.)

In the case of arguing about moral expertise in clinical ethics consultation (CEC), the logic of this debate has typically been the following: If ethics consultants are to claim a professional role in clinical ethics discussions, their position must be grounded on a solid and explicit foundation of expertise in moral issues. Like the cardiologist has undeniable expertise in the functions of the human heart, the ethics consultant should have comparable evidence of her expertise in human morality. Even though it has been argued that this reduction is absurd and the expertise that the ethicist has can never be truly compared to the highly specific expertise in a

[1] I acknowledge that the line between values and knowledge is blurred and theoretically more complex than is suggested here.

S. Saxén (✉)
University of Eastern Finland, Kuopio, Finland

J. C. Watson, L. K. Guidry-Grimes (eds.), *Moral Expertise*, Philosophy and Medicine 129, https://doi.org/10.1007/978-3-319-92759-6_16

medical subspecialty,[2] the underlying assumption is nevertheless the same: the ethicist must have at least some kind of recognizable form of moral expertise to validate her position (cf. Priaulx et al. 2016; Yoder 1998). In these debates, the hypothesis has also been that if this legitimation of moral expertise is not reached, the whole CEC practice hangs in the balance. Therefore, the question of moral expertise, with its empirical and theoretical validations, has naturally been very important to the legitimation of CEC as an institutional practice for healthcare.

In this essay, I claim that while the dynamics of arguing "for" and "against" moral expertise in CEC practice is an essential conceptual discussion for academics, it is often a debate that ultimately leads to a paralyzing and insolvable contradiction. At best, it gives credibility and legitimation to the ethicists to do their work; at worst, it jeopardizes efforts to work toward creating open social spaces for value discussions in all kinds of healthcare environments from administration to the bedside. I believe that the debate also indicates an "all-or-nothing" approach that categorically misses the point of why open acknowledgment of values is important in healthcare decision-making, policy and practice. I argue that such an approach also often treats the problem of moral expertise in CEC as taking place in a vacuum that fails to take social dynamics, including power, into account.

As a potential resolution to the problem of the dichotomy, I suggest that moral expertise in CEC practice could be understood as a locus of an inherent paradox. As a baseline for my argument, I maintain that moral expertise can be both defended and rejected with reasonable, rational arguments (as we have witnessed by now). By acknowledging that moral expertise is, inherently and indefinitely, a contestable concept, a third perspective can be envisaged. It is a perspective that takes into account the paradoxical nature of the concept of moral expertise, and strives to understand the politics entangled in and around it. It is also a perspective that attempts to make visible the power that is used around defining moral expertise in professional positions. It does not claim to eliminate power, but it aims to foster the kind of forms of power that are in tune with liberal and pluralistic values. Yet, I will claim that in order for CEC to embrace such a goal, it should categorically resist the overemphasizing of consensus in its professional rhetoric.

The idea of a constitutive paradox and its implications that I use in this essay can be tracked down to political theorist Chantal Mouffe's (2005 & 2013) work, especially to her concept of the "democratic paradox" and the theory of "agonistics." I will not delve into Mouffe's original thought in itself here very thoroughly, but in short, she constructs a model of political order—"agonistics"—that depicts the ineradicability of antagonisms in society. In her theory, Mouffe criticizes the "deliberative democracy" approach (such as Rawls and Habermas) for putting too much weight on consensus, and therefore in her view, intrinsically watering down the recognition that value divisions in pluralistic society are comprehensive and real.

[2] For example, Yoder (1998) argues that while a medical specialist embodies *specialist* expertise, expertise can also be *generalist* in its essence. Put simply, while the expertise of the specialist is defined by the depth of her knowledge, the expertise of the generalist is defined by the breadth of her knowledge.

She also claims that the theories that overemphasize the possibility of ultimate consensus categorically miss the most important element of democratic politics, which is seeing the open possibility of dynamic struggle *as a goal in itself.* I will use Mouffe's theory as a baseline and an inspiration in my effort to offer a perspective on the issue of moral expertise in CEC. I will build on the idea that CEC, as well, could embrace the open acknowledgment of dynamic struggle as a goal in itself. I will also connect these theoretical openings to some empirical insights flowing out of my previous qualitative interview study concerning the social construction of professional vision in CEC practice (Saxén 2016). The overall theoretical perspective of this essay can be placed under the umbrella of post-structuralism.

16.1 Mapping the Paradox

Having become familiar with CEC through the general literature and my interview study about the practice, the basic understanding I have accumulated about CEC appears, on a very general level, to point at two central ideas at once. The first idea is the recognition that *value pluralism*[3] is a central principle for healthcare decision-making.[4] The second is the understanding that *expertise* is a key element in fostering decision-making and social order in the conflicted circumstances that inevitably follow from the acknowledgment of pluralism. To build my argument about viewing moral expertise in CEC as a paradox, I start from presuming that these two constitutive constructions, or ideas, together form the basis for the practice of CEC.

The baseline of my argument in this chapter is that while both of these constructions intuitively make sense, they can be seen as being, in a deeper sense, incompatible. While the idea of pluralism acknowledges the lack of an objective perspective on moral matters, the idea of expertise presupposes that a certain version of knowledge serves as a framework to shape moral conversations. The tension between the two ideas is, therefore, profoundly a tension between inclusion and exclusion.

To make a rough simplification of a complex concept, *value pluralism* refers to accepting that everyone has a right to her own moral understanding. This means recognizing the fundamental equality of worldviews and the idea that no one worldview should take dominance over another. What is at stake, then, is the open legitimation of conflict and division in the medical setting (cf. Mouffe 2005, 19). This

[3] As Aulisio et al., for example, write in "Ethics Consultation—From Theory to Practice" (2003, 7): "[T]here is no particular privileged substantive moral view. (...) We are religious and nonreligious, utilitarians and Kantians, egoists and natural lawyers, atheists and theists, and we have a right to be so." Overall, I believe this statement to capture the spirit of CEC.

[4] The argument I make is based on a generalization of my empirical findings and what I find to be the ethos of CEC literature. One could ask the more in-depth philosophical question of what version of pluralism is the most productive baseline for healthcare decision-making. In order to focus sharply on my point of introducing a way to view moral expertise as a paradox, I will not address this topic in this essay. Therefore, the absence of a more detailed analysis of the concept of pluralism is a limitation of this text.

acknowledgment of pluralism and conflict also means that a deep foundation of the work of the ethicist is to recognize that diverse value systems have practical implications on medical decisions. For example, in practice, it means acknowledging that conceptions of what constitutes a good death are not based on medical judgments, but value judgments. Clinical ethicists, therefore, bring to the table the open recognition of how medicine and values overlap, and aim at making sense of these connections as well as making them explicit.

Expertise, on the other hand, here refers to a position of knowledge that can offer solutions for working with the pluralism by facilitating value uncertainties and conflicts. The idea of CEC expertise presumes a definitive base of knowledge and skill that the ethicist utilizes in her work. The claim of expertise is also fundamentally based on an act of exclusion: not everybody with moral values can be a moral expert. While the idea of expertise may not suggest moral authority per se, it does imply that something that the expert has learned through her education and experience enables her to bring a more sophisticated view to the negotiation table. The expert can, therefore, be expected to occupy a key position in shaping the direction that the conversation takes.

Consequently, it can be summarized that the ideas of pluralism and expertise seem to have the kind of inner logic that leads to a certain ideological tension—a clash between the inclusion of all views and an exclusion as to who can have the kind of knowledge that shapes and guides the discussion of such views. This inner tension, then, forms the core of what I call the paradox of moral expertise.

It should be noted at this point that my intention is not to trap CEC into a paradox as a form of critique toward CEC practice. Instead, on the contrary, my intention is to introduce the paradox *as a fruitful and productive baseline for CEC*. While I claim that there is a tension between the ideas of pluralism and moral expertise, this does not mean that both ideas were not valuable—that we should not try to embrace them both or that we would have to reject one idea in order to pursue the other. Following the ethos of Mouffe's theory of agonistics,[5] I suggest that it is possible to envisage a tension between two logics in a positive way, as a locus of a paradox rather than as a destructive contradiction.

As the paradox that arises out of the inner tension cannot be solved or closed, it forms an inherent struggle at the heart of CEC that must then be negotiated in and through social and political practices. I will next bring to attention the kind of social dynamics that aim at this negotiating and renegotiating of the relevant spaces for CEC in its professional and institutional communities and working contexts. I suggest that this negotiation appears as a rhetorical play that shapes the goals of CEC differently in different circumstances. I will call this social dynamic the "expertise game."

[5] In short, agonistics is a theory that accepts antagonisms and value divisions to be permanent, and presumes that conflict can only be disguised, but not overcome, by social manufacturing of consensus. The theory seeks to show how the existence of this conflict can be channeled in a positive way. "Adversaries fight against each other because they want their interpretation of the principles to become hegemonic, but they do not put into question the legitimacy of their opponent's right to fight for the victory of their position. This confrontation between adversaries is what constitutes the 'agonistic struggle' that is the very condition of a vibrant democracy" (Mouffe 2013, 7).

16.2 Dealing with the Paradox: Rationality, Neutrality, and the Consensus-Rhetoric

When the inherent tensions embedded in the concept of moral expertise in CEC are not openly acknowledged, two typical ways of dealing with the paradox can be roughly identified. The first one attempts to strive toward the perfect rational argument so convincing that it could be thought that the other side of the dichotomy would eventually vanish. As I have argued, this approach becomes eventually impossible, because both sides can be argued for rationally. A convincing rational argument for either side can temporarily eclipse the other pole of the dichotomy, but yet, cannot truly close the debate.

The second way to confront the struggle of the paradox is to deny it—often not straightforwardly, but by implying that eventually CEC does not attempt to legitimize any form of moral expertise after all, but instead only attempts to build consensus, or as I found in my interview study (Saxén 2016), to allow "neutral interaction". This leads to defining CEC expertise as the ability to create consensus—and not, as one might assume, the ability to argue about, define and understand *ethics* as a system of knowledge. Consequently, the "neutral interaction" rhetoric waters down the whole essence of claiming oneself an *ethicist*, instead of a *mediator*.

Both of these ways of dealing with the inner tensions of CEC start from the underlying assumption that the struggle in itself is insufferable. Therefore, the (invisible and implicit) assumption is that that the paradox must either be hidden from scrutiny or be rationally overcome. Given that the rational arguments are convincing from both sides, I will not go any deeper into exploring the arguments and their implications in this essay. Instead, I will pay attention to the emphasizing of consensus as the goal of clinical ethics consultation. While striving for consensus may be desirable in everyday life for social reasons, I argue that framing CEC fundamentally as a consensus-building effort may be a more dangerous path than is typically recognized. It could be argued that consensus as a primary goal does not do justice to recognizing pluralism and giving the value divisions a channel to be articulated and debated. This is because consensus categorically strives at reframing the inherent value divisions in such a way as to make them invisible. In other words, highlighting consensus constructs divisions in a way that, in a sense, (and put very roughly) makes the differences irrelevant. Therefore, the consensus-rhetoric obliterates the notion of pluralism, as it silences *the struggle* of values that can be argued to be the very condition of pluralism (cf. Mouffe 2005 & 2013).

Furthermore, underlining consensus as a central goal of CEC practice not only raises the question whether ethics is truly synonymous with consensus, but it also presumes that conflicts and antagonisms could eventually be overcome by putting the right professional practices in place. Put differently, it can turn CEC practice into a strategy of controlling divisions and, put in more radical terms, to colonize the moral space by establishing certain procedures through which antagonisms can be limited. Moreover, appealing to 'rationality' as a marker of consensus makes power imbalance invisible by assuming that everyone in the negotiation table have similar agencies and abilities to affect the outcome of the deliberation. (Mouffe 2005 & 2013; see also MacIntyre 1988.)

The core of my criticism toward consensus-building is, then, the realization that consensus eventually entails some form of exclusion: not all perspectives can be established in the compromise at once, but the solutions will arise out of a struggle of arguments. The problem I raise is not the struggle in itself, but the consensus-rhetoric that makes it invisible. Such rhetoric also typically abstracts the process of arriving at the consensus as a conversation between equals, which is not necessarily the case. As a consequence of these assumptions, all forms of social and power relations, as well as the subtle workings of deeply rooted cultural traditions and prejudices, tend to be therefore left out of the consensus-rhetoric. Constructing all participants as having equal starting points, therefore, creates an invisible smokescreen that hides the power structures that operate in such deliberations. What I am concerned about is, then, that the consensus-rhetoric "disguises the necessary frontiers and forms of exclusion behind the pretenses of 'neutrality'" (Mouffe 2005, 22). This is because to frame decisions as the outcome of purely rational deliberations is to make space for change and resistance practically impossible, since the "rational" and "pure" consensus silently illegitimates the forms of its challenge (cf. ibid., 32).

It is important at this point to specify that I do not view *power* in this context as an instrument of manipulation or coercion, but instead, as constituting the very identities and agencies that people bring to the negotiation table. Viewing the workings of power in this way arises out of poststructuralist philosophy, essentially Foucault's writings that (defined loosely) present power as something that is being diffused in discourse, knowledge and culturally hegemonic 'regimes of truth'[6] (1995 & 1998; see also Rabinow 1984). In other words, 'truth' or 'knowledge' are not viewed as something that exists outside of people's actions, but instead, truth is seen as "a thing of this world" (Rabinow 1984, 72)—a product of dynamic social action, manifesting through professional roles and expert statements in public discourse such as the media, science and education. In this way, Foucault claims that "power is exercised through the epistemes (underlying order) and discourses found in what passes as knowledge" (Allan 2007, 527). However, it is important to note that for Foucault (among other post-structuralist theorists) power does not appear solely as coercive and exclusive, but also as a productive force that gives shape to a working society. (See ibid.)

Based on this general idea of power as something that constitutes agencies, identities, and knowledge positions, experts in institutional hierarchies can be seen to have power in defining and controlling what counts as sayable and thinkable in the institutional circumstances even if the experts were not aware of this power or mak-

[6] "Each society has its regime of truth, its 'general politics' of truth: that is, the types of discourse which it accepts and makes function as true; the mechanisms and instances which enable one to distinguish true and false statements, the means by which each is sanctioned; the techniques and procedures accorded value in the acquisition of truth; the status of those who are charged with saying what counts as true." Foucault in Rabinow 1984, 73.

ing it explicit. Institutional roles therefore shape identity and agendas, reproduce forms of social hierarchy, and create social distance between people (Cribb and Gewirtz 2015, 14). To follow Mouffe (2005, 21), "we should not conceptualize power as an external relation taking place between two pre-constituted identities, but rather as constituting the identities themselves." This is also why power should be understood as something that can never be erased from such institutional environments and transactions.

I believe that while ethics consultants cannot detach themselves from such power-infused institutional processes, it is possible to envision an ideal in which power would be constituted in ways that are compatible with pluralism. This would require the ethics consultant to be aware of the subtle dynamics of who is doing the defining of ethics in different contexts and how, and to make the social process of negotiating the struggle visible. Equipped with analytical understanding of the moral landscape in a complex situation, the ethics consultant can be in a special vantage point to open space for the kind of voices that may otherwise remain silent—but only if the silent voices are not rendered invisible by assuming that they have an equal position in comparison to the more powerful voices in the first place. The ethics consultant has power to make the invisible visible, or the implicit explicit, which was well crystalized in my interview study by an interviewee who told me how she 'goes ahead and states the obvious'— and I quote—"*I will frequently be the person that says you know, maybe I am instigating here, but are you trying to say that you don't respect their beliefs*" (for the data quote, see Saxén 2016, 106). This kind of question is powerful as it challenges the social order of the situation by virtue of making the implicit explicit.

Yet, it should be noted that my criticism of the consensus-rhetoric in CEC does not attempt to make the claim that consensus would not be desirable or extremely relevant for managing difficult and conflictual situations. The ability of the ethicist to work between different worldviews and to find solutions in heated situations is, without a doubt, vitally invaluable and helpful for the everyday hospital life. However, it is important to distinguish this everyday consensus from a more fundamental idea of constructing consensus as professional ideal, and to recognize that building consensus is a secondary goal to the fostering of a vibrant climate of ethical discussions. Yet, it is a difficult task to differentiate between the two: the social demands of consensus-building, and the concept of consensus-building as a professional ideal. My intention here is to focus on the latter, even though the concepts no doubt are overlapping.

I acknowledge that suggesting that consensus building is at the heart of CEC practice in this way is a rough reduction and does not necessarily reveal what the ethicists actually do. I do not attempt to make the statement that ethicists would not see the difference between "consensus" and "ethics." However, I want to pay attention to the neutrality rhetoric and to point out its potential hazards for the goal of openly legitimating pluralism. It should also be noted that in my qualitative interview study of the professional vision in CEC practice, I found that creating consensus through mediation was not constructed as the primary aim of ethics consultation, even though it was mentioned in many contexts. The claim that the ethicist is simply a neutral arbiter (and, implicitly, not an expert in ethics after all), was mostly brought up in the interviews when inherent tensions in the ethicists' role came up implicitly or explicitly. Overall, in my study I found that the clinical ethicists adopt an intricate array of social positions in their hospital work field. Moving between the different expected social roles therefore creates a position of ideological tension for the ethics consultant. (Saxén 2016.)

More specifically, in the analysis of the interviews, I categorized consensus building, or neutral interaction, as a "bridging discourse" that was introduced as an attempt to mitigate inner tensions embedded in the professional vision of CEC. I found a clear tension in the discourses of how ethics consultation is carried through: for example, as to whether a consultation is the outcome of "deliberation" or the expert's "technique" (what I have called the tension between the collective and individual form of agency). However, I did not interpret consensus building to be a fundamental discourse in defining what ethics consultation is about. Discourses that I found to shape the professional vision of CEC were three discourses of order (managerial, emotional, and rational) and agency (exploration, technique, deliberation, and distancing). (See Saxén 2016 for examples and data quotes.) Only in addition to these discourses that I identified as constructing the base of ethics consultation, I interpreted "neutral communication" as a bridging discourse between the above-mentioned, more fundamental discourses, utilized against potential claims that the ethicists' role was illegitimate. These claims did not have to be articulate because I did not make them in my interviews; yet, occasionally during the interviews I found many of my interviewees talking as if viewing me as a potential contestant of their professional claims. I believe this reflected a more thorough social environment that appeared to place the ethicists into a position in which they were constantly shadowboxing against potential resistance. It has stuck with me how one of my interviewees crystallized these tensions well by stating, "*When you are the ethics consultant, you have to be constantly walking on eggshells*."

The management of the inner tensions embedded in CEC practice could be further labelled as the "expertise game" in which the ethics consultant simultaneously adapts a position of expertise in ethics as well as denies this expertise by claiming to be an expert in mediation and conflict resolution, instead of ethics. This observation comes close to what Dzur (2008, 218) calls the "liminality" of ethics consultation, meaning a position "which is neither this nor that, and yet is both" (ibid., original citation from Turner 1964, 7). Such dualistic accounts expose a socially dynamic "game" in which the ethicist is set to a position of fulfilling a range of expectations and must play her part differently in different kinds of situations. It is also important to acknowledge that this game involves other actors as well who have an interest in defining the parameters of ethics in their way. These actors may be hospital administrators and influential physicians, for example. The "expertise game" is, therefore, a dynamic social process in which norms and practices are tested, negotiated and legitimated in a world of social and ideological diversity.

I believe that grasping the nature of the "expertise game," and the tensions in which the ethicist is placed institutionally and socially, can potentially open a perspective on understanding ethics consultation more thoroughly, which helps to make the tensions that CEC practice faces more visible and tangible. It is my aim in this essay to construct a theoretical perspective for understanding the concept of moral expertise in CEC in a way that does not overlook the inner tensions of CEC, as well as the questions of power that surrounds CEC practice, but takes them into account.

16.3 Embracing the Paradox

A central starting point for my argument which has not been articulated very explicitly before is the understanding that the notion of moral expertise is not actually reserved only for clinical ethics consultants. The deeper question is, in my view, ultimately not whether or not the ethicists have moral expertise, but whether ethicists can actually challenge the more institutionalized, hegemonized and subtle forms of moral expertise in the hospital environment in which they work. In other words, even before ethics consultants were there at all, the hospital culture contained often indirect and typically invisible forms of practices that could also be called "moral expertise," such as certain conceptions of right and wrong knit tightly into the physician's profession, as well as the hospital institutions' structures of authority and decision-making (Freidson 1970 & 1988; Rosenberg 1999; Rothman 1991). Yet, these are forms of expertise that rather distort than divulge and clarify the claims of moral expertise. Due to their elusiveness, such forms of expertise are difficult to pin down, as they are typically not explicit, and they have usually been naturalized—put differently, appearing simply "normal"—within the existing conditions, its hierarchies, discourses, and distributions of authority. Especially because such forms of moral expertise are not made explicit, they can offer a strategic position to construct certain versions of ethics without facing open challenge. (For examples of how conceptions of ethics in healthcare professions can implicitly construct social hierarchy, see Saxén 2017.)

The difference between the moral expertise in CEC and the more traditional forms of moral authority in healthcare is therefore not that CEC brings a foundationally new concept of moral expertise into the healthcare organization; rather, CEC merely makes the dispute visible and open for argumentation. This openness may subject CEC practice to constant criticism, but it also contains a seed of wisdom that may be at least partly missing from the more traditional approaches to healthcare ethics. That seed is the *open, not implicit*, acknowledgment of values and the ways in which values may conflict in a world that is often exhaustively defined in economic, technological, and narrow professional terms.

Viewing the hospital as an institution shaped by acts of power and struggle creates the backdrop against which I see CEC taking its shape as a social practice that can foster space for pluralistic value discussions. What CEC can bring into open existence in the hospital is the acknowledgment of the *permanence of struggle* without attempting to bring this struggle to a closure. In this way, I view CEC as a practice that can serve to create, in Mouffe's terms, an agonistic space for healthcare decision-making—that is, to provide the hospital institution a social space "where the opponents are not enemies but adversaries among whom exists a conflictual consensus" (Mouffe 2013, xii). Understood this way, conflicts and confrontations are not an obstacle, but instead they indicate that democratic and pluralistic ideals are kept active and alive. This realization of the permanence of the struggle is, to me, what the idea of "keeping moral space open" means in its essence (quoting Walker 1993).

Viewed from this baseline of already existing struggle and negotiation about ethics, CEC appears as an emerging paradigm that challenges the more traditional discourses of healthcare ethics. The potential that I see for CEC is the institutionally embedded agency to create 'fragmentation' in the existing conceptions of 'truth'. Such fragmentation calls debate and vivid argumentation into being. Therefore, I view CEC practice as having potential to articulate the pluralism of values, as well as to challenge the existing hierarchies, in a way that creates open space for diversity. Viewed this way, CEC is seen to be caught in various processes of struggle, and making this struggle openly acknowledged appears as *the point* of CEC practice. Yet, ethics consultants are not above the struggle; they are inside it. Therefore, a real danger is that the ethicist might attempt to establish a position of actually being in control of the struggle, instead of making the struggle open and acknowledged. I view the concept of establishing a fixed sense of 'moral expertise' as a pathway to enabling such a dangerous position. This does not, yet, imply that the ethicist would not have knowledge or expertise in ethics altogether. Simply, it means being reflective and sensitive as to the concept of moral expertise and to the dangers that the concept contains potential to enable an agency of social control.

This perspective offers a way for envisioning the goal of CEC as articulating the existence of value pluralism[7] to all parties in the negotiation table. The goal of such articulation is to make possible to pay attention to various divisions and to create a possibility for *fragmenting the hegemonies*[8] that exist in any given social setting. Yet, to view the goal of CEC in this way is also to recognize that the struggle is never going to be closed. In order to keep the struggle open, in my view, it is crucial to understand moral expertise as a constitutive paradox. According to this view, moral expertise appears as a concept that holds an ideal that can be pursued, but not accomplished or closed. This means stretching toward moral expertise without ever actually reaching it. This is because the position of the moral expert is impossible to reach, since moral expertise is self-contradictory in its very essence—a conceptual impossibility, a paradox (cf. Mouffe 2005, 137). Indeed, to be a moral expert—as

[7] Again, a discussion of its own—which is out of the scope of this chapter—is what kind of understanding of pluralism is the most functional and justified to serve as the basis of CEC practice.

[8] A concept close to Foucault's "regimes of truth," hegemony refers to an established idea that has become so normalized in a given cultural setting that it is seldom openly questioned (Gramsci 1971 as cited by Fairclough 1992, 91–96). Many poststructuralist social research traditions, such as critical discourse analysis, presume that hegemonic social constructions that shape and constrain thinking in a given culture can be discovered by studying the use of language. While hegemonies present some culturally embedded ideas as normal, as "common sense," they silently marginalize other ways of thinking. (Ibid.)

understood in the sense of becoming a legitimated moral authority—would be to abolish the concept of recognizing authentic value pluralism, and therefore, to obliterate the whole basis for ethics consultation. Therefore, the expertise in CEC practice can only sustain as long as it is recognized that moral expertise cannot be reached—that is, only in so far as ethics consultants acknowledge the particularity and the limitation of their claims of moral expertise.

Acknowledging the paradox as the baseline of CEC and not attempting to close it may be the best way to foster the kind of profound value conversations CEC claims to enhance because open acknowledgment of the paradox would protect CEC practice against any attempts at establishing a closure of what moral expertise is. To acknowledge the inherent paradox is, therefore, to guarantee that the dynamics of open moral deliberation will be kept alive. This is crucial for recognizing that social space will always be ordered by certain versions of the truth, while leaving other narratives to cultural margins. The forceful recognition of the tendency of communities and institutions to establish certain versions of the truth can offer an open social space, "wiggle room," that is conducive to a broad acknowledgment of pluralism. Against this backdrop, the ethics consultant therefore could be thought of as being in a legitimated position for opening the discussion of pluralistic values and possibilities.

The professional challenge for CEC in my opinion is, thus, not to legitimize the concept of moral expertise, but to construct a claim of expertise strong enough to institute social space for facilitating value discussions, making them visible without actually permanently and fully defining moral expertise. The foundation of this claim is that without conscious efforts at making ethics visible, healthcare communities will develop value-laden and interest-ridden practices that, despite good will, may not necessarily benefit the patients and the public but the definers and the decision-makers. Making ethics visible may not reverse the impact of such influential dispositions, but it does cast light on existing decision-making practices and creates a due diligence process, categorically placing (at least to some extent) the burden of proof on the power holders. Therefore, CEC can construct social space for "a kind of interaction that invites something to happen, something which renders authority more self-conscious and responsibility clearer" (Walker 1993, 33). Yet, it is obviously important to recognize that the corruptive elements of power cut both ways, and that by legitimizing a position of "knowing ethics," ethics consultants themselves may attempt to establish an order that benefits their own professional aspirations.

In my view the deeper question about the legitimation of ethics consultation is, then, fundamentally not about moral expertise at all. It is eventually about the professional identity of clinical ethics consultants and the possibilities that identity entails—to put roughly, the question is whether ethics consultants become agents of manufacturing institutionally approved consensus, or whether ethics consultants are able to open spaces for a more pluralistic range of voices. In order for ethics consultation to open new horizons for the more traditional and profession-centered healthcare ethics ethos, it has to be clear that real alternatives are at stake.

Finally, it is reasonable to ask to what end, if any, these remarks bring us. Many critical writings on CEC are based on explicit or implicit assumptions that doing something differently will make CEC a better practice. While I have claimed that CEC should not fall into overemphasizing consensus, I want to resist the intuitive idea that this utopia would become closer by suggesting concrete changes. Instead, my endeavor has been to show that while the professional claims of CEC hold inner conflicts and tensions, these tensions are not detrimental to the goal of creating open space for moral discussions. I have attempted to argue that ethics consultants are in a special vantage point that has the potential to create awareness that is critical toward the clinical knowledge systems, rules and social hierarchies. All in all, my intention has been to demonstrate that making visible the moral realm of clinical decision-making is *in itself sufficient, because it opens up possibilities*. However, the task set for ethics consultants by these words is unquestionably difficult. How to actually use power in a way that is in tune with pluralism? How to define one's professional identity as someone who is both "knowing" and "not-knowing"? How, in other words, to embody a paradox? The essence of the challenge is well captured in a quote by Pakistani writer Raheel Farooq: "*Intelligence is to spot paradoxes. Wisdom is to live by them.*"

References

Allan, K. 2007. *The social Lens: An invitation to social and sociological theory*. London and New Delhi: Sage Publications.

Archad, D. 2011. Why moral philosophers are not and should not be moral experts. *Bioethics* 25 (3): 119–127.

Aulisio, M.P., R.M. Arnold, and S.J. Youngner. 2003. *Ethics consultation: From theory to practice*. Baltimore: Johns Hopkins University Press.

Cribb, A., and S. Gewirtz. 2015. *Professionalism*. Cambridge, UK: Polity Press.

Cross, B. 2015. Moral philosophy, moral expertise, and the argument from disagreement. *Bioethics* 30 (3): 188–194.

Crosthwaite, J. 1995. Moral expertise: A problem in the professional ethics of professional ethicists. *Bioethics* 9 (4): 362–379.

Dzur, A.W. 2008. *Democratic professionalism: Citizen participation and the reconstruction of professional ethics, identity, and practice*. University Park: Penn State University Press.

Fairclough, N. 1992. *Discourse and social change*. Cambridge UK: Polity Press.

Foucault, M. 1995. *Discipline and punish: The birth of a prison*. London: Penguin.

———. 1998. *The will to knowledge: History of sexuality*. Vol. 1. London: Penguin.

Freidson, E. 1970. *Professional dominance: The social structure of medical care*. New Brunswick: AldineTransaction.

———. 1988. *Profession of medicine: A study of the sociology of applied knowledge*. Chicago: University of Chicago Press.

Gordon, J.S. 2014. Moral philosophers are moral experts! A reply to David Archad. *Bioethics* 28 (4): 203–206.

Gramsci, A. 1971. Selections from the prison notebooks, ed. and Trans. Q. Hoare and G. Nowell Smith. London: Lawrence and Wishart.

Iltis, A.S., and L.M. Rasmussen. 2016. The "ethics" expertise in clinical ethics consultation. *The Journal of Medicine and Philosophy* 41 (4): 363–368.

MacIntyre, A. 1988. *Whose justice? Which rationality?* Notre Dame IN: University of Notre Dame Press.
Mouffe, C. 2005. *The democratic paradox*. London: Verso.
———. 2013. *Agonistics: Thinking the world politically*. London: Verso.
Noble, C.N. 1982. Ethics and experts. *The Hastings Center Report* 12 (3): 7–15.
Priaulx, N., M. Weinel, and A. Wrigley. 2016. Rethinking moral expertise. *Health Care Analysis* 24 (4): 393–406.
Rabinow, P., ed. 1984. *The Foucault reader: An introduction to Foucault's thought*. London: Penguin.
Rosenberg, C.E. 1999. Meanings, policies, and medicine: On the bioethics Enterprise and history. *Daedalus* 128 (4): 27–46.
Rothman, D.J. 1991. *Strangers at the bedside: A history of how law and bioethics transformed medical decision making*. New York: Basic Books.
Saxén, S. 2016. Untangling uncertainty: A study of the discourses shaping clinical ethics consultation as a professional practice. *Journal of Clinical Ethics* 27 (2): 99–110.
———. 2017. Same principles, different worlds: A critical discourse analysis of medical ethics and nursing ethics in Finnish professional texts. *HEC Forum* 30: 31. https://doi.org/10.1007/s10730-017-9329-0.
Shalit, R. 1997. When we were philosopher kings: The rise of the medical ethicist. *New Republic* April 28: 24–28.
Turner, V.W. 1964. *Betwixt and between: The liminal period in rites de passage*. Seattle: American Ethnological Society.
Walker, M.U. 1993. Keeping moral space open: New images of ethics consultation. *The Hastings Center Report* 23 (2): 33–40.
Yoder, S.D. 1998. The nature of ethical expertise. *The Hastings Center Report* 28 (6): 11–19.

Chapter 17
Building Clinical Ethics Expertise through Mentored Training at the Bedside

Evan G. DeRenzo

17.1 Introduction

This chapter is about building clinical ethics expertise by having an advance practice clinical ethicist (APCE) mentor a novice clinical ethicist (NCE) at the bedside. (The designation "advance practice" will be an interchangeable substitute for the term "expert" throughout this paper). Being at the bedside, that is by being present on work rounds and shadowing on clinical ethics consultations is how a NCE learns the substance of her work (the pronouns for he and she will be used interchangeably throughout this paper and will represent any and all gender identities, with no preference for any particular one). One learns about the work through academic training; one learns the work where the work takes place.

Before getting to the practice, itself, is a brief theoretical discussion. Not all training programs are alike. "Medical ethics is a field of scholarly inquiry that uses a wide variety of methods" (Sugarman and Sulmasy, 2010, pl. xi). This statement would go without saying if it weren't for the matter that unless one understands the philosophical context that guides a particular program's or hospital's practice, one may think that one method works just as well as any other. That is not the position this chapter takes.

Also, this chapter focuses attention on what the author sees as absent from most of today's discussions. That is, today many in the field avoid discussion of what we at the John J. Lynch, MD Center for Ethics at MedStar Washington Hospital Center (hereafter referred to as the Lynch Center) believe is required to become an APCE. For us, character matters. The chapter then moves to how, exactly, this bedside training is provided and how, through bedside training virtue can be taught.

E. G. DeRenzo (✉)
MedStar Washington Hospital Center, Washington, DC, USA
e-mail: evan.g.derenzo@medstar.net

J. C. Watson, L. K. Guidry-Grimes (eds.), *Moral Expertise*, Philosophy and Medicine 129, https://doi.org/10.1007/978-3-319-92759-6_17

The paper's conclusion looks to gaps and opportunities in bedside training of a NCE. The chapter ends with suggestions for how to make use of the opportunities and close the gaps.

17.2 Why Must Training of Clinical Ethicists Take Place at the Bedside? Going Where the Patients Are

Although Willie Sutton denied having said it, there is a wonderful, perhaps apocryphal story that when a reporter asked Sutton why he robbed banks he replied, "Because that's where the money is"' (Sutton and Linn 1976). Following this line of thinking, the answer to the question, 'Why train NCEs at the bedside?', is, "Because that's where the patients (and family and friends and clinician providers) are". Although this answer seems overly simplified, the bedside is where the complexities unfold.

The first bedside lesson is that medicine, at least in an acute care hospital, is the territory of the clinical providers, especially senior attending physicians. The bedside is where they practice. One has to be invited into their world.

That one sees this requirement for an invitation as mere etiquette is where the shaping of the NCE's character shaping begins. Rather than bristle at the need to wait to be invited, the process of respecting the special relationship of patient and physician begins to teach the NCE the humility with which her job should be faced. Although working with clinicians is part of what is so enticing about being a clinical ethicist, (it is great to work with smart people in the service of taking care of the sick and vulnerable), the NCE needs to appreciate – appreciation that only comes with close bedside observation - that even in today's busy hospital, there continues to be a special and privileged relationship between patient and physician. Entering usefully into this relationship takes skill and humility in the face of the patient's need to trust their physicians. Only once a physician trusts a CE enough to be invited into a hospital space previously the purview of only treating clinicians will that sense of trust be conveyed to the patient. And only then can a CE be of any real help.

This quality of humility ought to grow with time. Often NCEs come into the clinical ethics setting with high levels of education, giving one a false sense of mastery. But the NCE with the character appropriate to being an APCE quickly appreciates how little one knows and how long it will be 'til one achieves any semblance of mastery. Without this aspect of virtuous character, one is doomed to live in one's head, missing the other qualities that must be demonstrated to gain the kind of trust of the medical provider teams for them to allow and provide one access to their patients.

Without easy access to patients, it's quite impossible to learn about the ethical care of patients. This point is easily understood by thinking about the training of physicians and nurses. Would anyone want a physician or a nurse for oneself or one's loved one if that physician or nurse had never taken care of a patient? No, we all want to be taken care of by physicians and nurses who are, themselves, well-trained and experienced. For physicians and nurses that means many years of mentored and supervised training in the care of patients and their families and friends. It shouldn't be any different in the training of NCEs and our field fails the training of NCEs if this point is not well appreciated.

17.3 A Brief Philosophical and Historical Context

When training NCEs to become APCEs, it is to have them be assistive in a variety of ethically important ways in achieving good, at least improved, outcomes for patients, surrogates, and clinical providers. Good outcomes include optimized medical care while helping patients meet their own goals to the greatest degree medically reasonable. When patient and/or surrogate goals cannot be met, the term 'good outcomes' means to come as close as is medically reasonable while reducing and/or avoiding excess burden on patients and surrogates, cradling those who grieve, always taking opportunities to deepen ethical thinking in one's institution's providers and other hospital personnel, and help to dissolve lingering moral distress.

What is often missed in training NCEs, particularly if the mentoring APCE is not a clinical provider him or herself, is about what is required to produce good outcomes for individual patients, families and friends and/or improved outcomes for patient populations. There are two drivers of excellent medicine; 1. meeting the preferences of patients and surrogates to the greatest degree medically reasonable while 2. having the clinicians practice within appropriate standards of practice. Understanding of how these two drivers of moral medicine combine, intertwine and unfold in the course of any patient's care can only be learned if the NCE is closely observing of the intimacy of this process and discussing the fine points with a truly skilled APCE.

That is at least part of why this author diverges from others in the field relating to the methodology of mediation. (Here it is important to differentiate between the high utility in having CEs learn mediation skills. That is not what is meant here. Rather, this is a discussion of the utility of applying mediation as a legalistic process as the format for clinical ethics consultation). Mediation, as a methodological process calls for working towards consensus and compromise across parties. But as well-discussed many years ago (Moreno 1995), often, consensus and compromise result in parties ending up feeling as though something of their deeply held values have been lost.

We take, however, as our starting point that often it will be unrealistic, and perhaps harmful, to expect everyone to end up of one mind. Instead, we teach that a key lesson for the NCE is that persons of good judgment may disagree. To become an APCE one must not only learn to tolerate high levels of multi-layered disagreement, sometimes much of which is directed fully at the clinical ethicist, one must appreciate from where such disagreement comes and honor and uphold the moral right of persons to respectfully disagree. The space in which the CE works is often that of conflict and uncertainty. Under these conditions, bedside mentoring of the NCE assists the NCE to learn how to tolerate working constantly in this grey space.

But unlike the philosopher, whose job it is to debate virtually endlessly the great questions of all eternity, the job of the APCE is to assist in producing good outcomes for patients, families and clinical providers. And to do so in ways that are consistent with ethically justifiable outcomes, one must be well-grounded in the philosophical perspectives one brings to an ethics consult or to bedside rounds; understanding to which one inclines to and why.

17.4 Theories: To Which Does the Novice Cleave?

Since the earliest days of the field of bioethics there have been those who decried its very existence. Of these many critiques, perhaps the most eloquent, robust and influential comes from H. Tristram Engelhardt, Jr. (Minogue et al. 1997; Rasmussen et al. 2015). His elegant critique is essentially this: because Enlightenment philosophy has not produced a single, unified, content-filled theoretical or methodological structure upon which to build a secular morality, we are left with merely tolerance for a thin ethic of permission between moral strangers in the medical encounter. According to Engelhardt, because no single theory or methodology can be shaped into a one-size-fits-all, there can be no intellectually coherent ethics of bioethics. He grounds this claim in the demonstrably obvious fact of the diversity of values in our pluralistic society where there is often disagreement about what constitutes 'the good', where values conflicts abound and where at the bedside there is often intractable, irresolvable, discord.

Even if one wants to dismiss his theoretical challenge to the field because it produces such discomforting cognitive dissonance, one does so at the field's peril. Not so much because of the philosophical quicksand his message creates but even more because in the same breadth that Dr. Engelhardt's message is dismissed, it creeps into the most field-foundational spaces, perhaps uninvited, engendering theoretical confusion.

This confusion is seen in close examination of the field's foundational documents, the ASBH *Core Competences for Health Care Ethics Consultation* (1998; 2011). The casual reader of these documents may not pick up on how the *Core Competencies'* creationist Task Force members immediately shied away from explicitly addressing Engelhardt's challenge while fully embracing it in unarticulated ways. Perhaps it is just because we all accept that our field is growing up in a world of culture and values diversity, the creationist ASBH Task Force members, in the first edition of the *Core Competencies* stopped short of defining the good in clinical ethics consultation. Although they make note that "good character is important for optimal ethics consultation" (1998, p. 21), these creationist Task Force members pulled their punch by not overtly calling for character development as part of the process of becoming competent in the practice of clinical ethics. The 2nd edition removes all explicit discussion of virtues and character.

The problem for training the NCE these omissions cause, however, is that what is essentially a philosophical debate loses sight of the clinical nature of clinical ethics and totally misses the moral core of clinical medicine. While in an abstract philosophical discussion one can indefinitely debate what might become philosophical consensus about the good, this is not what moral medicine is about. Rather, moral medicine is about taking good care of patients. Once one understands the fundamental difference between the abstraction of philosophical debate and the practicalities of clinical medicine, and that it is the latter that is at the heart of moral medicine, than it becomes easier to figure out what might be the most ethically justifiable option or set of options for a particular patient at a particular point in that patient's care.

Pellegrino refutes Engelhardt's nihilist view of being able to make complex ethical recommendations for patients in a world of values diversity by laying claim to a thick philosophy of medicine that sees the goal of medicine as the helping and healing of patients (Pellegrino and Thomasma 1988; Englelhardt and Jotterand 2011). If one believes, as we at the Lynch Center do, that helping and healing patients is the morally correct goal or *telos* of medicine, embracing cultural and values diversity is just one aspect of acting in the patient's best interest. Relying only on Enlightenment Philosophy's rules and rights-based approaches to figuring out what ought to be done, for us, is simply not the place where the ethics of medicine stops. Rather, while subsuming important default rules, notions of duties and obligations, and justice concerns, we include attention to concerns about the character of those responsible for providing care as well as to that of the NCE.

This simple focus, simple only in the abstract and so complicated and difficult in the operationalization, is what we at the Lynch Center see as the ethically rightful philosophical perspective for medicine. That is all of medicine writ large, not just physicians' practice but that of nurses and respiratory therapists and hospital clinical social workers and clinical ethicists; all those who provide and are administratively responsible for patient care. Further, once this is understood by those we train, the methods for achieving the goal of helping and healing patients by working to act in their best interest, not one's own, can only be actualized in bedside training.

Finally, that we at the Lynch Center cleave to the ancient ethical norms of virtue for the helping and healing of patients does not mean that we ignore the great lessons of the Enlightenment. Expertise in clinical ethics is being able to intellectually encompass all that is important for working towards making decisions and recommendations that are in the best interests of patients, and then secondarily but not unimportantly, what is in the best interest of patient's families, friends and those who provide the actual care.

Certainly, we accept that attending to character development adds an additional layer of training to an otherwise difficult training task. But we do not think that one can, for example, expect clinical providers, patients and families to trust us if we do not aspire to be trustworthy persons. One can debate which virtues of character may be necessary, the most important and/or unnecessary. Nonetheless, to leave out of training NCEs how character matters is to our minds only addressing part of the job.

17.5 The Practice: Mentoring to Do What?

In Jim Childress'Forward to Fletcher et al. (1995), Childress opens with, "The field of bioethics is in creative ferment." (Childress 2005a, p. *v*). As far as this author is concerned, that applies now just as it applied then to (1) just what the responsibilities of a clinical ethicist are and (2) how those responsibilities ought to be implemented (Childress 2005b, p. v).

This ferment can be gleaned by reading the preface to the 2nd edition of Sugarman and Sulmasy's *Methods in Medical Ethics* (Sugarman and Sulmasy 2010). In it they say, "These methods derive from the humanities and the social sciences, including anthropology, economics, epidemiology, health services research, history, law, literature, medicine, nursing, philosophy, psychology, sociology, and theology" (p. xi). This academic and clinical wealth of diversity breeds ferment, which ought to continue to be eagerly welcomed by the field and explicitly acknowledged when teaching NCEs about methods and approaches to what it means to become an APCE.

One's appreciation for this diversity of discipline and preferred methodology is deepened if one reads all the editions of what can only be viewed as the most important scholarly project in modern medical ethics, the seven (7) editions of Tom L. Beauchamp and James F. Childress's *Principles of Biomedical Ethics* (first edition, 1989, 7th edition 2009). The early editions were the source of what evolved into being called Principlism, or colloquially the Georgetown mantra. This mantra was the result of Beauchamp's and Childress' focus on the biomedical principles of respect for autonomy, nonmaleficence, beneficence and justice. Although from the outset the authors made clear the ancient problem with principles is that there is no tried and true way to figure out which principle takes precedence over which under which sets of circumstances, that the authors always placed autonomy at the top of the list in a country where autonomy reigns supreme drove the field in the direction of autonomy taking the helm.

Over the 35 years spanning the 7 editions of *Principles of Biomedical Ethics*, as Principlism has lost its vice grip on the field, a new methodology has arisen as today's methodological buzz word, mediation (Fiester, this book; Dubler and Liebman 2011). Mediation, while derived theoretically from an important concern for increasing fairness and justice in clinical ethics consultation, looks more like the methods of Fletcher et al. (citation op cit) and Jonsen et al. (2006) than like the scholarship of Beauchamp and Childress. Nevertheless, mediation is sweeping the field. Consistent with the rule-based, legalistic approach that the field has preferred since it went down the Enlightenment philosophy path in the first place, mediation continues to exclude virtue theory and virtue ethics from its literature. Just as the Georgetown mantra was favored, now mediation is the mantra *de jour*. Unfortunately, like Principlism was, mediation is also the wrong mantra.

17.6 Learning to be Virtuous through Mentoring during Standard Work Rounds and Shadowing on Ethics Consults

There are many kinds of rounds in hospitals, such as sitting rounds or grand rounds. Many are appropriate and useful for APCEs', and often NCEs', participation. What is focused on here, however, are work rounds; the rounding on each patient, usually each morning, in which representatives and/or all members of the clinical team participate (DeRenzo et al. 2006; Messutta 2017). Additionally, at least in large, teaching hospitals, these rounds may include a social worker, a pharmacist or PharmD, a member of the hospital's palliative care group, and a psychologist or psychiatrist. These rounds are usually lead by the senior team clinicians; attending physician with fellow, senior residents and/or interns presenting; attending physician and Advance Practice Clinicians (APCs), ie physician assistant (PA) or nurse practitioner (NP) presenting; fellow with residents, interns and/or APCs presenting. The main reason we find joining work rounds so vital to successful hospital clinical ethics practice is because it gets us to the bedside of most, if not all, patients on a particular unit, hearing the patient information in real time.

Only at the bedside can an NCE learn to differentiate ethical weight among the cascade of clinical facts that are part of any acute hospitalized patient's care. It is only at the bedside that an NCE can come to understand how the medical facts of a case are perceived by the various members of the treating team. The process of rounds gives the clinical ethicist, both NCE and APCE, an appreciation of the subtleties of clinical team members' grasp (or not) of ethically salient medical and psycho/social factors. This learning environment for the well-mentored NCE provides patient-relevant ethics nuances that simply cannot be conveyed through reading or hearing about a case.

Another reason we find joining work rounds so useful is because it does double duty for training both house staff who are part of the patients' treating team and NCEs and any others who may be with us from the Lynch Center's side, such as university interns, visiting international scholars or our own accredited rotating residents. That is, the APCE joins rounds to teach; whether that is those who come with the APCE or it is teaching of the treating team. Both groups benefit educationally from the skilled participation of an APCE. That includes but is not limited to being able to judge well the timing of when an ethics question or comment is profitably interjected to advance patient care while simultaneously demonstrating respect for team time constraints.

As previously noted, the APCE will have already had to build trust between the APCE and members of the treating team, often including and perhaps first, the senior attending. To think that even in a generally ethics-welcoming hospital, clinical ethicists are always warmly welcomed into the world of doctors and nurses, where the hierarchy of medicine continues to be felt palpably when team rounds occur calls for a high level of trust across the physicians and nurses and the ethicists for the ethicists to have any ability to have any patient-specific or organizational impact. That trust may be most effectively built on rounds, one set of rounds at a time. For example, one organizational goal in clinical ethics is always to work towards greater democratization of moral discourse within this strict hierarchical system. Rounds is the perfect place to model such democratization. And doing so only increases the trust placed in the APCE by members of the treating team, particularly those who are not the senior attending but including senior attendings. Although joining work rounds is time consuming, we find it is worth the effort.

Here, again, we return to the importance of the APCE at any/every hospital to teach that the good of clinical ethics is contributing to the helping and healing of patients. That is, clinical ethicists ought to be focused on the operationalization of actions taken in and for the best interest of each individual patient and organizationally, to achieve improved patient outcomes. With this frame of reference one comes to appreciate how trust is the cornerstone of successful clinical ethics hospital practice. And there is simply no place or time better to build trust with such a wide array of senior physicians and other clinical providers than on work rounds.

That is only, however, if one is, and therefore can be perceived as, trustworthy. For this reason, work rounds are one of the best places for mentored NCEs to learn how to work on aspirational virtues such as trustworthiness and courage. For example, for team members to be comfortable enough to bring forward what might be considered by a senior attending as a 'touchy' ethical issue, the APCE will need to have the courage and skill to raise the issue in such a way as to reduce team tensions rather than increase them and without shining a spotlight on how the APCE came to appreciate that the issue exists. One hopes the APCE is successful. But even an APCE won't hit a home run every time. Perhaps on a particular case, mere mention of the issue angers the senior attending and thus initially exacerbates existing team tensions. If the APCE, also with courage and skill, acknowledges on the spot that s/he sees that mere mention of whatever the problem is makes the group uncomfortable, apologizes for any discomfort s/he may have just caused (without apologizing for whatever was said that seems, in the APCE's judgment to have needed to have been said out loud), and then holds one's ground, gently and sensitively but firmly and matter-of-factly returning to the substance of the issue until it is settled enough that the 'volume' comes back down and patient well-being is advanced at least a notch will speak volumes to the team about how an APCE can be of assistance to them in meeting their moral goal of taking the best care of patients they can.

Let's consider trust and trustworthiness building on rounds in the context of a published case (Fiester 2015). This case is introduced, accurately, as a classic scenario. But rather than having the moral distress on the team be off the Richter scale because the case percolated until the consult was called, think how differently it might have unfolded if there had been an APCE, with an NCE along, rounding with this team.

The case is described as one in which the nurse and physician want to withhold and withdraw life-extending technologies and shift to comfort measures only. The nurse is quoted as saying s/he believes they are 'torturing' the patient. The physician is reported to say that s/he is, "...appalled at being forced to pummel the patient...." The family, who are of a completely opposite view, are reported to believe withholding and/or withdrawing is 'murder' and that only God should be making life and death decisions. The patient is described as grimacing when turned and in obvious, continuous anguish. The case description is closed with a caution not to confuse one's own deeply held commitments and values for 'the universal good' (p. 21).

To usefully reframe this case, we need, for a moment, to return to this chapter's earlier brief philosophical and historical context-setting. Remember this author suggested first that the clinically relevant flaw is to confuse Engelhardt's accurate characterization of modern society as a sea of values diversity with that fact posing a seemingly insurmountable barrier to managing well such values diversity ethically in the hospital. Then this suggestion was followed by this author's assessment that the ASBH *Core Competencies'* 1st edition creationist task force members, without referencing Engelhardt, further confused the issue by substituting the millennial-spanning inability to arrive at an academically pristine definition of the good - what Fiester (2015) seems to be referring to in her conclusion about needing to express, "...a specific conception of the good' (p. 21), with ignoring Pellegrino's ethically rich philosophical theory of medicine that illuminates the practical definition of the good of medicine as the healing and helping of patients.

How such help is contributed by APCEs depends on the case and is well-taught on rounds. Continuing to analyze the Fiester (2015, op cited) case from the perspective of how it might be best used for teaching, developing trustworthiness in the NCE and trust across the APCE and the clinical team, the first thing that jumps out at the reader is how many ethically relevant medical facts are left out. Even if this were a new case to a rounding APCE, the distress is compelling and immediately identifiable. But to have the APCE demonstrate a stance of neutrality in the face of needing more facts in and of itself tends to slow the flow of the distress down, calming the waters a bit.

In just seeing an APCE get the necessary, morally salient medical facts in an orderly fashion, such as whether the patient has any reversible disease or only irreversible disease, what is the patient's mental status and how has this been assessed, and whether or not the family members who are in opposition to the recommended plan are family members who have been involved in the patient's care throughout or are they, or are some of them, coming to the situation more recently, is likely to help suspend some of the moral distress. These are just some of the facts that, as they are told to the APCE by the team, the APCE models for everyone a systematic approach to ethical analysis, building trust in everyone that the clinical ethicist, although who may have strong feelings and beliefs him or herself, behaves in a way that tells everyone that personal beliefs may be strong but personal beliefs need to be put aside so the ethically relevant facts of the case – and that means all ethically relevant facts, not just the strictly medical ones – need to be considered evenly in working towards decisions and/or recommendations that are best for this particular patient.

That is what we at the Lynch Center mean when we teach distinctions between objectivity and neutrality. We think it builds trust because others find us trustworthy when we admit openly that while there may not be objectivity, because we all have our conscious and unconscious biases, as clinical ethicists we work to come to a case neutral in our judgements about what might or might not be in this patient's best interests at this point in his/her disease course. That seems to us to be more accurate than to make any claim about whether or not we are ever likely to ferret out and then neutralize all our personal preferences and biases. But coming neutral to the facts of a case, and this is seen by the totality of the team best on rounds, builds in others a belief about the APCE as a person who honestly is trying to help them help their patient in an even handed way.

This case is an excellent demonstration of the differences between being hindered by calling on the parties to come to some consensus or compromise in their views of an abstract good versus working to figure out what might be the best practical good for helping and healing this patient. The actual need of the patient, family and clinicians in meeting this practical clinical goal is the heart of clinical ethics work. Rounds provides a highly effective venue to highlight the practice of moral medicine in a timely way.

Next in the discussion of the case with the team may include elucidating more facts about the words 'pain' and 'anguish.' It might take only a brief conversation to determine that, as it often is for ICU patients, this patient, although, evincing pain through grimacing, might not be conscious enough to be anguished. ICU patients who are on ventilators are usually so sedated that it might turn out to be difficult to sustain a fact-based concern that the patient is truly anguished. The anguish might be the nurses', a point at which support can easily be given the nurses by acknowledging the emotional burdens of the case for them, an acknowledgement that is ordinarily deeply appreciated. In these exchanges, the APCE models for all, including the NCE, the virtue of compassion for the clinicians, not just for the patient and his/her family. The team grasps quickly that the APCE's goals are aligned with the virtue-based goals of the nurses and other clinical providers, themselves.

As to the physician's feeling that s/he is being forced to 'pummel' the patient, this opens the way for the APCE to remind not only the senior attending but the physicians-in-training also that they are moral agents too, not merely the extension of the will of families and patients. The teaching point is the explanation, explicitly restated for the physician trainees, allowing the senior attending to save face if s/he has not yet been able to analyze her/his obligations in this way, that there are two main drivers of moral medicine as noted previously in this chapter; the first is enacting patient and/or surrogate preferences to the greatest degree medically reasonable and the second is that physicians and nurses must always practice within reasonable standards of practice, be they local hospital, regional and/or national standards of practice.

With this framework for professional obligations established, the APCE then may be best advised to return to ethically weighty fact findings. That is, s/he might then ask, "Do you (directed to all the treating team) believe that your patient is actively dying, ie so unstable that the patient will die in our hands or soon thereafter? Or do you think you can get this patient stable to discharge?" The ability to make ethically sound recommendations to the team hinges on the team physicians' and nurses' answers. This process ordinarily makes clear that proposed next steps for the care of this patient rest not on one's own or any of the team's own personal preferences.

Discussing the matter in this way demonstrates that the APCE is coming to the team's assessment of the clinical facts from a position neutral to patient outcome. Thus, for which ever answers the APCE receives from the team, those answers lead seamlessly into a teaching moment about where there appears to be national consensus about what is ethically permissible end-of-life care and where there are still areas of such deep values divides that end-of-life decisions in the hospital are most appropriately left in the hands of patients and/or surrogates, as long as what patients and/or surrogates want don't stretch standards of practice beyond reasonableness. True, figuring out what words such as 'stretch' and 'reasonableness' mean in the context of a particular patient, requires ethically sound judgments. But by working with the team, from a point of neutrality to the degree one can possibly recognize when neutrality is being infringed, shows the wisdom of the APCE and her commitment to ethically practiced medicine, no matter one's preferred care for oneself or one's own loved ones.

In the end, if the patient is so unstable as to be dying in the clinicians hands, and it is clear that removal of the life-extending technology of a ventilator is too much for the family to bear, ICU patients do die on vents. It may take longer but administrative time pressures can be resisted when it is ethically important to do so. The attending can be asked to suspend her/his concern about utilization of resources on the basis of the ethical obligation to care for the patient with whom one is in relationship. If, instead, the answer with team consensus was that the patient can be made stable to discharge then discharge may be the best outcome for this patient and family, as long as the patient can be kept comfortable. Deep but quick discussion on rounds often diminishes the professional distress these irreconcilable values conflicts produce. It does not take an inordinate amount of time to teach enough on rounds that many in this large group of clinical providers can glean new insights into how these conflicts of values can be managed in a way that preserves the family's trust in the treating team, the treating team can manage the patient within reasonable standards of practice, the treating team's moral distress can be reduced and the patient is well-managed while in the clinicians' care. Even if discharge with tracheostomy for breathing and a peg tube for feeding and fluids might not be either the preference for one's own loved ones of either most or all members of the treating team's or the personal preference of the APCE, all, including the NCE, have learned important lessons in ethically working through such an emotionally difficult case without having to compromise their diversified values away.

To the novice, a quiet day on rounds may seem like a huge waste of time. But when a case such as the one just discussed comes up, the lessons and trust building processes are learned in a way that can only be learned at the bedside. On other days, days where perhaps the APCE doesn't say anything at all, equally important but more subtle messages about patience and respect for clinician's precious time are broadcast and learned; learning that can only take place at the bedside. In either case, all members of the team become increasingly comfortable with each other and this familiarity makes it easier for team members to call ethics consultations in the future.

17.7 Strengthening the Ethical Climate of the Organization through Mentored Attention to the Virtues: Yes, We Can Learn to be Virtuous

Given that patient scenarios are essentially infinite but the space for this chapter is not, we move to the last section which is on how organizational ethics and bedside mentoring to the virtues are excellent ways to strengthen the moral climate of a hospital and how to learn to become ever more virtuous in the process. Here I start with a wonderful story that took place at a previous ASBH annual conference (2016 in Washington, DC). During our session on mentees and mentors, presented by one of this book's editors (Guidry-Grimes) and this chapter's author along with our mentee/mentor counterparts from Monte Fiore in NY, a question from the audience started with the disclaimer that, "Of course it's not our job to change culture….." to which my partner and I almost jumped up and replied in unison, "But of course it's our job to change culture…." The real "Of course…" of this little anecdote is that some clinical ethicists probably believe that it is not – or it has been made explicit in their contracts that it is not – their job to work on culture change. That is simply not the position taken in this chapter.

Like some from the early days of our field, we see strengthening the moral climate of the hospital as a critically important part of the job of our Lynch Center. Here, of the various virtues that apply and that we mentor to, this section's example is about courage. In what this author considers one of the most important papers of our new century, Hamric et al.'s *Must We Be Courageous*? (2015), they focus the field of medicine, and this includes clinical ethicists (DeRenzo 2015), on what it means to be courageous. This is absolutely must reading for several reasons, the one considered here is how Hamric et al. define courage in medicine. They state that,

> "Courageous action requires a difficult, painful, or dangerous situation. In order for an action to be deemed courageous, the stakes have to be high. There has to be a perceived personal cost to doing the right thing." (p. 34).

One of the examples they cite is of two Texas nurses who repeatedly complained about a physician's unsafe practices. For their efforts, they were fired and charged criminally. Only after a 3 year legal battle were they exonerated and the offending physician and hospital indicted. Certainly, these nurses demonstrated courage.

So let's return one last time to Fiester's case (2015) for an example of how a seeming lack of mentored understanding of what to do and what is called for, driven by the philosophy of Engelhardt and seemingly lacking an appreciation of the philosophy of Pellegrino, can result in a case that is clearly producing harm for the patient, family and providers. As explained earlier, the case is about an ICU patient who grimaces when turned. Although we don't know if the patient has underlying, progressive, terminal disease, it is implied. The physicians and nurses are at their wits' end because they want to shift to comfort measures only and are at odds with a family who is adamant that all life-extending measures that are technically feasible, seemingly regardless if such measures are clinically meaningful, be used and/or continued.

Although Fiester does not analyze the case but rather suggests using a post hoc check list to help the NCE assess his own feelings about the case, a novice ethicist in our Lynch Center would be assisted, in real time, to learn how to muster the courage to bring help and healing ala Pellegrino through demonstration of the ancient virtue of courage to the miseries of this case. When writing about this case earlier in the chapter, it was to present how the case could have played out if picked up on work rounds. Here, we are going to assume as seems the intent of the case as originally written, that it is being picked up by the clinical ethicist as a formally called clinical ethics consultation. Once picked up by the APCE, with the NCE, the APCE might start by talking first with the nurses. Next would likely be a page to the senior attending and then perhaps conversations with members of the house staff. Thereafter, perhaps, the APCE, along with at least a resident and one of the patient's nurses, might talk with the family ahead of asking social work to set up a family meeting with the full team and as many of the family as want to come. During a pre-family meeting the team decides that the only issues that ought to be addressed are the patient's seeming pain and that the life-extending technologies that are being applied now will be continued. It would be appropriate for the most senior physician in the room to start by explaining what the medical team sees – sticking only to medical facts, no conclusions being drawn - and ask the family what they see – making certain not to interrupt too quickly as is physician habit (Groopman 2008). In the meeting with the physicians taking their cues from what the family offers, cracks in the family front appear subtly. Only in person may these intra-family disagreements be seen and registered by the APCE and NCE. The meeting is ended with the family assured that at least for the time being all beneficial interventions the patient is presently receiving will be continued but not before the APCE makes clear that the hospital team will facilitate a transfer if the family wants it and finds a receiving physician and reminds the family that they can always go to court to obtain an injunction against us.

Once the meeting concludes but before the group breaks up, there is a 'de-brief' which includes the APCE offering the team the maxim of physician professionalism that only those procedures that are indicated ought to be provided, and assurance that the APCE will repeat this view point in her clinical ethics consultation note submitted into the electronic medical record. But let's say the APCE's reminder to the family that they have the right to go to court to force additional treatment has spooked the senior attending. Let's assume that telling her she has no ethical or regulatory obligation to do everything the family wants does not allay her fears. Being able to tell that senior attending that the case will be brought to the attention of the hospital's senior medical officer and then to have the senior medical officer contact this physician and repeat this maxim as the standard for the hospital broadcasts volumes and assists her in finding the ability to become more courageous about what she considers appropriate patient treatment. Even if support by hospital leadership and the hospital's legal department is necessary to comfort the senior attending physician and perhaps others on the team, this case certainly shines light on the kind of courage that might be needed for others to aspire to being able to muster when needed.

But both the Fiester (2015 op cit) case and the nursing case in Hamric et al. (2015 op cit) obscure, in their drama, what may be the inestimable contribution to the field the Hamric et al. paper makes in defining courage as called for in the every-day sense. That is, although the Texas nurses are clearly exercising courage, health care providers ought never be called on to demonstrate this kind of courage. It is excessive and merely an example of an intolerable moral climate in a hospital that obviously required overhaul. Rather, this author sees Hamric et al's real contribution the delineating of what courage is on a daily basis. Hamric et al. paint the picture that in every decision and/or recommendation about a patient, the self-reflective, ethically astute clinician knows one can be wrong. And in being wrong, one could hurt a patient. Nonetheless, making decisions and recommendations for patient after patient, day in, day out, while carrying the burden of the possibility of harming a patient or a family member; this is the true courage of patient care. This kind of courage extends to clinical ethicists, too. Developing the muscle memory needed to do one's work, whether a provider or a clinical ethicist, comes only after much well-mentored, bedside training.

The primary way this everyday courage helps clinical providers and clinical ethicists to learn to become ever more virtuous is that the repetitiveness of the demand on one's psyche habituates one to being ever more able to muster courage when needed. And this is not only true for the virtue of courage. Every time, for example, an APCE, or even an NCE, is called on to give an accounting of a stickie situation that grew worse regardless of multiple disciplines' efforts, including ethics', the individual providing the accounting is practicing being honest and trustworthy by presenting all party's role, even that of one's own. When one has to learn to hold one's tongue because saying something, even something that needs to be said, requires different timing, practicing patience allows one to learn – and to want to learn – how to become ever more patient. In sum, the qualities that allow the NCE to mature into an APCE are not only knowledge based but are also virtue based. The wise person learns that about the job of being a clinical ethicist and genuinely wants to get better and better as being that individual that others can depend on for assistance in the complex task of helping and healing patients.

17.8 Conclusion: Closing the Gaps, Squeezing Success out of the Opportunities

At this moment in the development of the field of clinical ethics, the gaps are still wide. While there are those who have set themselves out as authorities, there are still others who continue to argue against majority views. For example, this chapter demonstrates how some fundamentally disagree with the philosophical underpinnings of the field, believing that differing views on this starting point need to be uplifted, not ignored. The disagreements about the degree to which Enlightenment philosophy should be the primary theoretical underpinning of bioethics or whether there is room for those of us who cling to the inclusion of the more ancient norms of virtue and character as core and required qualities of the wise clinician are yet to be settled.

These theoretical distinctions lead to important differences among us about which methods lead to the best outcomes for patients, families and the physicians and other clinical providers who care for them. These differences lead to differences among us on what is necessary to develop credentialing and certification programs and who needs to be at the table in the development of these processes. There is, as yet, not even agreement that the field of clinical ethics should include responsibilities for organizational ethics concerns within one's own institution.

But just as these gaps pose threats to performance and progress of NCEs towards becoming APCEs or even being clear on what it means to be an APCE, at all, these gaps also present opportunities. Perhaps the greatest opportunity is the continuation of scholarship in the world of clinical ethics. Almost 20 years ago, Singer et al. (2001) called for more bedside teaching in clinical ethics, making the claim that such teaching "…is potentially the most effective and yet the least studied" (p. 4). Not much has changed since then. We need more research about patient outcomes with clinical ethicists applying different methods for CECs. We need more research about the influences on clinician training in ethics and about what influences on patient care processes having a clinical ethicist on rounds might produce. We need to start connecting all the quality data hospitals collect to the various activities clinical ethicists perform and pursue. We need much more research on how clinical ethicists can contribute to reduction of moral distress of physicians, nurses and other care providers such as social workers and respiratory therapists. And we need research and others' approaches to scholarly thinking about how clinical ethicists do or can assist in strengthening the moral fabric of a hospital.

This author looks forward to seeing such research performed and integrated into the many discussions in the field of bioethics broadly and especially relevant for those in the field who practice clinical ethics. Bedside mentoring of the next generation of clinical ethicists from novice to advance practice status is a responsibility we all share. Meeting this responsibility should be guided by broad, inclusive, collegial and scholarly thinking, publishing, and conversation.

Acknowledgement The author would like to thank Jack Schwartz, JD for his thoughtful and most helpful critique of an earlier draft of this chapter.

References

American Society for Bioethics and Humanities. 1998. *Core competences for health care ethics consultation.* 1st ed. Glenview, IL.

———. 2011. *Core competences for health care ethics consultation.* 2nd ed. Glenview, IL.

Beauchamp, Tom L., and James F. Childress. 2016. *Principles of biomedical ethics.* 1st edition in 1989, 7th edition in 2009. New York: Oxford University Press.

Childress, James F. 2005a. Foreword. In *Fletcher's Introduction to Clinical Ethics*, 1st. ed., ed. John C. Fletcher, Edward M. Spencer, Paul A. Lombardo and Mary Faith Marshall, p. v. Frederick: University Publishing Group.

———. 2005. Foreword. In *Fletcher's introduction to clinical ethics*, ed. John C. Fletcher, Edward M. Spencer, and Paul A. Lombardo, 3rd ed. Hagerstown: University Publishing Group.

DeRenzo, E.G. 2015. A clinical ethicist's thank you. *The Hastings Center Report*. Nov-Dec. 2015 45(6). Invited commentary on Hamric AB, Arras JD, ME Mohrmann. Must we be courageous? *The Hasting Center Report* May–June 2015, 45 (3): 33–40.

DeRenzo, E.G., J. Vinicky, B. Redman, J.J. Lynch, P. Panzarella, and S. Rizk. 2006. Rounding: A model for consultation and training whose time has come. *Cambridge Quarterly of Healthcare Ethics* 15 (2): 207–215.

Dubler, N.N., and C.B. Liebman. 2011. *Bioethics mediation: A guide to shaping shared solutions.* Nashville: Vanderbelt University Press.

Englelhardt, H.T., and F. Jotterand, eds. 2011. *The philosophy of medicine reborn: A Pellegrino reader.* Notre Dame: University of Notre Dame Press.

Fiester, A. 2015. Teaching nonauthoritarian clinical ethics: Using an inventory of bioethical positions. *Hastings Center Report.* 45 (2): 20–26.

Groopman, J. 2008. *How doctors think.* New York, NY: Houghton Mifflin Harcourt Publishing Co..

Hamric, A.B., J.D. Arras, and M.E. Mohrmann. 2015. Must we be courageous? *Hastings Center Report* 45, no. 3, 2015:33–40. DOI: 10–1002/hast. 449.

Jonsen, A.R., M. Siegler, and W.J. Winslade. 2006. *Clinical ethics: a practical approach to ethical decisions in clinical medicine.* First edition, Yr.....; 6th edition. New York, NY: McGraw Hill.

Messutta, Donna. 2017. Moral distress, ethical environment, and the embedded ethicist. *The Journal of Clinical Ethics* 28 (4. (Winter 2017): 318–324.

Minogue, B.P., G. Palmer-Fernandez, and J.E. Reagan, eds. 1997. *Reading Engelhardt: Essays on the thought of H. Tristram Engelhardt, Jr.* New York: Springer Science + Business Media, BV.

Moreno, J.D. 1995. *Deciding together: Bioethics and moral consensus.* New York, NY: Oxford University Press.

Pellegrino, E., and D.C. Thomasma. 1988. *For the patient's good: The restoration of beneficence in health care.* New York, NY: Oxford University Press.

Rasmussen, L.M., A.S. Iltis, and M.J. Cherry, eds. 2015. *At the foundations of bioethics and biopolitics: Critical essays on the thought of H. Tristram Engelhardt, Jr.* New York, NY: Springer.

Singer, P.A., E.D. Pellegrino, and M. Siegler. 2001. Clinical ethics revisited. *BMC Medical Ethics* 2 (1). http://www.biomedcentral.come/1472-6939/2/1.

Sugarman, J., and D.P. Sulmasy. 2010. *Methods in medical ethics.* Washington, DC: Georgetown University Press.

Sutton, W., and E. Linn. 1976. *Where the money was: The memoirs of a bank robber.* New York: Viking Press.

Zeitfracht Medien GmbH
Ferdinand-Jühlke-Straße 7
99095 Erfurt, Deutschland
produktsicherheit@kolibri360.de